Springer Water

The book series Springer Water comprises a broad portfolio of multi- and interdisciplinary scientific books, aiming at researchers, students, and everyone interested in water-related science. The series includes peer-reviewed monographs, edited volumes, textbooks, and conference proceedings. Its volumes combine all kinds of water-related research areas, such as: the movement, distribution and quality of freshwater; water resources; the quality and pollution of water and its influence on health; the water industry including drinking water, wastewater, and desalination services and technologies; water history; as well as water management and the governmental, political, developmental, and ethical aspects of water.

Etikala Balaji · Golla Veeraswamy ·
Prasad Mannala · Sughosh Madhav
Editors

Emerging Technologies for Water Supply, Conservation and Management

Springer

Editors
Etikala Balaji
Department of Geology
Sri Venkateswara University
Tirupati, Andhra Pradesh, India

Golla Veeraswamy
Department of Geology
Sri Venkateswara University
Tirupati, Andhra Pradesh, India

Prasad Mannala
Department of Geology
Central Tribal University of Andhra Pradesh
Vizianagaram, Andhra Pradesh, India

Sughosh Madhav
Department of Civil Engineering
Jamia Millia Islamia
New Delhi, Delhi, India

ISSN 2364-6934 ISSN 2364-8198 (electronic)
Springer Water
ISBN 978-3-031-35278-2 ISBN 978-3-031-35279-9 (eBook)
https://doi.org/10.1007/978-3-031-35279-9

This Springer imprint is published by the registered company Springer Nature Switzerland AG
The registered company address is: Gewerbestrasse 11, 6330 Cham, Switzerland

Preface

This book deals with emerging technologies for water supply, conservation and management using artificial intelligence, machine learning, remote sensing and GIS techniques. Moreover, it encompasses water literacy, groundwater resource management, conservation, morphometric characteristics, mapping, monitoring of water resources, urban water stress, identification of potential groundwater zones, management of coastal ecosystems, flood risk assessment, delineation of seawater intrusion, rainwater harvesting and recharge structures using advanced artificial intelligence, machine learning, deep learning, neural network, geophysical, and remote sensing and GIS. The book will be helpful to water resource planners, decision-makers, researchers, policymakers, NGOs and students in Hydrogeology, Geophysics, Hydrology, Remote Sensing and GIS, and Agriculture.

Tirupati, India — Etikala Balaji
Tirupati, India — Golla Veeraswamy
Vizianagaram, India — Prasad Mannala
New Delhi, India — Sughosh Madhav

Contents

Contributors

Sk Ajim Ali Department of Geography, Faculty of Science, Aligarh Muslim University, Aligarh, India

Rukhsar Anjum Department of Geography, Faculty of Science, Aligarh Muslim University, Aligarh, India

Arfan Arshad Department of Biosystems and Agricultural Engineering, Oklahoma State University, Stillwater, OK, USA

Rana Ammar Aslam Department of Structures and Environmental Engineering, University of Agriculture, Faisalabad, Pakistan

Uday Bhan Department of Petroleum Engineering & Earth Sciences, University of Petroleum and Energy Studies, Dehradun, Uttarakhand, India

Humberto Chaves Escola Superior Agrária de Beja, Instituto Politécnico de Beja, Beja, Portugal;
FibEnTech – Fiber Materials and Environmental Technologies, Covilhã, Portugal

Krupavathi Chinthala Department of Geology, Yogi Vemana University, Kadapa, Andhra Pradesh, India

Srilert Chotpantarat Department of Geology, Faculty of Science, Chulalongkorn University, Bangkok, Thailand;
Center of Excellence in Environmental Innovation and Management of Metals (EnvIMM), Environmental Research Institute, Chulalongkorn University (ERIC), Bangkok, Thailand

Sourav Choudhary Department of Water Resources Development and Management, Indian Institute of Technology Roorkee, Roorkee, India

Joystu Dutta Department of Environmental Science, UTD, Sant Gahira Guru University, Ambikapur, Sarguja (C.G.), India

Balaji Etikala Department of Geology, Sri Venkateswara University, Tirupati, AP, India

Ana Fernandes CBIOS, Escola de Ciências e Tecnologias da Saúde, Universidade Lusófona, Lisboa, Portugal

Margarida Figueiredo Departamento de Química e Bioquímica, Escola de Ciências e Tecnologia, Centro de Investigação em Educação e Psicologia, Universidade de Évora, Rua Romão Ramalho, Évora, Portugal

M. Suresh Gandhi Department of Geology, University of Madras, Chennai, India

Somenath Ganguly Department of Petroleum Engineering & Earth Sciences, University of Petroleum and Energy Studies, Dehradun, Uttarakhand, India

Annmaria K. George Department of Geology, University of Madras, Chennai, India

Harish Vijay Gudala Department of Geology, Yogi Vemana University, Kadapa, AP, India

Vijayakumar Gundala St. Ann's College of Engineering & Technology, Vetapalem, Chirala, AP, India

Avinash Pratap Gupta Department of Environmental Science, UTD, Sant Gahira Guru University, Ambikapur, Sarguja (C.G.), India

Murali Krishna Gurram Department of Geo-Engineering & RDT, College of Engineering, Andhra University, Visakhapatnam, AP, 530003 India

Jagriti Jain Department of Water Resources Development and Management, Indian Institute of Technology Roorkee, Roorkee, India

Ghulam Jeelani Department of Earth Sciences, University of Kashmir, Srinagar, India

Anvesh Jujjuvarapu Civil Engineering Department, St. Martin's Engineering College, Secunderabad, India

Shahbaz Nasir Khan Department of Structures and Environmental Engineering, University of Agriculture, Faisalabad, Pakistan

Deepak Khare Department of Water Resources Development and Management, Indian Institute of Technology Roorkee, Roorkee, India

Nooka Ratnam Kinthada Department of Geosciences, Adikavi Nannaya University, Rajamahendravaram, AP, India

Nanabhau Kudnar Department of Geography, C. J. Patel College, Tirora, Gondia, Maharashtra, India

Suhail A. Lone Department of Earth Sciences, University of Kashmir, Srinagar, India

Y. Mohana Prasada Rao Department of Geology, Sri Venkateswara University, Tirupati, AP, India

Francisco Munoz-Arriola School of Natural Resources, University of Nebraska Lincoln, Lincoln, USA;
Department of Biological Systems Engineering, University of Nebraska Lincoln, Lincoln, USA

P. Muthukumar Department of Geology, V.O. Chidambaram College, Thoothukudi, Tamil Nadu, India

Anusha Boya Nakkala Department of Geology, Yogi Vemana University, Kadapa, Andhra Pradesh, India

D. Naresh Kumar Civil Engineering Department, St. Martin's Engineering College, Secunderabad, India

José Neves Instituto Universitário de Ciências da Saúde, CESPU, Famalicão, Portugal;
Centro Algoritmi/LASI, Universidade do Minho, Braga, Portugal

Masilamani Palanisamy Department of Geography, School of Earth Science, Bharathidasan University, Tiruchirappalli, Tamil Nadu, India

Piyush Pandey Department of Environmental Science, UTD, Sant Gahira Guru University, Ambikapur, Sarguja (C.G.), India

Ravi Kumar Pappaka Department of Geology, Yogi Vemana University, Kadapa, Andhra Pradesh, India

Farhana Parvin Department of Geography, Faculty of Science, Aligarh Muslim University, Aligarh, India;
School of Liberal Arts, Noida International University, Noida, UP, India

Thilagaraj Periasamy Department of Geography, School of Earth Science, Bharathidasan University, Tiruchirappalli, Tamil Nadu, India

Santosh Murlidhar Pingale Hydrological Investigations Division, National Institute of Hydrology, Roorkee, India

M. Rajashekhar Department of Geology, Yogi Vemana University, Kadapa, AP, India

S. Selvam Department of Geology, V.O. Chidambaram College, Thoothukudi, Tamil Nadu, India

Srinivasa Gowd Somagouni Department of Geology, Yogi Vemana University, Kadapa, Andhra Pradesh, India

Vinoda Rao Tanikonda Department of Geology, Andhra University, Visakhapatnam, AP, India

Tarun Kumar Thakur Department of Environmental Science, Indira Gandhi National Tribal University, Amarkantak, M.P., India

Nguyen Ngoc Thanh Graduate School, Interdisciplinary Program in Environmental Science, Chulalongkorn University, Bangkok, Thailand;
University of Agriculture and Forestry, Hue University, Hue City, Thua Thien Hue, Vietnam

Vamsi Kalyan Veerla Civil Engineering Department, St. Martin's Engineering College, Secunderabad, India

Venkateswara Rao Vegala Department of Geo-Engineering & RDT, Andhra University, Visakhapatnam, AP, India

Thumati Venkateswara Rao Civil Engineering Department, VVIT, Nambur, AP, India

Henrique Vicente Centro Algoritmi/LASI, Universidade do Minho, Braga, Portugal;
Departamento de Química e Bioquímica, Escola de Ciências e Tecnologia, REQUIMTE/LAQV, Universidade de Évora, Évora, Portugal

Abdul Rehman Zahoor Department of Structures and Environmental Engineering, University of Agriculture, Faisalabad, Pakistan

Chapter 1
Assessment of Water Consumers Literacy

Ana Fernandes, Margarida Figueiredo, Humberto Chaves, José Neves, and Henrique Vicente

Abstract Nowadays, an increasing amount of water is used without the awareness that this resource is not inexhaustible. In fact, pollution, environmental degradation and/or climate change caused by human activities lead to the degradation of the quality of available water. In 2015, the United Nations warned about the risk of reaching a water deficit of 40%, in 2030, if consumption patterns are not changed. Indeed, population growth is one of the main causes for this deficit. The protection and

A. Fernandes
CBIOS, Escola de Ciências e Tecnologias da Saúde, Universidade Lusófona, Campo Grande 376, 1749-024 Lisboa, Portugal
e-mail: teresa.vila.fernandes@ulusofona.pt

M. Figueiredo
Departamento de Química e Bioquímica, Escola de Ciências e Tecnologia, Centro de Investigação em Educação e Psicologia, Universidade de Évora, Rua Romão Ramalho, 59, 7000-671 Évora, Portugal
e-mail: mtf@uevora.pt

H. Chaves
Escola Superior Agrária de Beja, Instituto Politécnico de Beja, Rua Pedro Soares, 7800-295 Beja, Portugal

FibEnTech – Fiber Materials and Environmental Technologies, Rua Marquês d'Ávila e Bolama, 6201-001 Covilhã, Portugal
e-mail: hc@ipbeja.pt

J. Neves
Instituto Universitário de Ciências da Saúde, CESPU, Rua José António Vidal, 81, 4760-409 Famalicão, Portugal
e-mail: jneves@di.uminho.pt

J. Neves · H. Vicente
Centro Algoritmi/LASI, Universidade do Minho, Campus de Gualtar, Rua da Universidade, 4710-057 Braga, Portugal

H. Vicente (✉)
Departamento de Química e Bioquímica, Escola de Ciências e Tecnologia, REQUIMTE/LAQV, Universidade de Évora, Rua Romão Ramalho, 59, 7000-671 Évora, Portugal
e-mail: hvicente@uevora.pt

E. Balaji et al. (eds.), *Emerging Technologies for Water Supply, Conservation and Management*, Springer Water, https://doi.org/10.1007/978-3-031-35279-9_1

sustainable consumption of water is one of the United Nations' Sustainable Development Goals, to ensure that the world's population has access to clean water, free from pollution and managed responsibly. Thus, governments and non-governmental organizations must promote the conscious and informed use of water by the population. It is mandatory that the population becomes aware of the need for efficient management of water resources, ensuring their quality and preventing their degradation, in order to not compromise/jeopardize their future availability. The knowledge of the population's literacy on water issues and on water quality—health interconnections is essential to design plans leading to the implementation of eco-sustainable practices. The goal of this research was to evaluate the literacy of water consumers and to establish a forecast model for water literacy managing. The collection of information was conducted through the inquiry by questionnaire technique and applied on a cohort encompassing 453 participants. The questionnaire includes three main dimensions (*Water Quality*, *Disease Prevention* and *Sustainability/Public Health Promotion*) and in each dimension, four competencies were evaluated (*obtain*, *understand*, *assess* and *apply* information regarding water consumption). The results obtained allow to assert that in the two first dimensions, the competence in which participants show more difficulty is *assess*. Regarding the *Sustainability/Public Health Promotion*, the participants show more difficulty in the competence *apply*. The model presented in this research, grounded on the connectionist paradigm, has shown great efficiency in the forecast of the target variable. The key contribution of the present research is to present an integrated and systematic approach that can give a contribution to the increase of water literacy, which allows the implementation of eco-sustainable practices.

Keywords Artificial intelligence · Artificial neural networks · Sustainable use of water · Water literacy assessment · Water management · Water quality

1.1 Introduction

In the mid-twentieth century, after the 2nd World War, the sustainable development model was based on a set of international and institutional agreements, based on two fundamental pillars, the economic and the social [50]. With the change in the living standards of a large part of the world's population, the global economy has grown, and average life expectancy has increased [72]. This fact, associated with high population growth, led to increasing on the exploitation of resources and in pollutant emissions [38, 39]. In the 1970s and 1980s, it began to recognize that pollution and the resulting environmental degradation were harming human well-being and sustainable development. Thus, a new approach that should integrate environmental factors became necessary [60]. The report of World Commission on Environment and Development, i.e., the Brundtland Report of 1987, is considered as the first step towards the current concept of sustainable development, defining sustainable development as "*the ability to make development sustainable to ensure that it meets the need of the present without compromising the ability of future generations to meet their own needs.*" [50]. According to this report, entities and policy makers are responsible for presenting truly integrative and holistic strategies for sustainable development, considering that the impact resulting from their decisions goes far beyond the immediate, lasting over the years [35]. In addition to the economic and social aspects included in the Brundtland Report, the United Nations Conference on Environment and Development [70] introduced the environment protection and restoration model. Indeed, the conclusions of this conference emphasized how social, economic, and environmental factors are mutually dependent and move forward simultaneously, and how an action in one sector involves efforts in the other sectors to be sustainable in the future. Moreover, the conclusions also highlighted the importance of incorporating economic, social, and environmental concerns in meeting current needs for the sustainability of human life on Earth and that such integrated attitude is practicable [70]. However, the combination of economic, social, and environmental factors requires new perceptions and/or changes on production and consuming habits, living and working habits, and on decision-making processes. This point of view was radical for its time, and it initiated a dynamic discussion in governments and between governments and citizens on how to guarantee sustainability for development.

In 2015, the United Nations proposed 17 Sustainable Development Goals, also known as the Global Goals. These goals were described in the 2030 Agenda for Sustainable Development and are deeply related, as the results attained in one of them influence the results of the others. The 2030 Agenda was the result of joined efforts made by governments and citizens around the world to create a new global model in order to put an end poverty, promote prosperity and well-being for all, protect the environment, and fight climate change. The 17 Sustainable Development Goals are [71]:

1. No poverty;
2. Zero hunger;
3. Good health and well-being;
4. Quality education;
5. Gender equality;
6. Clean water and sanitation;
7. Clean and affordable energy;
8. Decent work and economic growth;
9. Industry, innovation and infrastructure;
10. Reduced inequality;
11. Sustainable cities and communities;
12. Responsible consumption and production;
13. Climate action;
15. Life on earth;
14. Underwater life;
17. Partnerships for the goals
16. Peace, justice and strong institutions;

Goal 6 aims to ensure the availability and sustainable management of water and sanitation. According to the World Meteorological Organization, in sufficient quantity and quality, fresh water is crucial in all aspects of life and sustainable development. Indeed, safe drinking water and sanitation are basic human rights and are recognized by almost all countries. Water resources are intrinsically related to all dimensions of sustainable development (e.g., food security, health promotion, and poverty reduction). In addition, they are also indispensable to sustain economic growth in agriculture, industry, and energy generation, as well as to maintain healthy ecosystems" [78]. The agricultural sector is responsible for the use of about 70%

of water abstracted from rivers, lakes, and aquifers [44]. In fact, the intensification of agriculture requires an intensive use of water, leading to a significant increase in consumption. However, activities such as commerce, industry (particularly the food industry), and public supply also consume huge amounts of water. Furthermore, the issues related with the pollution of water resources, both surface and groundwater should be considered [53], and this problem is often caused by agricultural or agro-industrial activities [18]. In fact, 25% of European groundwater is of poor quality, being agricultural activity one of the main responsible for this poor quality [57]. Thus, it is mandatory to become aware of the need for efficient management of water resources, guarantee their quality and avoid both degradation and waste, aiming to ensure the future availability of water sources [47]. Thereby, the protection and sustainable consumption of water constitute one of the pillars of the United Nations Sustainable Development Goals (SDGs), ensuring that all population has access to clean, pollution-free, and responsibly managed water [71]. The main items of this goal, by 2030, are:

- Universal and equitable access to safe and affordable drinking water;
- Access to adequate and equitable sanitation and hygiene, and end open defecation;
- Improve water quality by reducing pollution, eliminating dumping and minimizing the release of hazardous chemicals and materials, halving the proportion of untreated wastewater and substantially increasing recycling and safe reuse globally;
- Increase water-use efficiency and ensure sustainable withdrawals and supply of freshwater to address water scarcity, and reduce the number of people suffering from water scarcity;
- Implement integrated water resources management at all levels, including transboundary cooperation;
- Protect and restore all water-related ecosystems;
- Expand international cooperation and capacity-building support to developing countries in water and sanitation related activities and programs; and
- Support and strengthen the participation of local communities in improving water and sanitation management.

Water quality is determined by a set of criteria and standards, which varies according to the objectives of its use [6]. The awareness of the need of conducting the quality control of the water from private abstractions has grown in Portugal. This fact is demonstrated by the increase of accredited entities for that purpose. Indeed, the maintenance of high levels of quality in water industry depends, to a large extent, on consumer demand. Consumers must be informed citizens, i.e., be aware of the consequences that water related issues can have on their daily lives, in particular on their health. Thus, it becomes necessary to assess consumer literacy in relation to issues like:

- The consequences for health resulting from the use of non-monitored water;
- How to get information about water quality;
- The need to consume monitored water;

- The need to ensure periodic water quality control;
- Human activities that can have repercussions on quality of water;
- The importance of accredited entities to conduct water analyses; and
- Behaviors that contribute to the sustainable use of water.

This work aims to evaluate the degree of literacy of water consumers and to generate a forecast model for water consumers literacy manage. This model also aims to be a learning system that facilitates water consumers literacy data analysis and contribute to identify possible improvements, preventing precipitate decision-making and the unnecessary use of resources, contributing for the sustainable use of water and global sustainability.

1.2 State of Art

The improvement of global management of water resources is paramount. According to the 2030 agenda for sustainable development of the United Nations, the entities that are in charge of water management should seek to provide clean water in the quantity and quality for all individuals and/or uses and manage it in a sustainable way. These entities should also provide sanitation, being responsible for wastewater management [71]. The governments of developed countries, despite the fact that the access to water is practically global [77], recognize that there is a need of a constant improvement in water resources management. These concerns include not only water quality, but also upgrade/replacement of infrastructures, impacts of climate change and water scarcity, just to name a few [61, 62]. The issues related with water resources governance can be addressed from distinct perspectives, which can encompass from water quality to the formulation of public policies, including integrated water resources management and the connection water and health. Several recent studies address the topics *related* with water governance [12, 15, 28, 58, 64, 68, 73, 76], integrated water resources management [16, 42], the supply-demand relationship [65, 75], water reuse [34, 67], issues related with water legislation [2, 4, 5, 11, 29, 66], water management in stressing conditions like extreme weather events [36], political tensions [33] or scarcity conditions [58, 68].

1.3 The Quality of Water for Human Consumption

A particular water abstraction is understood to be that which is carried out by an individual for his own use, regardless of its origin (well, hole, or other). Water from a particular source may be unfit for human consumption although its appearance is crystalline, transparent or has a pleasant taste. In fact, it may contain microorganisms or substances that are invisible to eyes, but harmful to health, causing diseases such as vomiting, gastroenteritis (diarrhea) or hepatitis A [10]. Conversely, the water supplied by the public distribution systems is, in general, quality water, duly controlled, ensuring all the necessary requirements for consumption. In other words, the presence of contaminating agents likely to pose a risk to health (e.g., pathogenic microorganisms and/or chemical contaminants) are controlled [7].

The waters of private abstraction should be analyzed by laboratories accredited for the test methods used, i.e., regularly audited laboratories where it is assured the quality of the analytical techniques used and the services provided. The technicians have the necessary skills to help consumers in the selection of the parameters to analyze, taking into account the water use. According to the regulatory authority of water services and waste in Portugal (i.e., The Water and Waste Services Regulation Authority) consumers are advised to analyze coliform Bacteria, Escherichia coli, Clostridium perfringens, Cryptosporidium, Enterococci, pH, iron, manganese, arsenic, nitrate, and pesticides [63].

1.4 Literacy

The term literacy has traditionally been interpreted as the capability to read and write. However, it has progressively been used in other more comprehensively contexts such as scientific literacy, health literacy, computer literacy, environmental literacy, political literacy. The term scientific literacy, commonly used in the United States of America, is synonymous with public understanding of science in Britain and scientific culture in France [21]. Taking into account the Latin root of the terms literacy and science, Branscomb [13] defined the concept of scientific literacy as the ability to read, write and understand systematized human knowledge. In fact, there are many interpretations and meanings of the term scientific literacy, which justifies the idea that this is a diffuse and ill-defined concept [14].

Although literacy is related to the ability to read and write, from a broader perspective, it is associated with knowledge, learning and education. Thus, a person can acquire knowledge, even without knowing how to read, through oral transmission or from life experience. However, when it comes to a discipline with its own body of knowledge there is a very close connection between knowledge and the ability to read and write. It is in this sense that [51] argue that (i) *science, as we know it, could never be what it is if it were not for the text on which it rests*; and that (ii) *given the dependence of science on the text, an individual who cannot read or write will be severely limited as regards the acquisition of scientific knowledge, learning and education.*

1.4.1 Literacy in Water Consumption

Nowadays, it is increasingly important for consumers to understand the importance and the need of quality control and water treatment. Thus, the evaluation of citizens' levels of literacy on these issues is of the utmost importance. In particular, with regard to water consumption, the promotion of citizens' literacy is essential. For example, knowing what steps to take to use water from private sources or to distinguish between distinct types of water (e.g., untreated water, treated water or drinking water) can become fundamental from a public health perspective, particularly in rural areas.

Fernandes et al. [23] point out the lack of studies on water literacy, regardless of the contexts (e.g., water quality, water-using policies, extreme climatic events), despite the relevance of this matter.

1.5 Artificial Neural Networks-Based Approach

Artificial Neural Networks (ANNs), also referred as connectionist paradigm, are artificial intelligence systems that can be used to solve complex problems, with non-linear relationships between inputs and outputs. ANNs are inspired by the living beings' nervous system and comprise at least two layers of interconnected artificial neurons or nodes operating in parallel [1]. The multilayer perceptron is regarded as the most widely used ANN architecture, where just forward connections occur. The relationships between the inputs and outputs are apprehended by adjusting the interconnection weights of the nodes through a feedback mechanism and an appropriate activation function. The optimum set of the interconnection weights' values is obtained through an iterative adjustment process during the training process [1]. In recent years, various studies have demonstrated the effectiveness of ANNs to solve complex problems in different backgrounds, such as health [3, 46, 69], water quality [19, 22] environment [31, 45], and industry [8, 56].

1.6 Materials and Methods

1.6.1 Research Design

Assessing consumer literacy regarding water quality control is crucial take into consideration the water quality—health interconnections. The present research was designed following the Water Consumption Literacy (WCL) model. This model considers three dimensions, i.e., *Water Quality*, *Disease Prevention* and *Sustainability/Public Health Promotion*. The assessment of each one was made through the competencies *obtain*, *understand*, *assess*, and *apply* information related to water issues [23]. *Obtain* deals with the aptness in seeking, finding, and accessing information. *Understand* is regards to the aptness in cognizing the information obtained. *Assess* is related to the aptness in interpreting, filtering, and evaluating information, whereas *apply* is associated with the aptness to use and share the information to make choices that contribute to sustainability and not compromise consumers health.

1.6.2 Data Collection

In order to meet the objectives defined above, the choice of a sufficiently versatile tool becomes necessary. Analyzing the strengths and weaknesses of each technique, the inquiry through questionnaire was adopt. The main reasons that support this decision are linked with its simplicity and versatility [17, 20, 49, 52].

A questionnaire (Annex A) was designed for this research to evaluate the lab customers' literacy regarding water quality thematic and its relationship with individual and public health as well as with sustainability. The questionnaire comprises two parts. The first one, with descriptive answers, contains issues that aims to characterize the cohort, whereas the last one uses a four levels Likert scale, and covers questions about the literacy of the lab customers regarding water thematic, encompassing the dimensions and competences mentioned to above (Table 1.1 and Annex A). The option for a Likert scale with four levels lies on the presumptive cohort size.

The questionnaire was validated following the methodology suggested by Bell [9]:

- The questionnaire was analyzed and evaluated by a set of specialists, composed by auditors, physicians, and nurses, who recommended some adjustments;
- The adjustments were incorporated into the questionnaire;
- The updated version of the questionnaire was then tested in a restricted set of customers to evaluate the validity and perceive problems in the understanding of questions; and
- The final version of the questionnaire was delivered to 468 customers.

A total of 453 questionnaires were received, corresponding to a return rate of 96.8%.

1.6.3 Location of Study

The research took place in the Municipality of Santiago do Cacém, Portugal, between September 2020 and December 2021, and was conducted at the Laboratory of Water Analysis. In addition to industry, agriculture is the predominant activity in the municipality. However, in the recent years, the municipality has been gaining relevance from the tourist point of view, namely through the promotion and enhancement of the natural heritage. In this municipality the underground abstractions are widely used for agriculture and/or animal feed, as well as for swimming pools, built by foreign residents next to their homes [54].

Table 1.1 Issues contained in the part II of the questionnaire categorized by *Dimensions (D)* and *Competencies (C)*

D	C		How easy would you say it is to …
Water quality	Obtain	Q1	… obtain information on locals where the water quality control can be carried out?
		Q2	… obtain information on the parameters to be considered when assessing the water quality?
		Q3	… obtain information on how to treat the water of private water abstractions?
		Q4	… obtain information on the meaning of the warning pictograms in water sources?
		Q5	… obtain information about the outcomes of water quality control of the public network of your residential area?
	Understand	Q6	… understand the importance of the water quality control?
		Q7	… understand the meaning/importance of each parameter used in water quality control?
		Q8	… understand the instructions to treat the water from private water abstractions?
		Q9	… understand the meaning of the warning pictograms next to the water sources?
		Q10	… understand the information about the outcomes of water quality control of the public network of your residential area?
	Assess	Q11	… assess the consequences of not linking your property to the public water network?
		Q12	… assess when the water of a private abstraction can be used for human consumption?
		Q13	… assess the need to carry out analyzes of water from private abstractions?
		Q14	… assess to the allowed uses of public fountain water based on warning pictograms?
		Q15	… assess the water quality of the public network in your residential area based on the results reported by the authorities?
	Apply	Q16	… apply the information in decision-making to connect your property to the public water network?
		Q17	… apply the information to decide on parameters to consider to control the quality of the waters from private abstractions?
		Q18	… apply the instructions for treatment of water from a private abstraction?
		Q19	… apply the instructions given by the warning pictograms next to the water sources?

(continued)

Table 1.1 (continued)

D	C		How easy would you say it is to …
Disease prevention	Obtain	Q20	… obtain information on the connections between the consumption of inappropriate water and health risks?
		Q21	… obtain information on disease symptoms and possible connections with inappropriate water use?
		Q22	… obtain information on the treatment of diseases caused by the consumption of inappropriate water?
		Q23	… obtain information on water quality parameters that are most relevant to health?
	Understand	Q24	… understand the health risks caused by the consumption of inappropriate water?
		Q25	… understand the information regarding the symptoms of diseases caused by the consumption of inappropriate water?
		Q26	… understand the health experts' indications to treat diseases caused by the consumption of inappropriate water?
		Q27	… understand the interrelations between the water quality parameters and health?
	Assess	Q28	… assess the health consequences of not linking your property to the public water supply?
		Q29	… assess whether a specific symptom is caused by the consumption of inappropriate water?
		Q30	… assess the need for specialized help in the face of symptoms that may be caused by the consumption use of inappropriate water?
		Q31	… assess which behaviors associated with the consumption of inappropriate water minimize the health risks?
	Apply	Q32	… apply the information about symptoms related with the consumption of inappropriate water in decision-making of seeking specialized help?
		Q33	… apply the guidelines of health experts to treat diseases caused by the consumption of inappropriate water?
		Q34	… adopt behaviors that avoid/minimize the health risks caused by the consumption of inappropriate water?
Sustainability/ public health promotion	Obtain	Q35	… obtain information on how to save water?
		Q36	… obtain information on water resources protection?
		Q37	… obtain information on how to reuse water?
		Q38	… obtain information on the legislation that regulates the water sector?

(continued)

Table 1.1 (continued)

D	C		How easy would you say it is to ...
	Understand	Q39	... understand the behaviors that must be adopted to save water?
		Q40	... understand the behaviors that must be adopted to protect water resources?
		Q41	... understand the behaviors that must be adopted to reuse water?
		Q42	... understand the legislation that regulates the water sector?
	Assess	Q43	... assess whether your everyday habits promote water saving?
		Q44	... assess the impact of human behaviors on water resources protection?
		Q45	... assess, in your daily routine, the situations in which it is feasible to reuse water?
		Q46	... assess whether legislation regulating water use contributes to its sustainable use?
	Apply	Q47	... change your daily habits aiming to save water?
		Q48	... advise family and/or neighbors on the behaviors to adopt to save water?
		Q49	... participate in community initiatives or activities on sustainable water use and protection of water resources?
		Q50	... change your daily habits aiming to reuse water?

1.6.4 Participants

This research involved an opportunity sample composed by 453 participants (lab customers), aged between 18 and 87 years old, with an average of 52 ± 34 years old. Regarding age distribution, 20.8% are under 25 years of age, 37.3% are aged in the interval 26–50 years of age, 28.7% in the range 51–70 years of age, whereas 13.2% are above 70 years of age. The percentage of male and female participants were 35.8% and 64.2%, respectively. In relation to the local of residence, 84.1% of participants came from rural zones while 15.9% were from urban areas. Relating to academic qualifications, 14.7% of the cohort finished basic education, 54.4% completed secondary education, 26.7% concluded a degree and 4.2% has a post-graduate level of education. With respect to the type of water to be analyzed, 48.1% of the lab customers declared to be water to consumption, 23.0% stated to be swimming pool water, 15.4% declared to be bathing water and 13.5% affirmed to be natural waters.

1.6.5 Qualitative Data Processing

The data gathered in the part II of the questionnaire is qualitative and expressed on a Likert scale with four levels. Aiming to convert the non-numerical into numerical one, the methodology suggested by Fernandes et al. [24] was followed. Thus, the answers to the k questions on a specific topic are detailed into circle of radius $1/\sqrt{\pi}$, divided into k sections, where each option of answer corresponds to a mark in the axis, as illustrated in the Water Consumers' Literacy Assessment section.

1.6.6 Artificial Neural Networks

In order to develop ANNs forecast models the WEKA software was used, retaining the parameters' default values [27, 32]. To ensure statistical significance of the results, thirty-six replicates were conducted in all tests. In each simulation, the data was randomly separated into two mutually exclusive partitions, i.e., the training set (with 67% of the available data), and the test set (with the remaining examples). The former set was used to generate the model, whereas the second one was used to investigate its generalization capability.

1.6.7 Ethical Aspects

This research was executed in conformity with the current legal regulations and was authorized by the responsible of the Laboratory of Water Analysis of Santiago do Cacém. All involved became aware of the research's aims and agrees to answer to the questionnaire.

1.7 Results and Discussion

1.7.1 Frequency of Response Analysis

Figure 1.1 presents the frequency of response to the questions that compose part II of the questionnaire. In this part the participants marked the alternative matching to their point of view, regarding the issues concerning each dimension (Table 1.1). The results regarding the former dimension (Q1–Q19) are displayed in Fig. 1.1a. A perusal of this figure shows that only for questions Q1 (*obtain* information on locals where to carry out water analysis) and Q6 (*understand* the relevance of carrying out periodic water quality control) the percentage of responses *very easy* or *easy* are higher than 50.0%. In the remaining questions the majority of participants chose the options that express difficulty, i.e., *difficult* or *very difficult*. It should be noted that the percentage of participants who select the *difficult/very difficult* options is higher than 82.5% in questions related to the highest-level skills such as *assess* and *apply*. The exception is question Q18 in which the percentage of answers expressing difficulty is 57.2%. It is interesting to highlight that in the items related with treatment of water from a private abstraction (Q3, Q8, Q13 and Q18) the competencies *obtain* (Q3) and *assess* (Q13) have more than 90% of answers indicating difficulty. In other words, the participants reveal difficulty in obtaining and assessing information, showing less difficulty in understanding and applying the information. Additionally, in the issues related to the parameters to consider in water quality control (Q2, Q7, Q12 and Q17) the results obtained demonstrate that more than 87.0% of participants reveal difficulties in all competences included in this research. Concerning the meaning of warning pictograms (Q4, Q9, Q14 and Q19) 54.5% of participants reveal difficulties in obtaining information (Q4). However, they reveal even more difficulty regarding *understand* (85.0%), *assess* (94.9%) and *apply* (84.6%) the information. With regard to the outcomes of water quality control of the public network (Q5, Q10 and Q15) the great majority of participants demonstrate difficulties in all competencies, i.e., *obtain* (84.1%), *understand* (89.6%) and *assess* (82.6%). These results are in accordance with the ones obtained in questions related with the parameters to consider in water quality control (Q2, Q7, Q12 and Q17). In fact, these results can indicate knowledge gaps on issues related to the parameters used in water quality control. A glance to the results obtained in this dimension shows that participants reveal a low literacy level regarding water quality issues, whatever the competence under review (i.e., *obtain, understand, assess,* or *apply*). The outcomes obtained in *Water Quality* dimension agree with those achieved by Gibson and Pieper [30], Maleckia et al. [48] and Knobeloch et al. [41]. According to these authors, most owners of private abstractions do not conduct water analysis even though their concerns with the environmental pollution problems do exist. Moreover, the results obtained in this research are in accordance with the ones obtained by Maleckia et al. [48], Fox et al. [26] and Kreutzwiser et al. [43]. These authors emphasize that the interpretation of water analysis reports is very hard for the majority of citizens.

Concerning *Disease Prevention*, the frequency of response (Fig. 1.1b) allows to state that only for questions Q26 (*understand* the health experts indications to treat diseases caused by the consumption of inappropriate water), Q33 (*apply* the guidelines of health experts to treat diseases caused by the consumption of inappropriate water), and Q34 (*adopt* behaviors that avoid/minimize the health risks caused by the consumption of inappropriate water) the percentage of responses *very easy* or *easy* are higher than 70.0%. However, it should also be emphasized that in questions Q20 (*obtain* information on the connections between the consumption of inappropriate water and health risks) and Q24 (*understand* the health risks caused by the consumption of inappropriate water) the percentage of responses *very easy* or *easy* was relatively high (above 40.0%). In the remaining questions most participants ticked the options *difficult* or *very difficult*. It also must be mentioned that the percentage of participants who select the *difficult/very difficult* options is higher than 83.0% in all questions related to the competence *assess*. Moreover, in question Q23

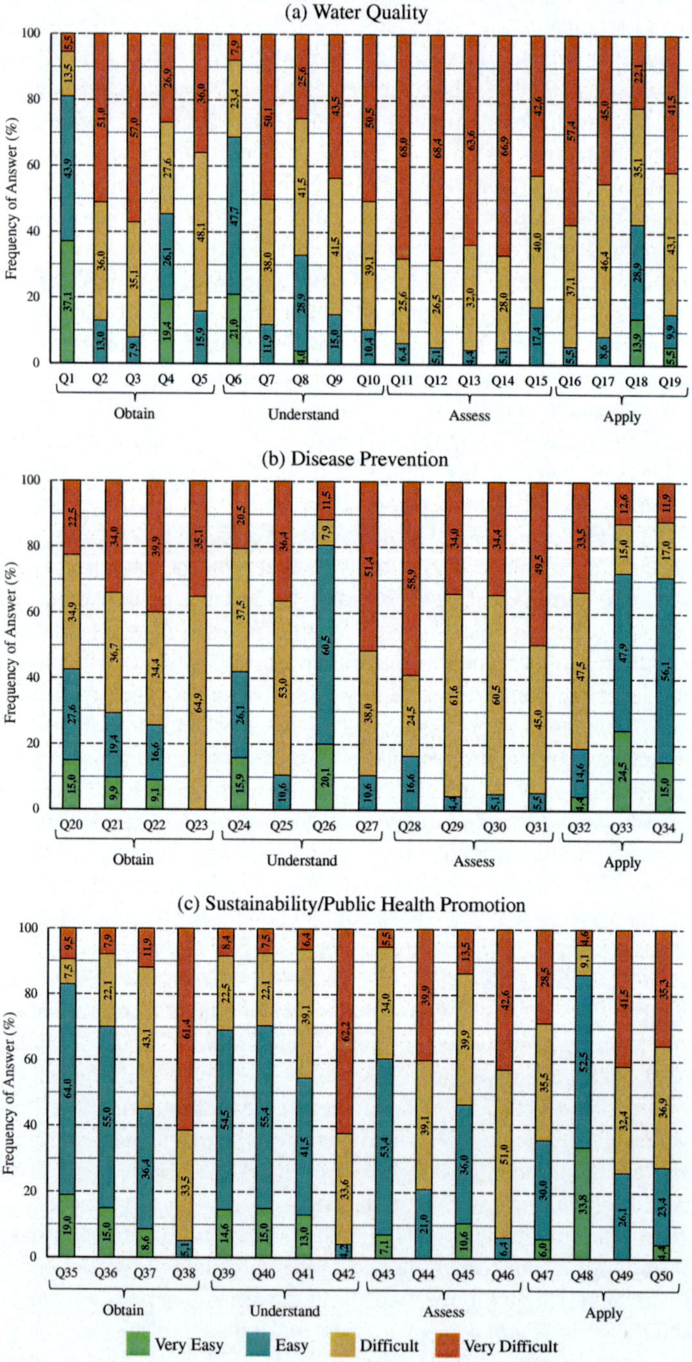

Fig. 1.1 Response frequency to the questions that compose the part II of the questionnaire

(*obtain* information on water quality parameters that are most relevant to health) all participants expressed difficulty.

An analysis of the frequency of response to the questions included in the dimension *Disease Prevention* allow to state that regarding the health risks that can arise from using inappropriate water (Q20, Q24, and Q28), more than 57.0% of participants reveal difficulty in obtaining (Q20) and understanding (Q24) information, expressing even more difficulty (83.4%) in assessing (Q28) information. Concerning the disease's signs related to the consumption of inappropriate water (Q21, Q25, Q29 and Q32) 70.7% of participants reveal difficulty in obtaining (Q21) information. However, they express even more difficulty in *understand* (89.4%), *assess* (95.6%), and *apply* (81.0%) the information. Concerning the treatment of diseases caused by the consumption of inappropriate water (Q22, Q26, Q30, and Q33) more than 72.0% of participants marked the options *very easy* or *easy* in the answers regarding the competencies *understand* (Q26) and *apply* (Q33). Conversely, 74.3% and 94.9% of participants reveal difficulty in the competencies *obtain* (Q22) and *assess* (Q30), respectively. In other words, the participants expressed difficulty in obtaining information and in assessing the need of specialized help, but in face of treatment instructions they reveal facility in understanding and applying the instructions. Finally, with regard to water quality parameters that are most relevant to health (Q23, Q27, Q31, and Q34), 71.1% of participants marked the options *very easy* or *easy* in the answers regarding the competence *apply* (Q34). On the contrary, more than 89.0% of the cohort reveal difficulty in *obtaining* (Q23), understanding (Q27), and *assessing* (Q31) information about this topic. Such as in the previous topic, these results show that the participants state to have difficulty in some competencies (i.e., *obtain, understand*, and *assess*), but when they know the behaviors that avoid/minimize the health risks they declare facility in their adoption.

Several studies address the assessment of the relationship between symptoms and the intake of untreated water and reveal lack of tests/treatment of the private water. The motives indicated by the holders are associated to the fact that the private water is pleasant to taste [25, 37, 59]. This is a major issue because some pollutants are tasteless, odorless and/or colorless. Additionally, the latency interval among the exposure and consequences makes the situation even more serious [55]. Consequently, assessing if a disease is associated to the consumption of untreated water is a hard task to be carried out by the people.

Concerning *Sustainability/Public Health Promotion*, the frequency of response (Fig. 1.1c) allows to state that only for questions Q38 (*obtain* information on the legislation that regulates the water sector), Q42 (*understand* the legislation that regulates the water sector), and Q46 (*assess* whether legislation regulating water use contributes to its sustainable use) the percentage of responses *difficult* or *very difficult* are higher than 90.0%. These results suggest that in the items related with legislation the participants reveal difficulty in obtaining, understanding, and assessing information. This fact can be justified considering the technical language used that is very specific and not accessible to the population in general. Conversely, in questions Q35, Q36, Q39, Q40, Q41, Q43, and Q48 the percentage of responses *very easy* or *easy* are higher than 50.0%. In the remaining questions the percentage of participants which marked the options *very easy* or *easy* range between 21.0% (Q44) and 46.6% (Q45).

The analysis of the answers to the questions that comprise the dimension *Sustainability/Public Health Promotion* topic by topic allow to state that concerning the topic related with water saving (Q35, Q39, Q43, Q47, and Q48) the participants expressed facility. A similar trend can also be found in issues related to the protection of water resources (Q36, Q40, Q44, and Q49), although in the competences *assess* and *apply* the percentage of participants revealing difficulty increase, 79.0% and 73.9%, respectively. Finally, regarding the topic related with water reuse (Q37, Q41, Q45, and Q50), about 50% of the participants reveal difficulty. This tendency is even more accentuated in the competence *apply*, where 72.2% of the participants marked the options *difficult* or *very difficult*. Thus, it is very important to sensitize the population to develop the appropriate behaviors regarding water reuse. Yu et al. [79] noted that the development of good practices relating water saving is more effective in populations with high water literacy levels. This population is more aware of issues related to water waste and water saving. The researchers state that the impact of consumer attitudes on water responsibility and ecological management is still small, mainly for financial reasons or time constraints. According to the researchers, the authorities should partially bear the costs related to water-saving devices and products, aiming to promote citizens' positive attitudes leading to water sustainability.

1.7.2 Water Consumers' Literacy Assessment

The quantification of non-numerical information concerning participant one is present in Fig. 1.2 and was conducted according to the methodology suggested by Fernandes et al. [24]. Aiming to demonstrate the procedure, the answers to the questions included in the competence *obtain* from the dimension *Water Quality*, given by participant one, were considered. Thus, the answers to the five questions included in the competence mentioned above was detailed into a circle of radius $1/\sqrt{\pi}$, divided into five sections, where each option of answer corresponds to a mark in the axis (Fig. 1.2).

Considering that answers to the questions Q1 and Q4 were *very easy* the corresponding areas were computed as $\frac{1}{5}\pi\left(\frac{4}{4\sqrt{\pi}}\right)^2 = 0.200$. The answers to the questions Q2 and Q5 were *easy* and the corresponding areas are $\frac{1}{5}\pi\left(\frac{3}{4\sqrt{\pi}}\right)^2 = 0.1125$. Finally, for the question Q3 the answer was *difficult*, and the corresponding area is $\frac{1}{5}\pi\left(\frac{2}{4\sqrt{\pi}}\right)^2 = 0.050$. The quantitative value corresponding to the answers of participant one to the question included in the competence *obtain*, from the dimension *Water Quality*, is the sum of the partial areas, i.e., 0.675 (Table 1.2). For the remaining competences and dimensions the process is analogous, and the values computed are presented in Table 1.2. The values shown in Table 1.2 were the input variables in the training of ANNs to estimate Water Consumers' Literacy.

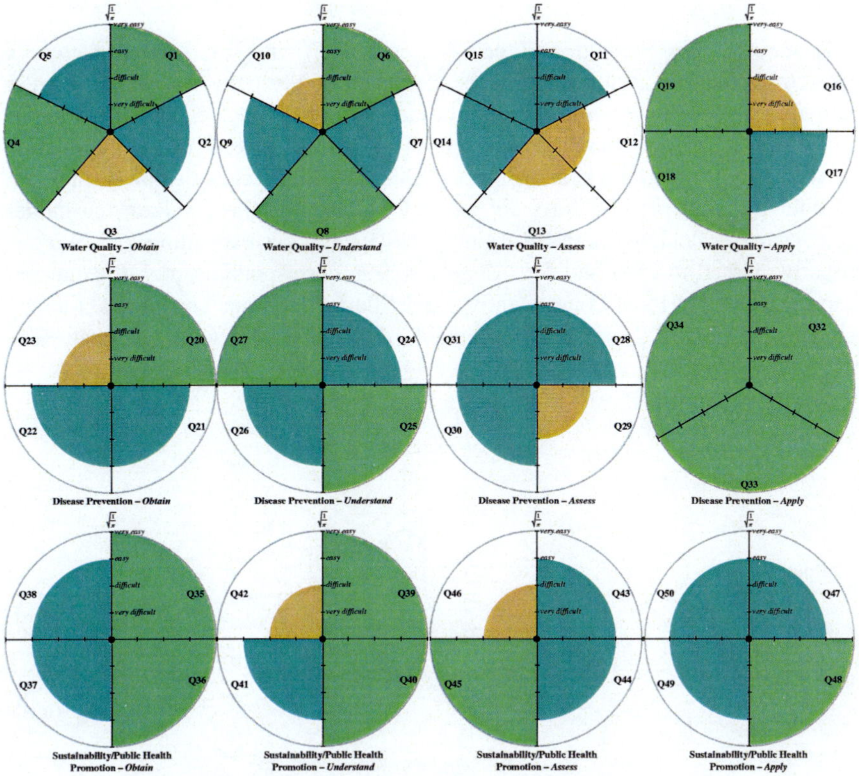

Fig. 1.2 A schematic representation of the quantification process of non-numerical information concerning participant one

Table 1.2 Excerpt of the data base used in water consumers' literacy evaluation

Dimension	Competence	Participant 1	Participant 2	…	Participant 453
Water quality	Obtain	0.675	0.450	…	0.238
	Understand	0.675	0.525		0.238
	Assess	0.438	0.238		0.162
	Apply	0.703	0.594		0.281
Disease prevention	Obtain	0.594	0.594	…	0.234
	Understand	0.781	0.469		0.359
	Assess	0.484	0.281		0.156
	Apply	1.0	0.708		0.458
Sustainability//public health prevention	Obtain	0.781	0.594	…	0.359
	Understand	0.703	0.547		0.359
	Assess	0.594	0.406		0.234
	Apply	0.672	0.547		0.281

To attain the best prediction of the water consumers' literacy, different networks were conceived and evaluated. The performance of the networks was evaluated computing the accuracy values, based on the matrix of coincidences [74]. Among the various network topologies evaluated, the 12-8-3-1 network (Fig. 1.3) was the one that demonstrated the best response (Table 1.3). The results displayed in Table 1.3 refer to an average of thirty-six replicates. Based on these values, the model accuracy was calculated both for training set (93.3%, corresponding to 279 cases properly classed in 299) and for test set (90.9%, corresponding to 140 examples properly labeled in 154). Consequently, the predictions of water consumers' literacy using the 12-8-3-1 ANN model are acceptable, attaining accuracies greater than 90%.

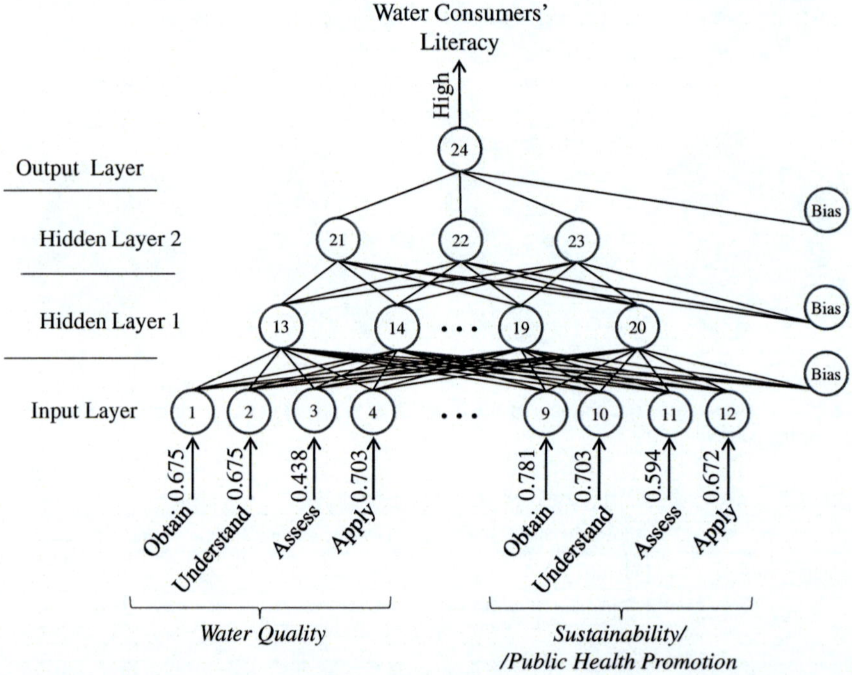

Fig. 1.3 A schematic representation of the 12-8-3-1 ANN model for water consumers' literacy evaluation

Table 1.3 Matrix of coincidences regarding 12-8-3-1 ANN model for water consumers' literacy evaluation

Predict	Training			Test		
Target	Low	Medium	High	Low	Medium	High
Low	169	8	2	87	4	1
Medium	5	70	3	4	35	2
High	0	2	40	0	3	18

Aiming to explore the influence of the inputs on output, the sensitivity based on variance was computed [40]. This analysis allows the computation of the Relative Importance (RI) of each input to infer about the effects of each competence and each dimension on water consumers' literacy. The results obtained indicate that the water consumers' literacy is more affected by the competence *assess* of the *Water Quality* dimension (RI = 0.20) and by the same competence of the *Disease Prevention* dimension (RI = 0.19), followed by the competence *apply* of the *Sustainability/ Public Health Promotion* dimension (RI = 0.12). These results are supported by the ones shown in Fig. 1.1. The competence *assess* of the dimensions *Water Quality* and *Disease Prevention* presents great frequencies of answers *difficult* and *very difficult* and, therefore, a slight change in these answers (questions Q11–Q15 and Q28–31) can have a strongly impact in the water consumers' literacy assessment.

According to the results presented in Table 1.3, only 13.9% of the participants reveal *high* levels of literacy regarding water issues, whereas the remaining reveal levels of literacy that are *medium* (26.3%) or *low* (59.8%). Among the participants who declare having higher education or post graduate education, 39.3% reveal *high* levels of literacy, whereas 60.7% show *medium* levels of literacy. The majority of participants who stated to have basic or secondary education show *low* levels of literacy (86.6%). Only 2.5% reveal *high* levels of literacy, whereas 10.9% show *medium* levels of literacy. A similar analysis considering the age groups demonstrates that 14.1% and 13.7% of participants that reveal *high* levels of literacy are under 50 years of age and above 50 years of age, respectively. Regarding the participants that reveal other levels of literacy, the results follow the same trend, i.e., 26.2% and 26.3% for *medium* levels of literacy and 59.7% and 60.0% for *low* levels of literacy, respectively.

1.8 Conclusions

Nowadays, the sustainable use of water is an emergent problem. However, this goal can become unattainable if the population is not led to adopt behaviors that can contribute to this purpose. To achieve such aim, it is crucial that the population becomes aware of issues related to the water problematic. Indeed, the adhesion of population to this kind of actions are more efficient if its knowledge on these themes is deeper. Thus, to know the water consumer's literacy on water issues it is essential to design strategies that lead to sustainable behaviors. The present work presents a methodology that can contribute to this goal and present a predictive model for water consumers' literacy management based on the ANN paradigm to computing. The proposed approach reveals an acceptable efficacy, attaining accuracies higher than 90%. The suggested methodology deals with data collected using the questionnaire survey technique, allowing the incorporation of various competences and dimensions. Moreover, the results permitted the recognition of the dimensions and competences in which the participants reveal more difficulties. Thus, the outcomes allow the planning and development of strategies to enhance water consumers' literacy, giving a contribute to the 6th United Nation Sustainable Development Goal, which aims to ensure the availability of water and its sustainable management. The results indicate reasonable levels of literacy in relation to the topics included in the dimension *Sustainability/Public Health Promotion*. Such fact can be justified considering that these topics have deserved greater attention/dissemination by the media. Conversely, the participants reveal very low levels of literacy in relation to other topics, associated with competence *assess* of the *Water Quality* and *Disease Prevention* dimensions. Indeed, the participants demonstrated difficulty when evaluating the usefulness of the solutions/information. In addition, this research led to the identification of other problematic issues related with water quality reports and with water legislation. These constraints are improvement opportunities and different actions can be put into practice, such as the publication of analytical results complemented with easy and perceptive descriptions. Moreover, training activities on water quality and water legislation (e.g., analytical periodicity, parameters to be considered, and allowed values) can be conducted. Regarding the problem of water scarcity, it is especially important to insist on publicity campaigns about the best practices related to water resources protection, saving water, and reusing water. The use of an opportunity sample can be considered the main limitation of this research as well as the reduced number of participants. In future developments the cohort should be extended to comprise general population aiming to validate the outcomes. The methodology followed, grounded on the connectionist paradigm, can be adopted by all types of organizations at different territorial contexts. However, the dimensions considered needs to be adapted to the nature of the organization. The use of this methodology allows the identification of potential improvements, allowing decision-makers to select the target topics that lead to the increase of the water consumers' literacy.

Acknowledgements This work received support from PT national funds (FCT/MCTES, Fundação para a Ciência e Tecnologia and Ministério da Ciência, Tecnologia e Ensino Superior) through the projects UIDB/50006/2020, UIDP/50006/2020, UIDB/04312/2020, UIDP/04312/2020 and UIDB/00319/2020.

Annex A

QUESTIONNAIRE ON WATER QUALITY AND ITS RELATIONS WITH HEALTH AND SUSTAINABILITY

SECTION I

RESPONDENT DATA

Mark the box or fill the blanks.

Gender Female ☐ Male ☐

Academic Qualifications	Basic Education	Secondary Education	Higher Education	Post Graduate Education
	☐	☐	☐	☐

Age ____________ **years old** **Locality** ____________

Type of Water to be analyzed

Water Consumption (reservoir; water network; tap water). ☐

Natural Waters (abstraction, lakes, rivers and groundwater). ☐

Bathing Water (fresh and salt water). ☐

Swimming Pool Water. ☐

QUESTIONNAIRE ON WATER QUALITY AND ITS RELATIONS WITH HEALTH AND SUSTAINABILITY

SECTION II

Using the scale *very easy*, *easy*, *difficult*, and *very difficult*, how easy would you say it is to …

WATER QUALITY

	Very Easy	Easy	Difficult	Very Difficult
Q1 ... obtain information on locals where the water quality control can be carried out?	☐	☐	☐	☐
Q2 ... obtain information on the parameters to be considered when assessing the water quality?	☐	☐	☐	☐
Q3 ... obtain information on how to treat the water of private water abstractions?	☐	☐	☐	☐
Q4 ... obtain information on the meaning of the warning pictograms in water sources?	☐	☐	☐	☐
Q5 ... obtain information about the outcomes of water quality control of the public network of your residential area?	☐	☐	☐	☐
Q6 ... understand the importance of the water quality control?	☐	☐	☐	☐
Q7 ... understand the meaning/importance of each parameter used in water quality control?	☐	☐	☐	☐
Q8 … understand the instructions to treat the water from private water abstractions?	☐	☐	☐	☐
Q9 ... understand the meaning of the warning pictograms next to the water sources?	☐	☐	☐	☐
Q10 ... understand the information about the outcomes of water quality control of the public network of your residential area?	☐	☐	☐	☐
Q11 ... assess the consequences of not linking your property to the public water network?	☐	☐	☐	☐
Q12 ... assess when the water of a private abstraction can be used for human consumption?	☐	☐	☐	☐
Q13 ... assess the need to carry out analyzes of water from private abstractions?	☐	☐	☐	☐
Q14... assess to the allowed uses of public fountain water based on warning pictograms?	☐	☐	☐	☐
Q15 ... assess the water quality of the public network in your residential area based on the results reported by the authorities?	☐	☐	☐	☐
Q16 … apply the information in decision-making to connect your property to the public water network?	☐	☐	☐	☐
Q17 ... apply the information to decide on parameters to consider to control the quality of the waters from private abstractions?	☐	☐	☐	☐
Q18 ... apply the instructions for treatment of water from a private abstraction?	☐	☐	☐	☐
Q19 ... apply the instructions given by the warning pictograms next to the water sources?	☐	☐	☐	☐

Questionnaire on Water Quality and its Relations with Health and Sustainability

Disease Prevention

	Very Easy	Easy	Difficult	Very Difficult
Q20 ... obtain information on the connections between the consumption of inappropriate water and health risks?	☐	☐	☐	☐
Q21 ... obtain information on disease symptoms and possible connections with inappropriate water use?	☐	☐	☐	☐
Q22 ... obtain information on the treatment of diseases caused by the consumption of inappropriate water?	☐	☐	☐	☐
Q23 ... obtain information on water quality parameters that are most relevant to health?	☐	☐	☐	☐
Q24 ... understand the health risks caused by the consumption of inappropriate water?	☐	☐	☐	☐
Q25 ... understand the information regarding the symptoms of diseases caused by the consumption of inappropriate water?	☐	☐	☐	☐
Q26 ... understand the health experts' indications to treat diseases caused by the consumption of inappropriate water?	☐	☐	☐	☐
Q27 ... understand the interrelations between the water quality parameters and health?	☐	☐	☐	☐
Q28 ... assess the health consequences of not linking your property to the public water supply?	☐	☐	☐	☐
Q29 ... assess whether a specific symptom is caused by the consumption of inappropriate water?	☐	☐	☐	☐
Q30 ... assess the need for specialized help in the face of symptoms that may be caused by the consumption of inappropriate water?	☐	☐	☐	☐
Q31 ... assess which behaviors associated with the consumption of inappropriate water minimize the health risks?	☐	☐	☐	☐
Q32 ... apply the information about symptoms related with the consumption of inappropriate water in decision-making of seeking specialized help?	☐	☐	☐	☐
Q33 ... apply the guidelines of health experts to treat diseases caused by the consumption of inappropriate water?	☐	☐	☐	☐
Q34 ... adopt behaviors that avoid/minimize the health risks caused by the consumption of inappropriate water?	☐	☐	☐	☐

Sustainability/Public Health Promotion

	Very Easy	Easy	Difficult	Very Difficult
Q35 ... obtain information on how to save water?	☐	☐	☐	☐
Q36 ... obtain information on water resources protection?	☐	☐	☐	☐
Q37 ... obtain information on how to reuse water?	☐	☐	☐	☐
Q38 ... obtain information on the legislation that regulates the water sector?	☐	☐	☐	☐
Q39... understand the behaviors that must be adopted to save water?	☐	☐	☐	☐
Q40 ... understand the behaviors that must be adopted to protect water resources?	☐	☐	☐	☐
Q41 ... understand the behaviors that must be adopted to reuse water?	☐	☐	☐	☐
Q42 ... understand the legislation that regulates the water sector?	☐	☐	☐	☐
Q43 ... assess whether your everyday habits promote water saving?	☐	☐	☐	☐

QUESTIONNAIRE ON WATER QUALITY AND ITS RELATIONS WITH HEALTH AND SUSTAINABILITY

	Very Easy	Easy	Difficult	Very Difficult
Q44 ... assess the impact of human behaviors on water resources protection?	☐	☐	☐	☐
Q45 ... assess, in your daily routine, the situations in which it is feasible to reuse water?	☐	☐	☐	☐
Q46 ... assess whether legislation regulating water use contributes to its sustainable use?	☐	☐	☐	☐
Q47 ... change your daily habits aiming to save water?	☐	☐	☐	☐
Q48 ... advise family and/or neighbors on the behaviors to adopt to save water?	☐	☐	☐	☐
Q49 ... participate in community initiatives or activities on sustainable water use and protection of water resources?	☐	☐	☐	☐
Q50 ... change your daily habits aiming to reuse water?	☐	☐	☐	☐

Thank you for your cooperation!

References

1. Aggarwal C (2018) Neural networks and deep learning: a textbook. Springer, Cham
2. Akinsete E, Apostolaki S, Chatzistamoulou N, Koundouri P, Tsani S (2019) The link between ecosystem services and human wellbeing in the implementation of the European water framework directive: assessing four river Basins in Europe. Water 11:508
3. Alshayeji M, Ellethy H, Abed S, Gupta R (2022) Computer-aided detection of breast cancer on the Wisconsin dataset: an artificial neural networks approach. Biomed Signal Process Control 71:103141
4. Apostolaki S, Akinsete E, Tsani S, Koundouri P, Pittis N, Levantis E (2019) Assessing the effectiveness of the WFD as a tool to address different levels of water scarcity based on two case studies of the Mediterranean region. Water 11:840
5. Apostolaki S, Koundouri P, Pittis N (2019) Using a systemic approach to address the requirement for integrated water resource management within the water framework directive. Sci Total Environ 679:70–79
6. Bagchi B (2013) Water in biological and chemical processes: from structure and dynamics to function. Cambridge University Press, New York
7. Baird RB, Eaton AD, Rice EW (2017) Standard methods for examination of water and wastewater, 23rd edn. American Public Health Association/Water Works Association/Water Environment Federation Edition, New York
8. Batista L, Marques C, Pires A, Minim L, Soares N, Vidigal M (2021) Artificial neural networks modeling of non-fat yogurt texture properties: effect of process conditions and food composition. Food Bioprod Process 126:164–174
9. Bell J (2010) Doing your research project: a guide for first-time researchers in education, health and social science, 5th edn. Open University Press, Maidenhead
10. Bell R, Morris S, Barrow J, Finney J, Harper C (2010) Water and life—The unique properties of H_2O. CRC Press, Boca Raton
11. Berardo A, Meyer M, Olivier T (2013) Adaptive governance and integrated water resources management in Argentina. Int J Water Gov 1:219–236
12. Biswas AK, Tortajada C (2010) Future water governance: problems and perspectives. Int J Water Resour Dev 26:129–139

13. Branscomb AW (1981) Knowing how to know. Sci Technol Hum Values 6:5–9
14. Champagne AB, Lovitts BE (1989) Scientific literacy: a concept in search of definition. In: Champagne AB, Lovitts BE, Callinger BJ (eds) This year in school science: scientific literacy. AAAS, Washington DC, pp 1–14
15. Chereni A (2007) The problem of institutional fit in integrated water resources management: a case of Zimbabwe's Mazowe catchment. Phys Chem Earth 32:1246–1256
16. Claassen M (2013) Integrated water resource management in South Africa. Int J Water Gov 1:323–338
17. Cohen L, Manion L, Morrison K (2017) Research methods in education, 8th edn. Routledge, New York
18. Costa JM, Oliveira M, Egipto RJ, Cid JF, Fragoso RA, Lopes CM, Duarte EN (2020) Water and wastewater management for sustainable viticulture and oenology in south Portugal—A review. J Viticult Enol 35(1):1–15
19. Dawood T, Elwakil E, Hector H, Delgado J (2021) Toward urban sustainability and clean potable water: prediction of water quality via artificial neural networks. J Clean Prod 291:125266
20. DeKetele J-M, Roegiers X (2016) Méthodologie du Recueil d'Informations: Fondements des Méthodes d'Observation, de Questionnaire, d'Interview et d'Études de Documents (Methodology of Information Gathering: Foundations of Observation, Questionnaire, Interview and Document Review Methods), 5th edn. DeBoeck Universite, Paris
21. Durant JR (1993) What is scientific literacy? In: Durant JR, Gregory J (eds) Science and culture in Europe. Science Museum, London, pp 129–137
22. Fernandes A, Chaves H, Lima R, Neves J, Vicente H (2020) Draw on artificial neural networks to assess and predict water quality. IOP Conf Ser: Earth Environ Sci 612:012028
23. Fernandes A, Figueiredo M, Neves J, Vicente H (2022) A conceptual model to assess the literacy of water consumers. Water Econ Policy 8:2250007
24. Fernandes A, Vicente H, Figueiredo M, Neves M, Neves J (2016) An evaluative model to assess the organizational efficiency in training corporations. In: Dang T, Wagner R, Küng J, Thoai N, Takizawa M, Neuhold E (eds) Future data and security engineering, Lecture notes in computer science, vol 10018. Springer, Cham, pp 415–428
25. Flanagan S, Spayd S, Procopio N, Chillrud S, Braman S, Zheng Y (2016) Arsenic in private well water part 1 of 3: impact of the New Jersey Private well testing act on household testing and mitigation behavior. Sci Total Environ 562:999–1009
26. Fox M, Nachman K, Anderson B, Lam J, Resnick B (2016) Meeting the public health challenge of protecting private wells: proceedings and recommendations from an expert panel workshop. Sci Total Environ 554:113–118
27. Frank E, Hall M, Witten IH (2022) The WEKA workbench. Online appendix for "data mining: practical machine learning tools and techniques" Morgan Kaufmann, 4th edn; 2016. Accessed 26 Oct 2022. https://www.cs.waikato.ac.nz/ml/weka/Witten_et_al_2016_appendix.pdf
28. García-Rubio MA, Tortajada C, González-Gómez F (2016) Privatising water utilities and user perception of tap water quality: evidence from Spanish urban water services. Water Resour Manag 30:315–329
29. Giakoumis T (2019) Improving the application of the integrated river basin management paradigm in the implementation of the EU water framework directive. Ph.D. thesis, Imperial College, London
30. Gibson J, Pieper K (2017) Strategies to improve private-well water quality: a North Carolina perspective. Environ Health Perspect 125:076001
31. Goudarzi G, Hopke P, Yazdani M (2021) Forecasting $PM_{2.5}$ concentration using artificial neural network and its health effects in Ahvaz, Iran. Chemosphere 283:131285
32. Hall M, Frank E, Holmes G, Pfahringer B, Reutemann P, Witten IH (2009) The WEKA data mining software: an update. SIGKDD Explor 11:10–18
33. Hartley K, Tortajada C, Biswas AK (2018) Political dynamics and water supply in Hong Kong. Environ Dev 27:107–117
34. Hartley K, Tortajada C, Biswas AK (2019) A formal model concerning policy strategies to build public acceptance of potable water reuse. J Environ Manag 250:109505

35. Harvard University (2022) Harvard university sustainability plan—Fiscal year 2015–2022. Harvard University, Harvard
36. Horne J, Tortajada C, Harrington L (2018) Achieving the sustainable development goals: improving water services in cities affected by extreme weather events. Int J Water Resour Dev 34:475–489
37. Imgrund K, Kreutzwiser R, de Loë R (2011) Influences on the water testing behaviors of private well owners. J Water Health 9:241–252
38. Intergovernmental Panel on Climate Change (2015) Climate Change 2014: mitigation of climate change. Cambridge University Press, Cambridge
39. Intergovernmental Panel on Climate Change (2021) Climate Change 2021: the physical science basis. Contribution of Working Group I to the Sixth Assessment Report of the Intergovernmental Panel on Climate Change, Cambridge University Press, Cambridge
40. Kewley R, Embrechts M, Breneman C (2000) Data strip mining for the virtual design of pharmaceuticals with neural networks. IEEE Trans Neural Netw 11:668–679
41. Knobeloch L, Gorski P, Christenson M, Anderson H (2013) Private drinking water quality in rural Wisconsin. J Environ Health 75:16–20
42. Kowalczak P, Maczak P, Slavikova L (2013) Institutional evolution in water management in the Czech Republic and Poland. Int J Water Gov 1:307–322
43. Kreutzwiser R, de Loë R, Imgrund K, Conboy M, Simpson H, Plummer R (2011) Understanding stewardship behaviour: factors facilitating and constraining private water well stewardship. J Environ Manag 92:1104–1114
44. Lamastra L, Suciu AN, Novelli E, Trevisan M (2013) A new approach to assessing the water footprint of wine: an Italian case study. Sci Total Environ 490:748–756
45. Li A, Qi M, Li W, Yu X, Yang L, Wang J, Li D (2022a) Prediction and verification of the effect of psoriasis on coronary heart disease based on artificial neural network. Heliyon 8:e10677
46. Li C, Zhang C, Yu T, Liu X, Yang Y, Hou Q, Yang Z, Ma X, Wang L (2022b) Use of artificial neural network to evaluate cadmium contamination in farmland soils in a karst area with naturally high background values. Environ Pollut 304:119234
47. Lundin M, Morrison GM (2002) A life cycle assessment based procedure for development of environmental sustainability indicators for urban water systems. Urban Water 4(2):145–152
48. Maleckia KM, Schultz AA, Severtson DJ, Anderson HA, VanDerslice JA (2017) Private-well stewardship among a general population based sample of private well-owners. Sci Total Environ 601–602:1533–1543
49. McMillan J, Schumacher S (2009) Research in education: evidence-based inquiry, 7th edn. Prentice Hall, New York
50. National Research Council (2011) Sustainability and the U.S. EPA. The National Academies Press, Washington
51. Norris S, Phillips L (2003) How literacy in its fundamental sense is central to scientific literacy. Sci Educ 87(2):224–240
52. Patton MQ (2015) Qualitative research and evaluation methods: integrating theory and practice, 4th edn. SAGE Publications Inc., Thousand Oaks
53. Paulo AA, Rosa RD, Pereira LS (2012) Climate trends and behaviour of drought indices based on precipitation and evapotranspiration in Portugal. Nat Hazard 12:1481–1491
54. Pordata (n.d.) PORDATA—Statistics on Portugal and Europe, Accessed 12 Oct 2022. https://www.pordata.pt/en
55. Rubin G, Amlôt R, Page L, Pearce J, Wessely S (2013) Assessing perceptions about hazardous substances (PATHS): the PATHS questionnaire. J Health Psychol 18:1100–1113
56. Salgado C, Dam R, Puertas E, Salgado W (2022) Calculation of volume fractions regardless scale deposition in the oil industry pipelines using feed-forward multilayer perceptron artificial neural network and MCNP6 code. Appl Radiat Isot 185:110215
57. Sihvonen S, Partanen J (2016) Implementing environmental considerations within product development practices: a survey on employees' perspectives. J Clean Prod 125:189–203
58. Stavenhagen M, Buurman J, Tortajada C (2018) Saving water in cities: assessing policies for residential water demand management in four cities in Europe. Cities 79:187–195

59. Stillo F, Gibson J (2017) Exposure to contaminated drinking water and health disparities in North Carolina. Am J Public Health 107:180–185
60. The Organization for Economic Cooperation and Development (2011) Towards green growth. Organization for Economic Cooperation and Development Publishing, Paris
61. The Organization for Economic Cooperation and Development (2015) OECD principles on water governance. Organization for Economic Cooperation and Development Publishing, Paris
62. The Organization for Economic Cooperation and Development (2018) Implementing the OECD principles on water Governance—Indicator framework and evolving practices. Organization for Economic Cooperation and Development Publishing, Paris
63. The Water and Waste Services Regulation Authority (2017) Use of private water abstractions for human consumption. Awareness Notebooks nº 4, The Water and Waste Services Regulation Authority Publishing, Lisbon
64. Tortajada C, Joshi YK (2013) Water resources management and governance as part of an overall framework for growth and development. Int J Water Gov 1:285–306
65. Tortajada C, Joshi YK (2013) Water demand management in Singapore: involving the public. Water Resour Manag 27:2729–2746
66. Tortajada C, Joshi YK (2014) Water quality management in Singapore: the role of institutions, laws and regulations. Hydrol Sci J 59:1763–1774
67. Tortajada C, Nambiar S (2019) Communications on technological innovations: potable water reuse. Water 11:251
68. Tortajada C, González-Gómez F, Biswas AK, Buurman J (2019) Water demand management strategies for water-scarce cities: the case of Spain. Sustain Cities Soc 45:649–656
69. Tumpa P, Kabir MA (2021) An artificial neural network based detection and classification of melanoma skin cancer using hybrid texture features. Sens Int 2:100128
70. United Nations (1993) Report of the United Nations Conference on environment and development—Rio de Janeiro, 1992, vol I. In: Resolutions adopted by the conference. United Nations Edition, New York
71. United Nations (n.d.) Transforming our world: the 2030 agenda for sustainable development. United Nations Publishing, New York
72. United Nations Development Programme (1999) Human development report. Oxford University Press, New York
73. van Buuren A, van Meerkerk I, Tortajada C (2019) Understanding emergent participation practices in water governance. Int J Water Resour Dev 35:367–382
74. Vilhena J, Vicente H, Martins MR, Grañeda J, Caldeira F, Gusmão R, Neves J, Neves J (2017) An artificial intelligence approach to thrombophilia risk. Int J Reliable Qual E-Healthc 6:48–68
75. Wang Y, Wan T, Tortajada C (2018) Water demand framework and water development: the case of China. Water 10:1860
76. Wegerich K, Warner J, Tortajada C (2014) Water sector governance: a return ticket to anarchy. Int J Water Gov 2:7–20
77. World Health Organization & United Nations Children's Fund (2017) Progress on drinking water, sanitation and hygiene: 2017 update and SDG baselines. World Health Organization and United Nations Children's Fund Publishing, Geneva
78. World Meteorological Organization (2021) State of climatic services—Water. World Meteorological Organization Edition, Geneva
79. Yu J-H, Lin H-H, Lo Y-C, Tseng K-C, Hsu C-H (2021) Measures to cope with the impact of climate change and drought in the island region: a study of the water literacy awareness, attitude, and behavior of the Taiwanese public. Water 13:1799

Chapter 2
Machine Learning Applications in Sustainable Water Resource Management: A Systematic Review

Rukhsar Anjum, Farhana Parvin, and Sk Ajim Ali

Abstract For all living things, water is one of the most important supplies. We require water for a variety of things, including food production, personal hygiene, electrical generation, fire control and most importantly, survival. Despite being a renewable resource, clean water shortage is a major problem in many regions of the world. Sustainable water management strategies and techniques are therefore more important than ever. Evolving digital technologies may encourage effective observation, management, optimization and foresee of freshwater use and pollution in line with the SDG for reasonable access to water and wise usage of nature driven resources. Evidently, smart water management has been accelerated by the use of sensors, machine learning, Internet of Things (IoT) and big data analytics. The goal of the current study was to identify scientific literatures that may contribute to the management of water resources. By conducting a comprehensive examination of the pertinent literature, this review aimed to educate researchers and practitioners about the benefits and drawbacks of machine learning and data analysis methodologies. The review found important research gaps in the application of cutting-edge solutions, which may result in water savings and more effective demand management.

Keywords Machine learning · Water resource management · Water conservation · Sustainability · Climate change · Systematic review

R. Anjum · F. Parvin (✉) · S. A. Ali (✉)
Department of Geography, Faculty of Science, Aligarh Muslim University, Aligarh 202002, India
e-mail: farhanaparvin93@gmail.com; fparvin@myamu.ac.in

S. A. Ali
e-mail: skajimali.saa@gmail.com; skajimali@myamu.ac.in

R. Anjum
e-mail: rukhsar.amu3@gmail.com

F. Parvin
School of Liberal Arts, Noida International University, Noida, UP 203201, India

E. Balaji et al. (eds.), *Emerging Technologies for Water Supply, Conservation and Management*, Springer Water, https://doi.org/10.1007/978-3-031-35279-9_2

Abbreviations

ANN	Artificial Neural Network
ANFIS	Adaptive Neuro-fuzzy Inference System
MLP	Multi-layer perceptron
ELM	Extreme Learning Machine
SVM	Support vector machine
MLR	Multiple Linear Regression
POAM	Predictive Ocean Atmosphere Model
BFtree	Best-First tree
AB	AdaBoost
MB	MultiBoosting
SVR	Support Vector Regression
NMFk	Non-negative Matrix Factorization with k-means clustering
SAE-ELM	Self-Adaptive Evolutionary Extreme Learning Machine
GA	Genetic algorithm
GWO	Gray Wolf Optimization
ICSOA-PPE	Improved Chicken Swarm Optimization Algorithm
RBF	Radial Basis Function
SELA	Shuffled Frog Leaping Algorithm
GWO	Grey Wolf Optimization
LR	Logistic Regression
RF	Random Forest
GMCR	Graph Model for Conflict Resolution
NSGA-II	Non-dominated Sorting Genetic Algorithm-II
MARS	Multivariate Adaptive Regression Spline
GEP	Gene-Expression Programming
SDG	Sustainable Development Goals
EPR	Evolutionary Polynomial Regression

2.1 Introduction

In order to maintain the environmental, economic and hydrological reliability of the water resources for the current society and the future perspectives, sustainable water resource management is devised and implemented [3, 22, 23]. Because of their strong potential for reasoning, flexibility, modeling and anticipating the water demand and size, artificial intelligence techniques are widely used in urban water resource planning [97]. For efficient environmental planning and management, water resources are regarded as a vital component in economic and social progress [13]. Water resources must be wisely distributed, used to determine the effects on the ecosystem and provide a unique challenge when developing ecological systems [47, 48, 53].

Climate change, urbanization and population expansion are all contributing factors to rising water demand, which must be properly managed for the diverse and complicated urban water resources utilizing current technology platforms [49, 68]. Progressive socio-economic improvements for environmental sustainability may be made possible by technology solutions that support the creation of a sustainable environment [8, 11]. These long-standing technical methods for managing water resources and adopting sustainable practices frequently correspond to conventional water governance [39].

Machine learning (ML) and data-driven modeling techniques have been developed in recent years to anticipate sustainable water resource management [63, 72, 86]. This kind of model may take a look at certain unforeseen hydrological words throughout the modeling process, in light of data-driven information, these terms might be seen as the hydrological phenomenon [78]. The accuracy of these ML approaches in modeling water management, however, was demonstrated to be outstanding [1, 64, 77, 86].

Accurate water resource modeling is the primary study field that impacts planning for water supply, water conservation, water management, water quality assessment etc. [5, 39, 65, 76, 80, 84]. It is scientifically established that this is quite challenging to forecast how sustainability and WRM would behave owing to the physical methods and natural modifications associated to the ground and surface water system [50, 99]. The need to increase the accuracy and dependability of hydrological variable forecast is received a lot of attention in hydrological applications [86]. Due to many physical phenomena, including patterns, arbitrariness or periodicity in model input and goal data, as well as natural randomness in general, ML techniques are often a method that may be used to mimic water related processes under various settings [66, 80]. Given this perspective, it may also be assumed that no one generic model outperforms all others when dealing with a variety of scenarios and perspective [3]. Due to these drawbacks, researchers favor researching and creating more robust and generic models in order to enhance performance utilizing existing historical data. Additionally, scientists must take into account the advantages of sophisticated and quickly developing computational capacity that can improve modeling approaches and threshold accurateness in predicting applications.

WRM has paid a lot of attention over the past two decades to the rationale for using machine learning algorithms to forecast river flow, public water supply, potential ground water zonation [5, 16, 26, 84]. Hydrological science's complicated challenges with missing data management and hydrological predictions have seen significant modifications using machine learning [95]. Numerous ML-based techniques are in use, including optimization algorithms, logical techniques, classification techniques, statistical learning techniques and likelihood techniques. The three machine learning subcategories of adaptive Neuro-fuzzy inference system (ANFIS) [42], support vector machine (SVM) and artificial neural networks (ANN) [36] are particularly popular in the hydrology and water management sector [66]. The current study focused on examining high impact factor journal publications that discuss various water sustainability issues in several global case studies. This study also aims to list the benefits and drawbacks of the models stated in various geographical locations. Table 2.1 describes the details of the literature studied in this study.

2.2 Conceptualization of Water Resource

Water resource refers to the fresh water resources available within a geographic region [10, 51]. It includes surface water, such as lakes and rivers and groundwater. Fresh water is a vital resource for all life on Earth [27, 33]. Humans use water for many purposes, including drinking, cooking, bathing, irrigation, manufacturing and generating electricity [31, 92]. In order to encounter the growing demand for water, it is important to manage this precious resource carefully [35, 38].

2.2.1 Water Resource Management (WRM)

The process of organizing, controlling, allocating and overseeing the best use of water resources is known as water resource management [34, 73]. It is a part of managing the water cycle. It is comparable to hydrology, but there is less frequent management because WRM mainly affects towns and districts here, whereas hydrology concerns States or the entire nation [90]. WRM is a subject that mainly emphases on methods for controlling water flow as they relate to hydrological elements along with river run off and sediment flow [52]. Sustainable water management is essential if the world's limited water resources are to be protected and preserved [20, 24]. Fresh water is necessary for almost all human needs. Therefore, WRM means minimizing losses in the reservoir and distribution hub while identifying strategies to use the current water as efficiently as feasible [18, 61]. WRM entails allotting water to various parties and creating priorities for requirements including drinking, industrial and agriculture [82]. The need for effective water usage at all levels has increased due to the rising focus on climate change and how it affects water supply [21, 74]. WRM will include the advanced level management of several watersheds or connected systems [55]. The Worldwide Water Partnership asserts that a synchronized progress and administration of water, land and other allied resources is the global approach to WRM [83, 91, 100]. The strategy aims to ensure the sustainability of vital ecosystems while maximizing the benefits to society and the economy.

2.2.2 Tools and Techniques of WRM

Water resource management is the process of planning, developing and optimizing water resources to meet the requirements of present-day and upcoming generations [44]. It includes the coordination of water resources across multiple sectors, including

Table 2.1 Information of the literature reviewed about the ML in this WRM study

References	Models used	Study area	Description
Abbot and Marohasy [1]	GCM, POAM, ANN	Queensland, Australia	Monthly and seasonal rainfall forecasting
Abernethy et al. [2]	XG Boost, RF, lasso LR	Flint, Michigan	Developed a model to calculate the amount of lead in people's blood
Ali et al. [3]	Indicators of hydrologic alteration (IHA)	Yangtze River, China	This study examined the effects of cascade dams placed upstream of the TGD in China on river flow regimes from 2003 to 2015
Avand et al. [5]	Bftree, AB, MB	Ysuj Dena, Iran	Potential groundwater mapping
Bazan-Krzywoszańska et al. [8]	Neuro fuzzy NN	Wilamowice Municipality, Silesia Province	Used for land use potential and risk assessment
Fleming et al. [26]	NMFk	Western US river	Used for the improvement of water supply forecast models interpretability
Gaya et al. [29]	ANFIS, MLR, ANN	Yamuna river, India	Used for assessing surface water quality assessment
Hmoud and Waselallah [37]	ANFIS, FFNN	666 different sources of rivers and lakes in India	Used for surface water quality assessment
Liu et al. [49]	ICSOA-PPE	Heilongjiang Province, China	Used to evaluate the environmental quality of surface water
Lin et al. [48]	Upwelling	AoShan Bay, China	Used for carbon sequestration in coastal Mari culture environments
Mohammadi et al. [64]	SFLA, ANFIS	Vu Gia Thu Bon river basin, Vietnam	Predicting river stream flow time series
Mohammadi et al. [63]	MLP-PSODE, SVM, RBF	Mahabad river, Iran	Used for the estimation of suspended sediment load
Msiza et al. [68]	ANN, SVR	Republic of South Africa	Used for the prediction of water demand
Nagel and Ptal [69]	MCAT	Pacific Northwest region, USA	Analyzed high risk hydropower station network across the same watershed
Najafzadeh et al. [70]	M5MT, EPR, GEP, and MARS	Karun River, Iran	Checked the reliability of WQI using remote sensing and machine learning models

(continued)

Table 2.1 (continued)

References	Models used	Study area	Description
Nourani et al. [71]	ANN	Ligvanchai watershed, Tabriz, Iran	Used for rainfall run off modeling
Sengorur et al. [78]	SOM, ANN	Melen River, Turkey	Used for the assessment of water quality
Sohrabi et al. [81]	GMCR+	Namak Lake basin, Iran	The paper focuses on water optimization based on strategic management and water value
Tikhamarine et al. [83]	GWO, SVR	Aswan High Dam, Egypt	Accuracy improvement for the monthly stream flow forecast
Wei [93]	Wavelet SVM	Tanshui River Basin, Taiwan	Used for the prediction of water level during Typhoon
Zhou et al. [103]	GA, NSGA-II	The Shihmen Reservoir, Taiwan	This study provided an overview of the prospects for small-hydropower generation in the context of sustainable development by investigating a comprehensive AI-based strategy

agriculture, industry, energy, urban development and the environment [94]. WRM must take into account the competing demands of different users, as well as the environmental impact of water development and use [7]. In recent years, there has been an increasing focus on sustainable WRM, which aims to balance the economic, environmental and social necessities of current and forthcoming generations [19, 30]. There are a variety of tools and techniques that is being used to manage water resources effectively, such as [14, 45, 67, 87].

Integrated Water Resources Management (IWRM): IWRM is a holistic approach to water resource administration that takes into account all aspects of water cycle management, from catchment to consumption. IWRM aims to optimize social, economic and environmental outcomes by considering all stakeholders in decision-making.

Water Efficiency: Water efficiency measures aim to reduce the amount of water lost or wasted through inefficient practices. This includes improving irrigation practices and using more efficient fixtures and appliances in homes and businesses.

Demand Management: Demand management strategies aim to reduce overall water demand through a variety of measures such as pricing reform, public education campaigns and restrictions on water use during periods of high demand.

Conservation: Conservation measures aim to reduce unnecessary or harmful uses of water resources.

Use of artificial intelligence: Since past two decades, the use of machine learning, deep learning and IoT is on the rise for their effective and precise results.

2.3 Methodology Adopted for Conducting Systematic Review

According to de Souza Melaré et al. [15], a systematic review is viewed as a research instrument that outlines the formal process and several methods used to perform the study, including the research problem, objectives and methodologies used for analysis. The following measures were used to perform the systematic analysis for this study:

- Designing (Preparation): Considering the essential concept, formulating the research process, search strategy, categorizing the search criteria, selecting platforms and identifying pertinent research questions.
- Execution: a pilot search (scientific publications in a database, choose and import citations and abstracts of similar research centered on the findings of the current review), the choice of criteria, the gathering and withdrawal of the data (extract the final complete publication).
- Review (Examination): thorough evaluation of the chosen publication and evaluation of outcomes.

As a result, originally, a variety of reputable and worldwide publications were taken into consideration while choosing papers that addressed various elements of SWRM (Table 2.1). For water resource management, published works were chosen throughout two decades (from 2002 to 2022). To highlight the problems and cutting-edge WRM approaches offered by various specialists, a collection of peer-reviewed works and their citation histories were compiled and examined in accordance with the standards. Various keywords, including "water resource management", "water conservation", "sustainability", "climate change and water", "AI and water", "sustainable water resource management" were explored in the Scopus® and Google Scholar databases for data extraction from scholarly journals. These two databases were chosen for their perks: Scopus is a large repository of peer-reviewed research papers from all disciplines conducted globally, whereas Google Scholar is a standalone platform that can browse thousands of academic research articles in a matter of seconds and that empirical studies recognized by Web of Science (WoS) can also be discovered by using Google Scholar [4].

Among the papers gathered from the aforementioned archives are original findings, review, full-length studies and pieces in press. The current study excluded any latest article published after August, 2022. The primary search resulted into a huge number of research paper hence, only those literatures were chosen which serves the similar area of interest. Later further filtration process was done by reading abstracts only. Finally 70 research items were selected for the full length assessments which are matched with the objective of the present study. Details of some significant literatures are highlighted above in Table 2.1.

2.4 Approaches and Modeling for Sustainable Water Resource via Machine Learning

By 2030, it is predicted that AI-enabled revolutions and results in the water industry would have a beneficial impact of USD 200 billion, or 0.04–0.2% of the world GDP [88]. Although this effect might seem insignificant in comparison to AI's impact on other sectors and SDGs, it will be crucial for protecting freshwater supplies, forests and seas. The important sectors in the water industry where AI is anticipated to have an impact are given below, along with some sample AI applications. While the possibilities, foresight and policy recommendations may be utilized throughout the policy development process, the applications and micro level case studies can be utilized to assess the present status of technology adoption.

2.4.1 Maintenance Planning for Water Infrastructure

The fourth industrial revolution, also known as maintenance 4.0 in the water industry, is being driven by AI [1]. The detection of non-revenue water and the upkeep of water infrastructure will be done through the use of smart sensor-physical systems rather than manual techniques in this revolution. The AI-IoT market, recently esteemed at USD 4 billion, is anticipated to rise to USD 15 billion by 2024, with an estimated yearly growth rate of 20% [46]. This planned growth and investment will help with the long-term supply of water, the protection of watersheds and water sources and disaster preparedness in addition to the renovation and replacement of ageing water and wastewater infrastructure.

An efficient and long-lasting water distribution management system was established in Singapore by the Public Utilities Board using AI and smart sensors. The system utilizes data from intelligent sensors that AI has referred to discover leaks, failures and preventative maintenance in the utility system along with other real-time monitoring [85]. HydroIQ was created a home-level solution in Kenya to track water pressure, water quality and leaks and faults. The technology combines artificial intelligence (AI), along with intelligent sensors and payment facility, allowing home clients to only pay for the amount used [40]. A US-based startup called Fracta is assisting water utility management organizations to implement AI-based approaches to better manage water infrastructure. By utilizing ML algorithms to determine the probability of failure, their solutions assist in evaluating the conditions and dangers associated with drinking water distribution mains. The technology facilitates decision-makers to evaluate water infrastructure and decide on repairs and replacements with knowledge. The business wants to address the $1 trillion water main issue in North America (Water Intelligence). AI is used by Smart Cover Systems, a business located in the USA, to monitor water and wastewater infrastructure. The system continually monitors, collects and transmits data through satellite communication when combined with IoT technologies. With the use of AI-based trend analysis, it can assess obstructions, find storm water penetration and deliver real-time maintenance information (Water Intelligence).

2.4.2 *Forecasting Water Demand and Use*

Facility managers may track water demand and use as well as evaluate the functioning of the water system in real-time thanks to AI-IoT empowered water management systems [6]. A latest version of water supervision strategies can offer short- and long-term projections by thoroughly addressing information on water demand and usage utilizing learning technologies. Effective reservoir management relies on the use of short-term forecasts as well as the accompanying infrastructure. Future projections are used to build and upgrade water infrastructure [54]. The Ministry of Economy and Competitiveness of Spain has organized an ANN based technique for forecasting water demands. The system provides short-term daily estimates of water irrigation needs using Genetic Algorithms (GA) and ANN. The Bayesian framework is also utilized. When compared to models that are not AI-based, the system's prediction accuracy is 11% higher [1].

An integrated structure for controlling and safeguarding the urban water supply has been implemented in South East Queensland, Australia. The grid employs AI to estimate water demand and consumption for residential users, water treatment facilities, desalination plants, reservoirs and dams and water pumping stations. The smart grid annually manages 400 million m^3 of water and up to USD 7.6 billion of water infrastructure [88]. The Metropolitan areas of Southern California utilizes ANN-based model to predict how economic expansion and population would affect water demand [1]. These projections are employed to manage water supply and advocate for water conservation measures. In addition to handling the estimate for 26 stores and 18 million customers, the model also manage the supply from local resources [89]. The largest water and wastewater service provider in the UK, Thames Water, employs AI to forecast the number of homes in water resource zones, population growth and water usage [86].

Aside from this, one of the prevalent tasks to effective management of water resources, reasonable water supply and consistent water supply is the issue of losses related to non-revenue water [6, 54]. Physical losses, commercial losses and non-revenue water are the main causes of non-revenue water. Physical losses include transmission leaks, storage leaks, water tank overflows and metering service linking leaks. Commercial losses include fraudulent manipulation, systemic bugs, unofficial consumption and data supervision mistakes [6, 54]. In industrialized nations, smart water grids often utilize physiochemical measures (to detect pH, optical characteristics, temperature, electrical conductivity, turbidity, TSS, water flow, chlorine concentration and microbiological readings (to identify toxicity) to monitor water quality [12].

Smart water meters are specifically used to successfully reduce non-revenue water expenses, prevent water waste and address issues with stress variations, leaks, damages, intake rate, unapproachability to client and their consumption data, issues with service quality etc. These characteristics are tracked online, detected in real time and attached to microcontrollers in order to process and analyze the water quality records utilizing event detection and predicting systems [12]. The resident water administration is then promptly informed of the site and severity of the identified abnormalities so that they can take the necessary precautionary action. The constant water sampling and hydraulic model adjustment using the real-time sensed data is an alternative to the pricey, boring and laborious classical calibration approach [57, 86].

Charge and billing concerns, uncontrolled water losses and other problems may be remedied by using clever pressure controlling techniques, smart step testing as well as other cunning methods [54]. Smart water grids comprehensively incorporate these computerized meters, analytic tools and sensor array for smart control and effective water administration to guarantee that top graded water supply is consistently and reliably supplied only when and where it is desired [17, 54, 62].

2.4.3 Observing Dams and Water Reservoirs

Around 7,320 water reservoirs and associated dams exist in the world, with a combined storage capacity of 6700 km^3 [88]. There are currently 3,700 additional major dams under construction or planned, largely in developing nations [102]. The building of water reservoirs is increasingly utilizing AI-based approaches for operations and decision-making. This development is most obvious in South America and Asia [54, 59]. Modern AI algorithms are being used to forecast indicators for water resources such as precipitation, sediment, river discharge, quality of the water, water level and evapotranspiration. Each of these constituents is predicted with fluctuating degrees of precision depending on the recorded data supplied and the AI method utilized [103].

18% of the dams in the USA are classified as high risk potential. 2,000 of these high-risk dams require rehabilitation, which is expected to cost $20 billion [70]. The Columbia Water Center is integrating artificial intelligence, geospatial data along with climate models to determine which dams are most at risk and to guide the repair and retiring process [60]. Deep learning techniques were used to examine climate data from 1970 to 2019 in order to determine whether patterns of humidity and wind circulation suggest rain. 95% of the time, the deep learning algorithm can correctly forecast whether it will rain or not. This rain projection is the result of combining altitude maps, dam heights and storage capacity and runoff equations [25, 75].

2.4.4 Water Quality Monitoring and Reporting

One of the biggest problems in the 21st century is poor water quality, since one in nine individual acquires their drinking water from questionable sources [96]. Additionally, 90% of the sewage that is released into aquatic bodies in underdeveloped nations is untreated [89]. Disruptive technologies must be deployed to solve these problems relating to water quality [9, 56]. Artificial intelligence is being utilized to recognize remote sensing satellite imageries for geographically distributed watersheds and water bodies in cases when it is not practical to deploy devices to obtain water samples [81]. This technique allows the detection of changes in the quality of water over time. Typical monitoring metrics in this group comprise heavy metal concentration, total suspended particles and sea surface temperature.

Improvements in AI-based pattern detection and sensor image quality allow for the quick identification of bacterial pollutants in water. AI-based solutions are being given as a substitute to manual and labor-intensive approaches of mapping color-based indications, which are typically used to analyse the quantity of specific poisons, diseases and infections [86]. Hmoud and Waselallah [37] confirmed a plan to create an effective system for monitoring drinking water to maintain a sustainable and hospitable environment using ANFIS and FFNN and K nearest. [29] showed the traditional linear model for predicting the Water Quality Index at Palla station of the Yamuna River, India, for this work he used ANFIS, ANN and MLR techniques. The SOM-ANNs were used to assess the associations between high-low flow periods and water-quality indicators in Melen River, Turkey. This was done in order to measure differences in river water quality parameters between high and low flow periods [79]. Through the use of contaminant isolation techniques and smart flood mitigation strategies, smart valves are used to stop pollutant infiltration and stop contaminated water from blending with pure water [54].

Since smart water metering and water quality monitoring are currently insufficient in the majority of African communities, the bulk of water pollution instances are typically discovered and informed by resident consumers who have already been exposed to the risk [6]. Water adulteration happens more frequently in the water supply system than in the treatment facilities, as noted by Ayemba [6] and Jallé et al. [43]. The reasons of adulteration in the water distribution system include colorization, microbiological entry, sanitizing expiration, pressure differentials, biofilms, stagnant water, biochemical anomalies, industrial accidents, diagnosis error and acts of sabotage [12].

2.4.5 *Use of AI in Assessing Water Quality Using Remote Sensing*

Remote sensing (RS) brings high geospatial and temporal resolution water quality data for large numbers of water bodies at once [29]. It aids in the valuation of ecological concerns and related health risks by examining alterations in water chemistry and spotting various dangerous algal blooms. GIS and Remote Sensing technologies are well adapted for analyzing, controlling, measuring and monitoring water availability in order to identify an appropriate methodology for sustainable WRM that can provide enough fresh water for life on Earth [58]. AI is currently being quickly developed and utilized to a wide range of remote sensing applications. Among many AI models, supervised and semi-supervised algorithms are most commonly used and require a big amount of training data, particularly for deep learning approaches [3, 48]. The possibilities by using remote sensing data to contribute significantly to commercial and human activities are extremely appealing since they acquire a good amount of information across wide spatial areas, allowing for precise identification of natural characteristics, diverse materials, and physical objects on earth [5]. McGrath and Gohl [58] assessed the impact of meteorological mapping on flood susceptibility using RS and ML. Najafzadeh et al. [70] checked the reliability of WQI using RS and machine learning models. Various scholars used AI along with RS to identify water bodies automatically and monitoring their water quality [41, 98]. Soon the modified AI algorithms along with latest advanced remote sensing satellite imageries will become more useful to identify environmental issues in more precise way.

2.5 Challenges in WRM

Natural processes such as precipitation, surface runoff and ground water levels refresh our supply of freshwater. Working with people is WRM's biggest challenge [32, 101]. Human activities have contaminated freshwater lakes and rivers [28]. If management of water resources is not done to make them climate resilient, future disputes over access to fresh water may occur. Since water is a necessity for life, life on this planet would not be imaginable without it. Drought is currently triggering significant water scarcity across the globe and drought is assumed to be a consequence of climate change. Many rivers are being besieged, drying up and becoming poisoned. The aquifer level significantly declines as a result of managing the ground water reservoir in a way that allows for excessive water pumping at a rate that is faster than the rate of recharging. The life of a reservoir is constrained to a finite amount of time by utilizing operational rules for designing and managing dams and reservoir systems that optimize the advantage from water use without addressing the accumulation of sediments. Managing a surface water storage facility that permits a pumping rate for water that is excessive compared to the rate of recharging. It is more crucial than ever to consider whether future generations will have access to enough freshwater to survive. What would appear to be an option for people actually causes enormous problems for nature, beginning with our natural ability to control how much water we use.

2.6 Review Findings and Research Gap

There is a growing body of researches on WRM but there are still many gaps in our knowledge:

- One major research gap is the lack of reliable data on water resources. Although a lot of work was done on WRM using machine learning, still there are many places which are not covered for the assessment.
- Our understanding of the social and economic factors that influence WRM.
- Our understanding of the hydrological and ecological processes which affect water resources. The development of effective policies and practices for WRM.
- Globally, researchers, environmentalists and biologists are increasingly issuing warnings that climate change would significantly modify the earth's hydrological regime and hydrological cycle, influencing the availability of groundwater and surface water to a new degree. Researchers need to focus on the impact of climate change on water resources.

Therefore, the existing literatures show that majority of water sustainability related works were done on developed and developing nations and very less focus is given to the third world nations. Those nations which were covered by the researchers, the study were mostly concentrated into selected regions only.

2.7 Conclusion

It is encouraged to develop integrated management approaches and multi-service strategies. Water conservation programs, quality monitoring activities and quality control techniques shouldn't be restricted to treatment facilities instead they should be scientifically assimilated network-wide to accomplish complete quality control for the water supply network.

This smart system will further augment the functionality of the smart water meters by offering a strong, dependable and resilient water supervision system that can fulfill the present and upcoming water needs of the towns and societies. The current trend indicates that block chain, the Internet of Things and artificial intelligence will soon be the dominant technologies. The many benefits that the current AI era has to offer must be utilized by the underdeveloped countries. However, the amalgamation of AI and IoT showed the potential to transform water resource management in order to achieve SDG as anticipated by the UN in 2035, via groundbreaking, effective and scalable solutions using these two technologies. The sustainable water management crisis afflicting the community cannot be solved in a one-size-fits-all manner. This paper suggests that future scholars use recently created optimization techniques to enhance the functionality of ML models. In this study, a number of instances were used to demonstrate the prediction, regression and classification proficiencies of ML in connection to SWRM modeling challenges. These incidents also illustrated the need for caution while using ML because of the possibility of over-fitting problems due to its non-linear character.

2.8 Recommendations

The outcomes of the literature review reveal that ML has several uses in computational hydrology, particularly in WRM. Upcoming academics can build their research on this strategy to produce cutting-edge machine learning and new hybrid systems that can handle the complexity of hydrological forecasts.

Making ML an effective tool for managing water resources, the machine learning model may provide more precise runoff simulation predictions, determine the state of the aquifer, observe surface and ground water quality, etc.

Future prospective scientists can use hybrid-based models utilizing hydrological and ML models in order to take advantage of the benefits of ML-based and physical-based models for diverse WRM assessments.

Acknowledgements We thankfully acknowledge the reviewers and editorial team for their valuable time, productive comments and suggestions for improving the overall quality of the manuscript.

Ethical Approval NA

Consent to Participate NA

Consent to Publish NA

Authors Contributions **Rukhsar Anjum**—Conceptualization, literature search, methodology development, Writing- original draft, Formal analysis; **Farhana Parvin** and **Sk Ajim Ali**—Formal analysis; Data curation, Writing, Visualization, Review and editing.

Competing Interests There are no conflicts of interest to declare.

References

1. Abbot J, Marohasy J (2012) Application of artificial neural networks to rainfall forecasting in Queensland, Australia. Adv Atmos Sci 29(4):717–730
2. Abernethy J, Chojnacki A, Farahi A, Schwartz E, Webb J (2018) Activeremediation: the search for lead pipes in flint, michigan. In Proceedings of the 24th ACM SIGKDD international conference on knowledge discovery & data mining, pp 5–14

3. Ali R, Kuriqi A, Abubaker S, Kisi O (2019) Hydrologic alteration at the upper and middle part of the Yangtze river, China: towards sustainable water resource management under increasing water exploitation. Sustainability 11(19):5176
4. Ali SA, Parvin F (2022) Examining challenges and multi-strategic approaches in waste management during the COVID-19 pandemic: a systematic review. Waste Manag Res 0734242X221079303
5. Avand M, Janizadeh S, Tien Bui D, Pham VH, Ngo PTT, Nhu VH (2020) A tree-based intelligence ensemble approach for spatial prediction of potential groundwater. Int J Digit Earth 13(12):1408–1429
6. Ayemba D (2017) Why Africa needs water metering. https://constructionreviewonline.com/2017/04/africa-needs-water-metering/. Accessed 13 Oct 2022
7. Badham J, Elsawah S, Guillaume JH, Hamilton SH, Hunt RJ, Jakeman AJ et al (2019) Effective modeling for integrated water resource management: a guide to contextual practices by phases and steps and future opportunities. Environ Modell Softw 116:40–56
8. Bazan-Krzywoszańska A, Mrówczyńska M, Tront S (2019) GIS technology, 3D models and mathematical models as a tool for assessing development capabilities of flood risk land to make arrangements of municipal planning documents. J Ecol Eng 20(1)
9. Bublitz M, Oetomo F, Sahu S, Kuang A, Fadrique L, Velmovitsky P et al (2019) Disruptive technologies for environment and health research: an overview of artificial intelligence, blockchain and internet of things. Int J Environ Res Publ Health 16(20):3847
10. Burak S, Margat J (2016) Water management in the Mediterranean region: concepts and policies. Water Resour Manag 30(15):5779–5797
11. Cazorla-Montero A, de los Rios-Carmenado I, Pasten JI (2019) Sustainable development planning: master's based on a project-based learning approach. Sustainability 11(22):6384
12. Cloete NA, Malekian R, Nair L (2016) Design of smart sensors for real-time water quality monitoring. IEEE Access 4:3975–3990
13. Cosgrove WJ, Loucks DP (2015) Water management: current and future challenges and research directions. Water Resour Res 51(6):4823–4839
14. Crini G, Lichtfouse E (2019) Advantages and disadvantages of techniques used for wastewater treatment. Environ Chem Lett 17(1):145–155
15. de Souza Melaré AV, González SM, Faceli K, Casadei V (2017) Technologies and decision support systems to aid solid-waste management: a systematic review. Waste Manag 59:567–584
16. Díaz-Alcaide S, Martínez-Santos P (2019) Advances in groundwater potential mapping. Hydrogeol J 27(7):2307–2324
17. Dogo EM, Salami AF, Nwulu NI, Aigbavboa CO (2019) Blockchain and internet of things-based technologies for intelligent water management system. In: Artificial intelligence in IoT. Springer, Cham, pp 129–150
18. dos Santos Amorim JMB, Bezerra SDTM, Silva MM, de Sousa LCO (2020) Multicriteria decision support for selection of alternatives directed to integrated urban water management. Water Resour Manag 34(13):4253–4269
19. Durán-Sánchez A, Álvarez-García J, del Río-Rama MDLC (2018) Sustainable water resources management: a bibliometric overview. Water 10(9):1191
20. EL-Nwsany RI, Maarouf I, Abd el-Aal W (2019) Water management as a vital factor for a sustainable school. Alex Eng J 58(1):303–313
21. Erfani T, Pachos K, Harou JJ (2018) Real-options water supply planning: Multistage scenario trees for adaptive and flexible capacity expansion under probabilistic climate change uncertainty. Water Resour Res 54(7):5069–5087
22. Etikala B, Adimalla N, Madhav S, Somagouni SG, Keshava Kiran Kumar PL (2021) Salinity problems in groundwater and management strategies in arid and semi-arid regions. In: Groundwater geochemistry: pollution and remediation methods, pp 42–56
23. Etikala B, Madhav S, Somagouni SG (2022) Urban water systems: an overview. Curr Direct Water Scarcity Res 6:1–19

24. Everard M (2019) A socio-ecological framework supporting catchment-scale water resource stewardship. Environ Sci Policy 91:50–59
25. Falga R, Wang C (2022) The rise of Indian summer monsoon precipitation extremes and its correlation with long-term changes of climate and anthropogenic factors. Sci Rep 12(1):1–11
26. Fleming SW, Vesselinov VV, Goodbody AG (2021) Augmenting geophysical interpretation of data-driven operational water supply forecast modeling for a western US river using a hybrid machine learning approach. J Hydrol 597:126327
27. Flint RW (2004) The sustainable development of water resources. Water Resour Update 127:41–51
28. Gao Y, Zhou F, Ciais P, Miao C, Yang T, Jia Y et al (2020) Human activities aggravate nitrogen-deposition pollution to inland water over China. Natl Sci Rev 7(2):430–440
29. Gaya MS, Abba SI, Aliyu MA, Tukur AI, Saleh MA, Esmaili P, Wahab NA (2020) Estimation of water quality index using artificial intelligence approaches and multi-linear regression. IAES Int J Artif Intell 9(1):126
30. Gazzola P, Pavione E, Pezzetti R, Grechi D (2020) Trends in the fashion industry. The perception of sustainability and circular economy: a gender/generation quantitative approach. Sustainability 12(7):2809
31. Gleick PH (2003) Water use. Ann Rev Environ Resour 28(1):275–314
32. Goldin J, Mokomela R, Kanyerere T, Villholth KG (2021) Diamonds on the soles of their feet: groundwater monitoring in the Hout Catchment, South Africa. J Educ Sustain Dev 15(1):25–50
33. Gregory KB, Vidic RD, Dzombak DA (2011) Water management challenges associated with the production of shale gas by hydraulic fracturing. Elements 7(3):181–186
34. Hajkowicz S, Collins K (2007) A review of multiple criteria analysis for water resource planning and management. Water Resour Manag 21(9):1553–1566
35. Hanjra MA, Blackwell J, Carr G, Zhang F, Jackson TM (2012) Wastewater irrigation and environmental health: implications for water governance and public policy. Int J Hyg Environ Health 215(3):255–269
36. Haykin S, Network N (2004) A comprehensive foundation. Neural Netw 2(2004):41
37. Hmoud Al-Adhaileh M, WaselallahAlsaade F (2021) Modelling and prediction of water quality by using artificial intelligence. Sustainability 13(8):4259
38. Hossain MZ (2015) Water: the most precious resource of our life. Glob J Adv Res 2(9):1436–1445
39. How ML, Cheah SM, Chan YJ, Khor AC, Say EMP (2020) Artificial intelligence-enhanced decision support for informing global sustainable development: a human-centric AI-thinking approach. Information 11(1):39
40. HydroIQ Africa (2020) HyrdoIQ. https://www.hydroiq.africa. Accessed 15 Oct 2020
41. Ighalo JO, Adeniyi AG, Marques G (2021) Artificial intelligence for surface water quality monitoring and assessment: a systematic literature analysis. Model Earth Syst Environ 7(2):669–681
42. Jain A, Kumar AM (2007) Hybrid neural network models for hydrologic time series forecasting. Appl Soft Comput 7(2):585–592
43. Jallé LC, Désille D, Burkhardt G (2013) Urban stormwater management in developing countries. http://documents.irevues.inist.fr/bitstream/handle/2042/51221/1A2P07-005LEJ.pdf?sequence=1. Accessed 13 Oct 2022
44. Janga Reddy M, Nagesh Kumar D (2020) Evolutionary algorithms, swarm intelligence methods and their applications in water resources engineering: a state-of-the-art review. H2Open J 3(1):135–188
45. Kehrein P, Van Loosdrecht M, Osseweijer P, Garfí M, Dewulf J, Posada J (2020) A critical review of resource recovery from municipal wastewater treatment plants–market supply potentials, technologies and bottlenecks. Environ Sci: Water Res Technol 6(4):877–910
46. Lasi H, Fettke P, Kemper HG, Feld T, Hoffmann M (2014) Industry 4.0. Bus Inf Syst Eng 6(4):239–242

47. Li Y, Yang W, Shen X, Yuan G, Wang J (2019) Water environment management and performance evaluation in central China: a research based on comprehensive evaluation system. Water 11(12):2472
48. Lin T, Fan W, Xiao C, Yao Z, Zhang Z, Zhao R et al (2019) Energy management and operational planning of an ecological engineering for carbon sequestration in coastal Mariculture environments in China. Sustainability 11(11):3162
49. Liu D, Liu C, Fu Q, Li T, Khan MI, Cui S, Faiz MA (2018) Projection pursuit evaluation model of regional surface water environment based on improved chicken swarm optimization algorithm. Water Resour Manag 32(4):1325–1342
50. Loucks DP (2000) Sustainable water resources management. Water Int 25(1):3–10
51. Lubell M, Lippert L (2011) Integrated regional water management: a study of collaboration or water politics-as-usual in California, USA. Int Rev Adm Sci 77(1):76–100
52. Lv T, Wang L, Xie H, Zhang X, Zhang Y (2021) Evolutionary overview of water resource management (1990–2019) based on a bibliometric analysis in web of science. Eco Inform 61:101218
53. Macias-Fauria M, Jepson P, Zimov N, Malhi Y (2020) Pleistocene arctic megafaunal ecological engineering as a natural climate solution? Philos Trans R Soc B 375(1794):20190122
54. Martyusheva O (2014) Smart water grid. Plan B Technical Report, M.Sc thesis, Department of Civil and Environmental Engineering, Colorado State University. http://www.engr.colostate.edu/~pierre/ce_old/resume/Theses%20and%20Dissertations/Martyusheva,Olga_PlanB_TechnicalReport.pdf. Accessed 13 Oct 2022
55. Mashaly AF, Fernald AG (2020) Identifying capabilities and potentials of system dynamics in hydrology and water resources as a promising modeling approach for water management. Water 12(5):1432
56. Maurya PK, Ali SA, Alharbi RS, Yadav KK, Alfaisal FM, Ahmad A et al (2021) Impacts of land use change on water quality index in the upper Ganges River near Haridwar, Uttarakhand: a GIS-based analysis. Water 13(24):3572
57. Mazumder RK, Salman AM, Li Y, Yu X (2018) Performance evaluation of water distribution systems and asset management. J Infrastruct Syst 24(3):03118001
58. McGrath H, Gohl PN (2022) Accessing the impact of meteorological variables on machine learning flood susceptibility mapping. Remote Sens 14(7):1656
59. Mehmood H, Liao D, Mahadeo K (2020a) A review of artificial intelligence applications to achieve water-related sustainable development goals. In: 2020 IEEE/ITU international conference on artificial intelligence for good (AI4G). IEEE, pp 135–141
60. Mehmood H, Mukkavilli SK, Weber I, Koshio A, Meechaiya C, Piman T et al (2020b) Strategic foresight to applications of artificial intelligence to achieve water-related sustainable development goals. UNU-INWEH Report Series, Issue 09. United Nations University Institute for Water, Environment and Health, Hamilton, Canada. https://doi.org/10.53328/LOTC2968
61. Mirdashtvan M, Najafinejad A, Malekian A, Sa'doddin A (2021) Sustainable water supply and demand management in semi-arid regions: optimizing water resources allocation based on RCPs scenarios. Water Resour Manag 35(15):5307–5324
62. Mohamed N, Al-Jaroodi J, Lazarova-Molnar S, Jawhar I (2021) Applications of integrated IOT-fog-cloud systems to smart cities: a survey. Electronics 10(23):2918
63. Mohammadi B, Guan Y, Moazenzadeh R, Safari MJS (2021a) Implementation of hybrid particle swarm optimization-differential evolution algorithms coupled with multi-layer perceptron for suspended sediment load estimation. Catena 198:105024
64. Mohammadi B, Linh NTT, Pham QB, Ahmed AN, Vojteková J, Guan Y et al (2020b) Adaptive neuro-fuzzy inference system coupled with shuffled frog leaping algorithm for predicting river streamflow time series. Hydrol Sci J 65(10):1738–1751
65. Mohammadi B (2021c) A review on the applications of machine learning for runoff modeling. Sustain Water Resour Manag 7(6):1–11
66. Mojiri A, Zhou JL, Ohashi A, Ozaki N, Kindaichi T (2019) Comprehensive review of polycyclic aromatic hydrocarbons in water sources, their effects and treatments. Sci Total Environ 696:133971

67. Mrówczyńska M, Sztubecka M, Skiba M, Bazan-Krzywoszańska A, Bejga P (2019) The use of artificial intelligence as a tool supporting sustainable development local policy. Sustainability 11(15):4199
68. Msiza IS, Nelwamondo FV, Marwala T (2008) Water demand prediction using artificial neural networks and support vector regression
69. Nagel AC, Ptak T (2022) Approaching obsolescence? A multi-criteria analysis of high-risk dams in the United States Pacific Northwest. Int J Water Resour Dev 38(2):217–241
70. Najafzadeh M, Homaei F, Farhadi H (2021) Reliability assessment of water quality index based on guidelines of national sanitation foundation in natural streams: integration of remote sensing and data-driven models. Artif Intell Rev 54(6):4619–4651
71. Nourani V, Komasi M, Mano A (2009) A multivariate ANN-wavelet approach for rainfall–runoff modeling. Water Resour Manag 23(14):2877–2894
72. Palansooriya KN, Yang Y, Tsang YF, Sarkar B, Hou D, Cao X et al (2020) Occurrence of contaminants in drinking water sources and the potential of biochar for water quality improvement: a review. Crit Rev Environ Sci Technol 50(6):549–611
73. Pearson LJ, Coggan A, Proctor W, Smith TF (2010) A sustainable decision support framework for urban water management. Water Resour Manag 24(2):363–376
74. Pörtner HO, Roberts DC, Adams H, Adler C, Aldunce P, Ali E et al (2022) Climate change 2022: impacts, adaptation and vulnerability. In: IPCC sixth assessment report
75. Rajna G (2018) AI locate risky dams. https://vixra.org/abs/1808.0594. Accessed 15 Oct 2022
76. Sang YF (2013) A review on the applications of wavelet transform in hydrology time series analysis. Atmos Res 122:8–15
77. Sawaf MBA, Kawanisi K, Jlilati MN, Xiao C, Bahreinimotlagh M (2021) Extent of detection of hidden relationships among different hydrological variables during floods using data-driven models. Environ Monitor Assess 193(11):1–14
78. Sengorur B, Koklu R, Ates A (2015) Water quality assessment using artificial intelligence techniques: SOM and ANN—A case study of Melen River Turkey. Water Qual Expo Health 7(4):469–490
79. Sharafati A, Khazaei MR, Nashwan MS, Al-Ansari N, Yaseen ZM, Shahid S (2020) Assessing the uncertainty associated with flood features due to variability of rainfall and hydrological parameters. Adv Civil Eng
80. Sheffield J, Wood EF, Pan M, Beck H, Coccia G, Serrat-Capdevila A, Verbist K (2018) Satellite remote sensing for water resources management: potential for supporting sustainable development in data-poor regions. Water Resour Res 54(12):9724–9758
81. Sohrabi M, AhaniAmineh ZB, Niksokhan MH, Zanjanian H (2022) A framework for optimal water allocation considering water value, strategic management and conflict resolution. Environ Dev Sustain 1–32
82. Su Y, Gao W, Guan D, Zuo TA (2020) Achieving urban water security: a review of water management approach from technology perspective. Water Resour Manag 34(13):4163–4179
83. Tikhamarine Y, Souag-Gamane D, Ahmed AN, Kisi O, El-Shafie A (2020) Improving artificial intelligence models accuracy for monthly streamflow forecasting using grey Wolf optimization (GWO) algorithm. J Hydrol 582:124435
84. Tortajada C, Joshi YK, Biswas AK (2013) The Singapore water story: sustainable development in an urban city state. Routledge
85. Tung TM, Yaseen ZM (2020) A survey on river water quality modelling using artificial intelligence models: 2000–2020. J Hydrol 585:124670
86. Tyralis H, Papacharalampous G, Langousis A (2019) A brief review of random forests for water scientists and practitioners and their recent history in water resources. Water 11(5):910
87. UN (2015) Sustainable development goals: 17 goals to transform our world. www.un.org/sustainabledevelopment/. Accessed 13 Oct 2022
88. United Nations World Water Assessment Programme (WWAP) (2015) The United Nations World Water Development Report 2015: water for a sustainable world. UNESCO, Paris
89. Uprety M, Ochoa-Tocachi BF, Paul JD, Regmi S, Buytaert W (2019) Improving water resources management using participatory monitoring in a remote mountainous region of Nepal. J Hydrol: Region Stud 23:100604

90. Vieira EDO (2020) Integrated water resources management: theoretical concepts, basis, responsibilities and challenges of IWRM. In: Integrated water resource management. Springer, Cham, pp 1–12
91. Ward FA, Michelsen A (2002) The economic value of water in agriculture: concepts and policy applications. Water policy 4(5):423–446
92. Water Intelligence. Using AI to diagnose water consumption patterns. https://wint.ai/using-ai-to-diagnosewater-consumption-patterns/. Accessed 15 Oct 2022
93. Wei CC (2012) Wavelet kernel support vector machines forecasting techniques: case study on water-level predictions during typhoons. Expert Syst Appl 39(5):5189–5199
94. Wei F, Zhang X, Xu J, Bing J, Pan G (2020) Simulation of water resource allocation for sustainable urban development: an integrated optimization approach. J Clean Prod 273:122537
95. Wen X, Feng Q, Deo RC, Wu M, Yin Z, Yang L, Singh VP (2019) Two-phase extreme learning machines integrated with the complete ensemble empirical mode decomposition with adaptive noise algorithm for multi-scale runoff prediction problems. J Hydrol 570:167–184
96. World Health Organization (2017) Progress on drinking water. Sanitation and Hygiene, pp 1–66
97. Xiang X, Li Q, Khan S, Khalaf OI (2021) Urban water resource management for sustainable environment planning using artificial intelligence techniques. Environ Impact Assess Rev 86:106515
98. Yang L, Driscol J, Sarigai S, Wu Q, Lippitt CD, Morgan M (2022) Towards synoptic water monitoring systems: a review of AI methods for automating water body detection and water quality monitoring using remote sensing. Sensors 22(6):2416
99. Yaseen ZM, Sulaiman SO, Deo RC, Chau KW (2019) An enhanced extreme learning machine model for river flow forecasting: state-of-the-art, practical applications in water resource engineering area and future research direction. J Hydrol 569:387–408
100. Yeleliere E, Cobbina SJ, Duwiejuah AB (2018) Review of Ghana's water resources: the quality and management with particular focus on freshwater resources. Appl Water Sci 8(3):1–12
101. Zahid A (2022) Groundwater resources, challenges and governance in the Bengal Basin, Bangladesh. In: Bangladesh geosciences and resources potential. CRC Press, pp 399–436
102. Zarfl C, Lumsdon AE, Berlekamp J, Tydecks L, Tockner K (2015) A global boom in hydropower dam construction. Aquatic Sci 77(1):161–170
103. Zhou Y, Chang LC, Uen TS, Guo S, Xu CY, Chang FJ (2019) Prospect for small-hydropower installation settled upon optimal water allocation: an action to stimulate synergies of water-food-energy nexus. Appl Energy 238:668–682

Chapter 3
Remote Sensing and Machine Learning Applications for the Assessment of Urban Water Stress: A Review

Jagriti Jain, Sourav Choudhary, Francisco Munoz-Arriola, and Deepak Khare

Abstract Water stress is a critical factor and depends on the balance between water demands and supplies at any given time and locations. The continuous expansion of urban areas attributed to rural immigration to urban centers and population growth, exacerbated by a changing climate, has broken the balance between water supplies and demands, making cities more water insecure. The use of remote sensing (RS) products and machine learning (ML) analytics for the assessment of water stress areas has increased in the past twenty years. This paper reviews scientific and technological evidence published in the past years in the intersection of RS, ML, and water stress. We explore how RS and ML are shaping the current and future needs for research and innovation for water stress assessment. This review focuses on the contrasting sources of water stress various water stress when water surpluses and deficits are present, and how indicators have incorporated the use of RS and ML to identify temporal and geospatial attributions, scales, and the metrics. It has been found that metrics such as rainfall, population, runoff, drainage network have been diversely and extensively are used in different case studies which plays a major role. For the water quality assessments, the parameters of salinity, pH, dissolved oxygen, suspended solids, and ammoniacal nitrogen, sediment load has been utilized. ML techniques such as ANN, XGBoost, SVM, CNN, ANFIS, RF have been implemented in the

J. Jain (✉) · S. Choudhary · D. Khare
Water Resources Development and Management, Indian Institute of Technology, Roorkee, India
e-mail: jagjain8@gmail.com

S. Choudhary
e-mail: schoudhary@wr.iitr.ac.in

D. Khare
e-mail: deepak.khare@wr.iitr.ac.in

F. Munoz-Arriola
School of Natural Resources, University of Nebraska Lincoln, Lincoln, USA

Department of Biological Systems Engineering, University of Nebraska Lincoln, Lincoln, USA

F. Munoz-Arriola
e-mail: fmunoz@unl.edu

E. Balaji et al. (eds.), *Emerging Technologies for Water Supply, Conservation and Management*, Springer Water, https://doi.org/10.1007/978-3-031-35279-9_3

previous literature. These techniques have been found useful with the requirement of exploitation of other methods as well. The use of knowledge graphs can be promoted which can help in the integration of the various parameters and help in defining water scarcity as one entity. With the development and improving upon the urban water stress management can help in micro level planning at the city level to adapt to the water scarcity.

Keywords Remote sensing · Machine learning · Urban water stress · Water scarcity · Climate change

List of Abbreviations

ANN	Artificial Neural Network
CART	Classification and Regression Trees
FDA	Flexible Discriminant Analysis
FFNN	Feed Forward Neural Network
IPCC	International Panel on Climate Change
MARS	Multivariate Adaptive Regression Splines
ML	Machine learning
NFP	Network Flow Programming Model
RS	Remote sensing
SDG	Sustainable Development Goals
SPEI	Standardized Precipitation Evapotranspiration Index
SVM	Support Vector Machines
UNICEF	United Nations International Children's Emergency Fund
WWDR	World Water Development Report

3.1 Introduction

With the increase in the occurrence of the extreme events the quantity and quality of the water gets affected and leaving the areas water scarce. The Sustainable Development Goals (SDGs) have established desired objectives for the consumption of freshwater responsibly (SDG 12) and for universal and equitable clean water availability (SDG 6) by 2030. By 2050, 68% of the world's population will reside in cities, according to certain forecasts [1]. The inability to obtain enough water is a significant obstacle to achieving all the Sustainable Development Goal 6. Cities that struggle to provide residents with enough water are known as water scarce cities [25]. As an indicator of water scarcity, water stress has been used to track changes in water supply-demand in urban areas. In many cities worldwide water demands have increased due to rapid population growth and industrial expansion. The efficient operation and management of an urban water distribution system depends on the accuracy of water demand estimates, and the sustainable planning of regional water supply systems depends on the prediction of future urban water usage [89]. Complex connections between human and environmental systems operating at different sizes put the management of urban water needs in danger because of water stress and shortage [28, 75]. At watershed scale, determination of water quantity and quality has been implemented by [32], which discussed on the drought and flooded seasons by using SWAT model and remote sensing products. But with the use of remote sensing products and in the data scarce regions the machine learning techniques and to integrate the various parameters are getting a good ground which is also demonstrated by Kumar et al. [39] and Sarzaeim et al. [70]. The use of knowledge graphs which integrates the different parameters and can help in defining water scarcity as one entity. Etikala et al. [20, 21] has highlighted the importance of the quantity and quality of the water which is altered by the urban runoff.

The study focuses on those non-urban-to-peri urban areas that will be transformed into urban areas in the future. One of the main causes of people migrating to areas with a lack of water is the changing of land use. The International Panel on Climate Change [31] examines how migration is growing because of droughts, water shortages, and coastal floods as well as how climate change affects the regional and local people [57]. The minimum daily water requirement for an individual is 50 L, according to the United Nations World Water Development Report [82]. Any variations from this value, whether in developed or developing nations, results from problems with accessibility, availability, and water quality. According to the United Nations International Children's Emergency Fund (UNICEF) about two-thirds of the world's population will be water-scarce for at least one month in any given year. It has been estimated that around 700 million will be displaced by the year 2030 and almost half of the population will be dwelling in the water scare regions by early 2025 (https://www.unicef.org/wash/water-scarcity). According to projections, India will be most negatively impacted by the development of the urban population that lacks access to water (increase of 153–422 million people). It is anticipated that there will be an increase in the number of significant cities affected by water scarcity from 193 to 284, including 10–20 megacities. More than two thirds of water-scarce cities can reduce water scarcity by investing in infrastructure, however large-scale water scarcity solutions may have major environmental trade-offs that must be considered. The key concept and definitions of water stress have been discussed in the past year as the characteristics of quantity and quality characteristics. Definition 1: "Water stress occurs when the amount of water needed during a given time exceeds the amount that is available, or when poor quality makes it difficult to utilize. Definition 2: Freshwater resources suffer from water stress, which affects both quality and quantity (in terms of eutrophication, saline intrusion, organic matter pollution, etc.) such as over-exploitation of aquifers and dry rivers. Definition 3: "The capacity, or lack thereof, to satisfy freshwater demand on a human and ecological scale" [78]. In water-stressed areas, the quantity and quality of drinking water can act as a catalyst for conflict and unrest in vulnerable people. Urban water scarcity has been found to exacerbate existing difficult circumstances and endanger the region's development.

Water shortages result from the deterioration in the water quantity and quality. The mail idea of conducting this review is to manifest the idea of remote sensing (RS) and machine learning (ML) methods in (1) to establish the diagnostics and prognostics of water scarcity is constrained by the constrains in collecting water-supplies data and the specificity of water demands which will be based on the various indices that have been developed. (2) to the access to observations (while their performance and application is in continuous progress). (3) to fosters the creation of spatiotemporally sound products (i.e., higher temporal and spatial resolution, predictive products, and products for better process-understanding). In this paper we discuss the various applications of the remote sensing (RS) and machine learning (ML) for the urban water stress assessment. Machine learning and remote sensing will also help in the prediction of the missing data and providing an accurate assessment of results. In Fig. 3.1 the pictorial representation of the machine learning model is provided.

Mainly the focus is on physical water stress management, the key indicators used and the methods of the assessment that have been used until now. Physical water stress means there is not enough water (in terms of quantity and quality) to deal with the demand of an ecosystem [78]. This paper targets the audience involving urban water demand and supply managers, urban and regional planners, and research professionals working on developing management practices of the urban water stress. Figure 3.2: show the results what the queries searched in Scopus: (*urban AND water AND stress*) AND (*urban AND water AND scarcity OR machine AND learning OR remote and sensing)* executed for the years from 2000 to 2022. This query retrieved

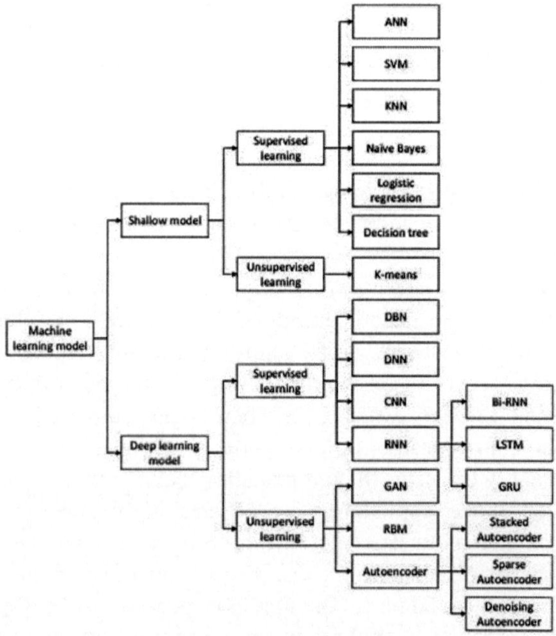

Fig. 3.1 Machine learning models can be utilised for the urban water stress

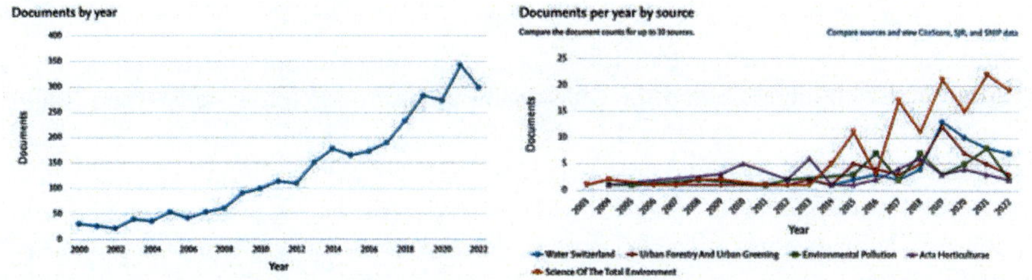

Fig. 3.2 Trend of research in urban water stress as results by SCOPUS search

around 3048 documents such that either articles title, key or abstract satisfies these criteria (Fig. 3.1). United states, China and India has been focusing on the area of the water scarcity. When further searched for the specifics of the remote sensing and machine learning techniques together are used which account for around 308 documents but with the individual search it did not yield good results. Therefore, final selection was made based on after reading the abstract and the conclusion. This chapter is divided into three sections: key indices and indicators for the water quantity and quality; application of remote sensing and machine learning in the water quantity and quality assessment and conclusion.

3.2 Key Indices and Indicators for the Water Quantity and Water Quality

Some of the key indices that are being used in previous literature are: Pfister's water quantity stress index (a log-function of the proportion of blue water supplies to water withdrawal and others are used to determine physical water stress; Falkenmark indicator (annual runoff per person); Criticality ratio (the ratio of withdrawal to blue water resources); Green water scarcity (the amount of per-person available green water resources) and blue water scarcity (the proportion of the blue water footprint to the blue water availability (difference of run-off and environmental flow requirements) [78].

A potent tool for facilitating talks among diverse participants are indicators. Several water indexes, but only a few of them concentrate on the management of urban water stress. For the urban waters, Van Leeuwen et al. [76] presented the City Blueprint sustainability strategy, which contains metrics that are governed by municipal organizations. Rather than information at the municipal level, it relies on information at the national level. For the long-term sustainability of urban water, Arcadis [3] created the Sustainability City Water Index, a standardized metric utilizing global data at varying spatial resolutions and public water utilities. It offers details about the waters' efficacy, endurance, and health. Milman and Short [49] created a Water Provision Resilience Index that considered how the water supply has changed over time. Demographics, facilities, services offered, financial resources, quality of water, and administration are all factors that affect access to potential clean safe waters. In 2013, Carden and Armitage created the Sustainability Index for Integrated Urban Water Management, a composite index that divides data into societal, socioeconomic, environmental, and organizational categories [14]. Jensen and Wu [33] developed a new index for the subcategories of water available resources, access, risks associated with water use, and administrative ability to manage water supplies. Water and environment protection were also considered in the Green City Index [73], City Resilience Index [4], and SDEWES Index [37].

3.3 Application of Remote Sensing and Machine Learning for Urban Water Quantity

The reasons for the transition of the water quantity in any urban system is based on the events that are occurring in the region such as floods and droughts which all the supply of the water availability in the system. Long-term transformations, alterations in the climate, shifting in population, and increases in urbanization, will probably lead to an increase in risk of flooding in the future cities which alters the underlying assumptions of flood risk analysis and management [23], and necessitating the development of new risk analysis tools [48]. It's crucial to comprehend the circumstances that contribute to flooding to learn to foresee them and use new techniques to lessen their consequences on metropolitan areas. Locations and events that cause floods include catchment-level flooding, river flooding, atmospheric processes, and water accumulating in metropolitan areas which are prone to [47]. One of the biggest obstacles to sustainability of the urban areas prone to flooding events is enhancing one's ability to withstand natural calamities [16]. Pielke [63] discussed that urban river—flooding can inflict significant amounts of damage, and while there may be a correlation between a hydrological features and destructive floods, understanding an area's hydrological features may not always imply knowledge of its exposure to devastating floods. Given that urban landscapes typically respond to heavy rainfall more quickly than natural surfaces, this knowledge is crucial for hazard-mitigation design in urban locations [68]. Shamseldin [71] highlighted that detention ponds, permeable concrete, soak aways, green spaces, or river training works involving the construction of dams and levees should be assessed and applied by understanding the risks and the responses of the area of the flood. For instance, [54] evaluated the entire Pattani basin, which includes two dams for managing water: Pattani Dam (diversion-type), and Bang Lang Dam (hydroelectric plant,) to anticipate floods for the city of Pattani lies in the southern part of Thailand. It is well known that the Pattani Dam's rapid flooding overflows into the city, causing the most frequent floods.

The experts have agreed that a plan for floods the region's management must include both structural and non-structural measures, such as technological advancements for monitoring drainage network data and expanding sensor coverage and frequency. The ML techniques used in each of these procedures can help in overcoming the obstacles for the sustainable development. Noymanee et al. [54] investigated five different machine learning techniques based on the available ground and the remotely sensed data sets which provided information about the hydrology of the region, the structural measures that are taken such as dams, the network of drains, and the technological aspects of the dams to provide an explanation for why catastrophic floods occur and to predict when flood peaks will occur in metropolitan areas. The five methods investigated were an Artificial Neural Network (ANN), choice forest regression, Bayesian linear regression, boosted decision tree regression, both of which are comparable to random forest analysis, and linear regression. Results from the Bayesian linear regression performed better which the least error and with the observations showing maximum correlations. This method's success may have been attributed to the fact that it was guided by probability distributions derived from historical data. Integration of decision support systems with prediction models is frequently informative for understanding and managing risks of urban floods beyond solely hydrological concerns. For instance, one study on Athens, Greece, [69] integrated a hydrological model with a network flow programming model (NFP) for demand management and a Feed Forward Neural Network (FFNN) to simulate the water supply network. Given hydrological inputs, the NFP simulates and improves how a water supply system function. According to [51], FFNNs are the simplest type of ANN and are well-suited for multi-model coupling since information passes put nodes in a forward direction from the output nodes to the hidden layer. In this instance, the NFP used synthetic data that was long enough to capture the risk associated with each insurance. Following that, the NFP's penalty functions were chosen to match the operational rules with various levels of risk acceptance. This procedure produced a sizable training data set that was collected over a considerable amount of time and then fed into the FFNN. Damage happening to assets goes beyond the flooded region is considered an indirect flood effect. After a significant flooding event, these assets may have effects that persist for many days or weeks or years, whether they are physical, fiscal, sociological, or ecological [15]. Simulations based on multiple agents have been used to measure the magnitude of these effects. To evaluate the consequences of individual activities on the system, agent-based models replicate the behaviors and autonomous agents, which could be communities, and their interactions. In a study in Tokyo, Japan, agent-based simulation was used in conjunction with reinforcement learning, which rewards software systems for actions taken to optimize their overall benefit, maximizing both individual and corporate post-disaster recovery. Through comparison to statistical techniques and imprecise empirical models, this study demonstrated increased indirect damage reduction ability and make provision for mitigation. Two studies were used by [86] to show the impact of reinforcement learning using agent-based models in aiding recovery decision-making following a flood disaster. The case that uses the ML approach fared better than the other example and had a faster rate of recovery.

A long-term absence of precipitation that can continue for diverse spatial and temporal scales, from micro to regional, may span several years or even decades

[27]. Drought can be made worse by changes in extreme heat, a lack of soil moisture, feedbacks from the land to the atmosphere, in the sea surface temperature, air circulation, and anthropogenic impacts including altering land use and vegetation and increasing water needs [17, 18, 34]. Droughts are severe weather risks that have an influence on human health, agriculture, the economy, the environment, and water supplies [27]. Huntingford et al. [30] estimated the that the actual budget of the dryness over the past two decades was in the hundreds of billions. Pagán et al. [58] and Pendergrass et al. [62] have highlighted that droughts are expected to last longer and be more intense in a warmer climate. It can be improved by working on the ability to predict events in real time and creating early warning systems are essential for strategic planning and risk management related to drought. A detailed understanding of the physical systems that generate drought is required to better drought forecasting [77]. To comprehend the complex physical mechanisms that result in the extremely low moisture levels of drought, numerous investigations have been carried out. To predicting droughts, scientists have used dynamical methods such as simulations of climate and hydrological models, statistical models that make use of a variety hybrid frameworks, drought indicators, and classifiers [19, 27, 82]. The application of ML approaches to increase drought predictability has increased during the past ten years [27]. Rahmati et al. [64] for the Australia's southeast part assessed how well 6 different ML algorithms worked at predicting the danger of agricultural drought by using RF, multivariate adaptive regression splines (MARS), flexible discriminant analysis (FDA), classification and regression trees (CART), BRT, and Support Vector Machines (SVM). They found that of the models, RF had the accuracy and predication capacity. For instance, studies that anticipate dryness are increasingly using random forest ML algorithms. It has been found that the RF performed better for the Standardized Precipitation Index (SPI) among all techniques [40]. Similar study was performed on Shipra River in India [61]. It is possible to develop different decision tree types together as a "forest" in a computing system to create Tree based analysis extensions known as random forests. They maintain flexibility in the choice of analytical techniques while providing complex predictors interactions are exceptionally accurately classified and characterized [2]. In comparison to conventional linear regression models, random forests additionally have the capacity to address the problems of overfitting and multicollinearity [38]. Park et al. [60] by utilizing 16 drought variables derived from remote sensed in both dry and subtropical cities in the United States, employed boosted regression techniques, RF, and Cubist Machine learning algorithms (rule-based modelling trees in which the terminal leaves contain linear regression models) to detect agriculture and climatic drought. Zaniolo et al. [87] contributed to the FRIDA (Framework for Index-based Drought Analysis) structure for the automatic analysis of droughts by applying an ML parameter selection algorithm for the determination for the basin -drought indices. The core of the method is the W-QEISS process, it finds Pareto an efficient group of parameters using an inter evolutionary process. By computing and combining all the pertinent data about the system's water cycle that was found using the feature selection algorithm. As a result, it produced a methodology that can determine an indicator that acts as a stand-in for basin's dry seasons. Forecasting droughts has also been done using ANN-ML

methods [9, 50]. Belayneh et al. [10] utilised a multiple data processing techniques such as SVR, ANN and bootstrap ensemble techniques for Ethiopia. By creating a sequence of models that are focused on training scenarios that were not previously successfully predicted, boosting approaches enhance the performance of an algorithm. In comparison to either ANN or SVR alone, the researchers discovered that coupled models performed better and offered more reliable SPI predictions. Model accuracy issues, computation performance issues, and local minima pitfalls are all potential drawbacks of ANN models.

As a result, compared to other ML methods such as SVM, ANN, and RF, there has been an increase in the quick execution and better prediction accuracy by the XGBoost [22, 88]. Zhang et al. [88] analyzed the effectiveness of XGBoost classifier with ANN and a standard statistical model for the Shaanxi Province of China to predict the Standardized Precipitation Evapotranspiration Index (SPEI). It was found to be faster and more accurate. It has been seen that different models perform differently when estimating the risk of various risks.

3.4 Application of Remote Sensing and Machine Learning with the Water Quality

The global water supply, ecosystems, infrastructure, and human well-being are all suffering of the reduction in quality of water in both surface and ground water [36]. The administration of water distribution systems and watershed frequently requires precise and reliable methods for evaluating quality of the water and foreseeing future contamination of water [13]. It is difficult to develop reliable and timely predictions of water quality. The conventional methods analysis and forecast water quality characteristics using water quality models. Most of these models are quantitative models of the physical processes that govern how contaminants are transferred from ground inputs into ecosystem of water [67]. Water quality models are valuable for simulating circumstances, but they can only offer one piece of information that approximates reality in an imprecise way [35]. This is since (1) there are several interconnected multi-domain activities (2) a significant number of underlying mechanisms that may have an impact on water quality are yet unknown. Procedures using sophisticated water quality modelling are frequently requires time, expensive and labour intensive [53]. The new developing data-driven strategies for predicting quality of water usually rely on a huge amount data from many sources [36]. Beran and Piasecki [11] have used the data sources are the National Water Information System (NWIS) online resource from the United States Geological Survey (USGS) and the STORET Data Warehouse from the United States Environmental Protection Agency (USEPA). These evaluations often consider the combined impact of several water quality indicators, including salinity, pH, dissolved oxygen, suspended solids, and ammoniacal nitrogen (NH_3–N). As watersheds are changeable and affected by the basin dynamics, parameters of quality of water may vary between basins. Bui et al. [13] indicated that using conventional methods to calculate Water Quality Index (WQI) takes time and results in errors when estimating the sub-indices. Researchers have used numerous ML strategies to overcome these constraints and enhance water quality analysis and prediction [36, 53], and a few hybrid systems [79]. Next, are some of these strategies can be used. ANN models were used to forecast river and coastal water quality in India and Singapore by Singh et al. [74] and Palani et al. [59]. García-Alba et al. [24] created an ANN model, to evaluate bathing water quality in estuaries. They estimate Es.coli concentrations that are comparable to those estimated by process-based models. Lu and Ma [45] recommended integrating two ML models to improve water quality forecasting: hybrid XGBoost model outperformed the random forest model for several factors such as temperature, fluorescent dissolved organic matters specific conductance prediction, pH values, dissolved oxygen (DO), turbidity. But the two methods' aggregate efficiency was the best for maximizing the creation of a water quality index.

A convolutional neural network (CNN), an ANN with a convolutional activation function, and the long short-term memory (LSTM) model, which includes feedback in addition to feedforward networks, were applied to the two water quality variables DO and Chlorophyll in the Small Prespa Lake in Greece by Barzegar et al. [8]. Where the LSTM outperformed the CNN model for DO and Chl-a prediction. Li et al. [41] and Lu and Ma [45], both used similar effective methods that involved. Isn northern Iran [13], used—random forests and its three variants as well as 12 other algorithm combinations. Total solids had the least impact while faecal coliform contents had the most impact. Read et al. [66] merged theoretical and ML approaches to improve predictions of factors linked to water quality that are governed by physical laws. The study provided a hybrid modelling framework called Process-Guided Deep Learning (PGDL) for predicting depth-specific lake water temperature, which is a crucial indicator of water quality. The researchers showed through the use case that incorporating scientific knowledge into deep learning methods has the potential to improve predictions of many crucial environmental factors. Other causes of water quality degradation are the sedimentation and soil erosion such as agriculture and urbanization. Without sufficient prevention, sedimentation and eroding in urban settings can cause structural damage (such as harm to utility supply chain network, fences, and gates), a rise in water treatment costs, and other undesirable outcomes [55]. Earlier, numerous types for siltation and erosion have been used Merritt et al. [46]. Liang et al. [42] proposed a comparison study showing that physical based models can be supplemented by the data driven data-driven models. There are numerous studies being conducted right now that use different ML techniques to address distinct problems with sediment study. For modelling the sediment transport ANN, Adaptive fuzzy interference system and M5 model trees are used [12, 26, 56, 81, 86]. The prediction of the sediment load can be done by RF, Genetic Algorithms, and unsupervised algorithms [44, 84, 85]. The soil erosion prediction uses Tree-based machine learning techniques, ANN and SVM [29, 52, 65]. The impacts of sediment on urban infrastructure are studied by RF, Adaptive-Network-based Fuzzy Inference System (ANFIS) [5, 6]. A list of tables is compiled for the literature that is being used in this article as Table 3.1.

Table 3.1 A consolidated list of ML and RS methods discussed

Topic	Machine learning	References
Water quantity	ANN, linear RF, linear Bayesian regression, boosted decision tree and forest regression	Noymanee et al. [54]
	Reinforcement learning with agent-based models	Yang et al. [86]
	CART, BRT, random forests, MARS, FDA, SVM	Rahmati et al. [65]
	CART, random forests	Kuswanto and Naufal [40]
	XG Boost	Zhang et al. (2019)
	W-QEISS	Zaniolo et al. [87]
	LSTM	Shi et al. [72]
Water quality	ANN	Tayfur (2002), Bhattacharya et al. [12], Lin and Montazeri Namin [43], Yang et al. [85]
	M5 model trees	Onderka et al. [56], Goyal [26]
	Adaptive-network-based fuzzy inference system (ANFIS)	Lin and Montazeri Namin [43], Wieprecht et al. [80], Bakhtyar et al. [7]
	XG Boost, RF	Lu and Ma [45]

3.5 Conclusion

It can be concluded there aren't any standardized methods for figuring out the environmental flows that affect water quality, and people don't know about the return flows following water usage at different levels, such domestic and industrial, should be kept to a minimum. When viewed from the standpoint of water scarcity, there has been a problem with the standardisation of terminology and words, particularly the water quality. The security of the water supply should also be improved. The emphasis should be on water-sensitive cities that provide for both quantity and quality of water while fostering water resilience [78]. The creation of resilient and long-term urban water systems which can handle future and current climate changes, water shortages, and growing demands must be a part of any strategy for resolving the urban water problem. In addition to the infrastructure, control systems, delivery mechanism, industry-supportive climate, and more general aspects like fee structures all play a role in how long services will be available to customers. Water shortage management will require collaboration and communication between partners from a range of industries [25]. A planned framework for the conduct of the actionable measures and supporting water sustainability and security must be in the place. This can be aided with the help of the remotely sensed and the machine learning techniques in an integrated and cyclic manner. It has been seen that many ML algorithms have been used in past for the analysis of the risk to the water quantity and water quality in the urban water systems. It has also been discussed which models can improve upon the predication of the parameters used. There is requirement of the work in coupling the remote sensing technology and the machine learning specifically for the urban water quantity and quality assessment are still underrepresented. It has been found that there was a lack of the inclusion of elements such as political, governance, socio-economic challenges much be kept in mind while working with these models. The developed models can be further incorporated within the indexes and improve upon the urban water accessibility to all in a sustainable an efficient manner. The indexes created have limitations in terms of an information that is available, the claims that were made, and the surrogate data, particularly when looking at country specific data for the regions, which results in a significant amount of variation in the geographical distribution. Data losses have occurred when many indexes are mixed, which causes information losses. The dashboard technique, which shows all the factors and allows for group collaboration, can be used to overcome issue. Also, the use of knowledge graphs can make the assessment of the water scarcity in the urban systems better by the integration of the various parameters. These problems about water shortages can be resolved by integrating indigenous knowledge, synchronizing institution actions, scheduling innovation, managing for predicted climate risks, especially expanding environmental amenities, and enhancing climate science awareness.

References

1. Aivazidou E, Banias G, Lampridi M, Vasileiadis G, Anagnostis A, Papageorgiou E, Bochtis D (2021) Smart technologies for sustainable water management: an urban analysis. Sustainability (Switzerland)
2. Allen MR, Zaidi SMA, Chandola V, Morton AM, Brelsford CM, McManamay RA, KC B, Sanyal J, Stewart RN, Bhaduri BL (2018) A survey of analytical methods for inclusion in a new energy-water nexus knowledge discovery framework. Big Earth Data 2:197–227
3. Arcadis (2016) Sustainable cities water index. Retrieved from https://www.arcadis.com/media/4/6/2/%7B462EFA0A-4278-49DF-9943-C067182CA682%7DArcadis_Sustainable_Cities_Water_Index-Web.pdf
4. ARUP (2014) City resilience framework: ARUP & the Rockefeller Foundation. http://publications.arup.com/publications/%20c/%20city_resilience_framework
5. Azamathulla HM, Ab. Ghani A, Fei SY (2012) ANFIS-based approach for predicting sediment transport in clean sewer. Appl Soft Comput J 12:1227–1230
6. Azamathulla HMD, Ghani AA (2011) ANFIS-based approach for predicting the scour depth at culvert outlets. J Pipeline Syst Eng Pract 2:35–40
7. Bakhtyar R, Bakhtiary AY, Ghaheri A (2008) Application of neuro-fuzzy approach in prediction of runup in swash zone. App Ocean Res 30(1):17–27
8. Barzegar R, Aalami MT, Adamowski J (2020) Short-term water quality variable prediction using a hybrid CNN–LSTM deep learning model. Stoch Env Res Risk Assess 34:415–433
9. Belayneh A, Adamowski J, Khalil B, Ozga-Zielinski B (2014) Long-term SPI drought forecasting in the Awash River Basin in Ethiopia using wavelet neural networks and wavelet support vector regression models. J Hydrol (Amst) 508:418–429
10. Belayneh A, Adamowski J, Khalil B, Quilty J (2016) Coupling machine learning methods with wavelet transforms and the bootstrap and boosting ensemble approaches for drought prediction. Atmos Res 172–173:37–47

11. Beran B, Piasecki M (2008) Availability and coverage of hydrologic data in the US geological survey national water information system (NWIS) and US environmental protection agency storage and retrieval system (STORET). Earth Sci Inform 1:119–129
12. Bhattacharya B, Price RK, Solomatine DP (n.d.) Machine learning approach to modeling sediment transport
13. Bui DT, Khosravi K, Tiefenbacher J, Nguyen H, Kazakis N (2020) Improving prediction of water quality indices using novel hybrid machine-learning algorithms. Sci Total Environ 721
14. Carden K, Armitage NP (2013) Assessing urban water sustainability in South Africa—Not just performance measurement. Water SA 39:345–350
15. Castello L, Bayley PB, Fabré NN, Batista VS (2019) Flooding effects on abundance of an exploited, long-lived fish population in river-floodplains of the Amazon. Rev Fish Biol Fish 29:487–500
16. Chang F-J, Hsu K, Chang L-C (n.d.) Flood forecasting using machine learning methods
17. Cook ER, Seager R, Cane MA, Stahle DW (2007) North American drought: reconstructions, causes, and consequences. Earth Sci Rev 81:93–134
18. Dai A (2011) Drought under global warming: a review. Wiley Interdiscip Rev Clim Change
19. Dutra E, Pozzi W, Wetterhall F, di Giuseppe F, Magnusson L, Naumann G, Barbosa P, Vogt J, Pappenberger F (2014) Global meteorological drought-Part 2: seasonal forecasts. Hydrol Earth Syst Sci 18:2669–2678
20. Etikala B, Golla V, Li P, Renati SR (2019) Deciphering groundwater potential zones using MIF technique and GIS: a study from Tirupati area, Chittoor District, Andhra Pradesh, India. HydroResearch 1:1–7
21. Etikala B, Madhav S, Srinivasa Gowd S (2022) Urban water systems: an overview. Curr Direct Water Scarcity Res 6:1–19
22. Fan J, Wang X, Wu L, Zhou H, Zhang F, Yu X, Lu X, Xiang Y (2018) Comparison of support vector machine and extreme gradient boosting for predicting daily global solar radiation using temperature and precipitation in humid subtropical climates: a case study in China. Energy Convers Manag 164:102–111
23. Gangrade S, Kao SC, Dullo TT, Kalyanapu AJ, Preston BL (2019) Ensemble-based flood vulnerability assessment for probable maximum flood in a changing environment. J Hydrol (Amst) 576:342–355
24. García-Alba J, Bárcena JF, Ugarteburu C, García A (2019) Artificial neural networks as emulators of process-based models to analyse bathing water quality in estuaries. Water Res 150:283–295
25. Gaya S, Ward F, Alvarez-Sala J, Agberemi B (n.d.) Urban water scarcity guidance note preventing day zero
26. Goyal MK (2014) Modeling of sediment yield prediction using M5 model tree algorithm and wavelet regression. Water Resour Manag 28:1991–2003
27. Hao Z, Singh VP, Xia Y (2018) Seasonal drought prediction: advances, challenges, and future prospects. Rev Geophys 56:108–141
28. House-Peters LA, Chang H (2011) Urban water demand modeling: Review of concepts, methods, and organizing principles. Water Res Res 47(5):1–15
29. Hosseinalizadeh M, Kariminejad N, Chen W, Pourghasemi HR, Alinejad M, Mohammadian Behbahani A, Tiefenbacher JP (2019) Gully headcut susceptibility modeling using functional trees, naïve Bayes tree, and random forest models. Geoderma 342:1–11
30. Huntingford C, Jeffers ES, Bonsall MB, Christensen HM, Lees T, Yang H (2019) Machine learning and artificial intelligence to aid climate change research and preparedness. Environ Res Lett 14
31. IPCC (2007) Is the assessment report N4. Please refer also on the IPCC AR5 (2014) and IPCC AR6 (2022)
32. Jaimes-Correa JC, Muñoz-Arriola F, Bartelt-Hunt S (2022) Modeling water quantity and quality nonlinearities for watershed adaptability to hydroclimate extremes in agricultural landscapes. Hydrology 9

33. Jensen O, Wu H (2018) Urban water security indicators: development and pilot. Environ Sci Policy 83:33–45
34. Kam J, Sheffield J, Wood EF (2014) A multiscale analysis of drought and pluvial mechanisms for the southeastern United States. J Geophys Res 119:7348–7367
35. Kebede S, Kebede S (2013) Geochemistry and water quality. Groundw Ethiop: Featur, Numbers Oppor 163–185
36. Khan Y, See CS (2016) Predicting and analyzing water quality using machine learning: a comprehensive model. In: 2016 IEEE long island systems, applications and technology conference, LISAT 2016. Institute of Electrical and Electronics Engineers Inc.
37. Kılkış, Ş. (2018). Sustainable Development of Energy, Water and Environment Systems (SDEWES) Index for policy learning in cities. Int J Innov Sustain Dev 12(1–2):87–134
38. Konapala G, Mishra A (2020) Quantifying climate and catchment control on hydrological drought in the continental United States. Water Resour Res 56
39. Kumar A, Ramsankaran RAAJ, Brocca L, Muñoz-Arriola F (2021) A simple machine learning approach to model real-time streamflow using satellite inputs: demonstration in a data scarce catchment. J Hydrol (Amst) 595.
40. Kuswanto H, Naufal A (2019) Evaluation of performance of drought prediction in Indonesia based on TRMM and MERRA-2 using machine learning methods. MethodsX 6:1238–1251
41. Li C, Li Z, Wu J, Zhu L, Yue J (2018) A hybrid model for dissolved oxygen prediction in aquaculture based on multi-scale features. Inf Process Agricult 5:11–20
42. Liang J, Li W, Bradford SA, Šimůnek J (2019) Physics-informed data-driven models to predict surface runoff water quantity and quality in agricultural fields. Water (Switzerland) 11.
43. Lin B, Montazeri Namin M (2005) Modelling suspended sediment transport using an integrated numerical and ANNs model. J Hydraul Res 43:302–310
44. López-Tarazón JA, Batalla RJ, Vericat D, Francke T (2012) The sediment budget of a highly dynamic mesoscale catchment: the River Isábena. Geomorphology 138:15–28
45. Lu H, Ma X (2020) Hybrid decision tree-based machine learning models for short-term water quality prediction. Chemosphere 249
46. Merritt WS, Letcher RA, Jakeman AJ (2003) A review of erosion and sediment transport models. In: Environmental modelling and software. Elsevier BV, pp 761–799
47. Merz B, Hall J, Disse M, Schumann A (2010) Natural hazards and earth system sciences fluvial flood risk management in a changing world, Hazards Earth Syst Sci
48. Milly PCD, Betancourt J, Falkenmark M, Hirsch RM, Kundzewicz ZW, Lettenmaier DP, Stouffer RJ (2008) Climate change: stationarity is dead: whither water management? Science (1979)
49. Milman A, Short A (2008) Incorporating resilience into sustainability indicators: an example for the urban water sector. Glob Environ Chang 18:758–767
50. Morid S, Smakhtin V, Bagherzadeh K (2007) Drought forecasting using artificial neural networks and time series of drought indices. Int J Climatol 27:2103–2111
51. Mosavi A, Ozturk P, Chau KW (2018) Flood prediction using machine learning models: literature review. Water (Switzerland)
52. Mustafa MRU, Sholagberu AT, Yusof KW, Hashim AM, Khan MWA, Shahbaz M (2018) SVM-based geospatial prediction of soil erosion under static and dynamic conditioning factors. In: MATEC web of conferences. EDP Sciences
53. Najah Ahmed A, Binti Othman F, Abdulmohsin Afan H, Khaleel Ibrahim R, Ming Fai C, Shabbir Hossain M, Ehteram M, Elshafie A (2019) Machine learning methods for better water quality prediction. J Hydrol (Amst) 578
54. Noymanee J, Nikitin NO, Kalyuzhnaya AV (2017) Urban pluvial flood forecasting using open data with machine learning techniques in Pattani Basin. In: Procedia computer science. Elsevier B.V., pp 288–297
55. National Resources Conservation Service (NRCS) (2008) Web soil survey. Data from survey. Accesssed 12 Nov 2008
56. Onderka M, Krein A, Wrede S, Martínez-Carreras N, Hoffmann L (2012) Dynamics of storm-driven suspended sediments in a headwater catchment described by multivariable modeling. J Soils Sediments 12:620–635

57. Pachauri RK, Reisinger A, Bernstein L (2007) Intergovernmental panel on climate change. Core writing team, intergovernmental panel on climate change, n.d. Climate change: synthesis report
58. Pagán BR, Ashfaq M, Rastogi D, Kendall DR, Kao SC, Naz BS, Mei R, Pal JS (2016) Extreme hydrological changes in the southwestern US drive reductions in water supply to Southern California by mid century. Environ Res Lett 11
59. Palani S, Liong SY, Tkalich P (2008) An ANN application for water quality forecasting. Mar Pollut Bull 56:1586–1597
60. Park S, Im J, Jang E, Rhee J (2016) Drought assessment and monitoring through blending of multi-sensor indices using machine learning approaches for different climate regions. Agric For Meteorol 216:157–169
61. Patidar R, Santosh, Pingale M, Khare D (n.d.) An integration of geospatial and machine learning techniques for mapping groundwater potential: a case study of the Shipra river basin, India
62. Pendergrass AG, Meehl GA, Pulwarty R, Hobbins M, Hoell A, AghaKouchak A, Bonfils CJW, Gallant AJE, Hoerling M, Hoffmann D, Kaatz L, Lehner F, Llewellyn D, Mote P, Neale RB, Overpeck JT, Sheffield A, Stahl K, Svoboda M, Wheeler MC, Wood AW, Woodhouse CA (2020) Flash droughts present a new challenge for subseasonal-to-seasonal prediction. Nat Clim Chang
63. Pielke Jr RA (2000) Policy history of the US global change research program: Part II. Legislative process. Glob Environ Chang 10(2):133–144
64. Rahmati O, Falah F, Dayal KS, Deo RC, Mohammadi F, Biggs T, Moghaddam DD, Naghibi SA, Bui DT (2020) Machine learning approaches for spatial modeling of agricultural droughts in the south-east region of Queensland Australia. Sci Total Environ 699
65. Rahmati O, Tahmasebipour N, Haghizadeh A, Pourghasemi HR, Feizizadeh B (2017) Evaluation of different machine learning models for predicting and mapping the susceptibility of gully erosion. Geomorphology 298:118–137
66. Read JS, Jia X, Willard J, Appling AP, Zwart JA, Oliver SK, Karpatne A, Hansen GJA, Hanson PC, Watkins W, Steinbach M, Kumar V (2019) Process-guided deep learning predictions of lake water temperature. Water Resour Res 55:9173–9190
67. Refsgaard JC, Henriksen HJ (2004) Modelling guidelines—Terminology and guiding principles. Adv Water Resour 27:71–82
68. Rodriguez F, Andrieu H, Creutin JD (2003) Surface runoff in urban catchments: morphological identification of unit hydrographs from urban databanks. J Hydrol (Amst) 283:146–168
69. Rozos E (2019) Machine learning, urbanwater resources management and operating policy. Resources 8
70. Sarzaeim P, Muñoz-Arriola F, Jarquín D (2022) Climate and genetic data enhancement using deep learning analytics to improve maize yield predictability. J Exp Bot 73:5336–5354
71. Shamseldin AY (2010) Artificial neural network model for river flow forecasting in a developing country. J Hydroinf 12:22–35
72. Shi X, Chen Z, Wang H, Yeung DY, Wong W, Woo WC (2015) Convolutional LSTM network: A machine learning approach for precipitation nowcasting. Adv Neural Inf Process Syst 28
73. Siemens A (2012) The green city index. Siemens AG
74. Singh KP, Basant A, Malik A, Jain G (2009) Artificial neural network modeling of the river water quality-A case study. Ecol Modell 220:888–895
75. Tiwari M, Adamowski J, Adamowski K (2015) Water demand forecasting using extreme learning machines. J Water Land Dev 28(1):37
76. van Leeuwen CJ, Frijns J, van Wezel A, van de Ven FHM (2012) City blueprints: 24 indicators to assess the sustainability of the urban water cycle. Water Resour Manag 26:2177–2197
77. van Loon AF (2015) Hydrological drought explained. Wiley Interdiscip Rev Water 2:359–392
78. Wang D, Hubacek K, Shan Y, Gerbens-Leenes W, Liu J (2021) A review of water stress and water footprint accounting. Water (Switzerland)
79. Wang X, Zhang F, Ding J (2017) Evaluation of water quality based on a machine learning algorithm and water quality index for the Ebinur Lake watershed, China. Sci Rep 7

80. Wieprecht S, Tolossa HG, Yang CT (2013) A neuro-fuzzy-based modelling approach for sediment transport computation. Hydrol Sci J 58:587–599
81. Wood EF, Schubert SD, Wood AW, Peters-Lidard CD, Mo KC, Mariotti A, Pulwarty RS (2015) Prospects for advancing drought understanding, monitoring, and prediction. J Hydrometeorol 16:1636–1657
82. WWDR (2019) World water development report (UNESCO World Water Assessment Programme). The United Nations World Water Development Report 2019: Leaving No One Behind. Paris, UNESCO
83. Xu H, Demir I, Koylu C, Muste M (2019) A web-based geovisual analytics platform for identifying potential contributors to culvert sedimentation. Sci Total Environ 692:806–817
84. Yadav A, Chatterjee S, Equeenuddin SM (2018) Suspended sediment yield estimation using genetic algorithm-based artificial intelligence models: case study of Mahanadi River, India. Hydrol Sci J 63:1162–1182
85. Yang CT, Marsooli R, Aalami MT (2009) Evaluation of total load sediment transport formulas using ANN. Int J Sedim Res 24:274–286
86. Yang S, Ogawa Y, Ikeuchi K, Akiyama Y, Shibasaki R (2019) Firm-level behavior control after large-scale urban flooding using multi-agent deep reinforcement learning. In: Proceedings of the 2nd ACM SIGSPATIAL international workshop on geospatial simulation, GeoSim 2019. Association for Computing Machinery, Inc, pp 24–27
87. Zaniolo M, Giuliani M, Castelletti AF, Pulido-Velazquez M (2018) Automatic design of basin-specific drought indexes for highly regulated water systems. Hydrol Earth Syst Sci 22:2409–2424
88. Zhang R, Chen ZY, Xu LJ, Ou CQ (2019) Meteorological drought forecasting based on a statistical model with machine learning techniques in Shaanxi province, China. Sci Total Environ 665:338–346
89. Zhou S, McMahon T, Walton A, Lewis J (2002) Forecasting operational demand for an urban water supply zone. J Hydrol 259(1–4):189–202

Chapter 4
Role of Artificial Intelligence in Water Conservation with Special Reference to India

Piyush Pandey, Avinash Pratap Gupta, Joystu Dutta, and Tarun Kumar Thakur

Abstract All life on earth depends on water as a resource. The availability of water is dwindling over time. Water is a renewable resource that, if not managed, will eventually run out of resources. The artificial intelligence can play a very important role in water conservation, especially in developing countries such as India. The employment of artificial intelligence in modelling approaches produces answers for linear, non-linear, and other systems that are close to the real data, as shown by a number of studies in the literature. For the following water variables such as rainfall-runoff, evaporation and evapotranspiration, water quality components, streamflow, sediment, and changes in dams or lakes levels, we looked at the state-of-the-art and advancements in artificial intelligence modelling. This article investigates the function of artificial intelligence in water conservation in order to achieve the goal of serving as a reference for the general public. It accomplishes this by fully introducing this notion, its import, and relevant application scenarios. The study has also offered a number of interesting research topics and models for the water variable variables.

Keywords Evapotranspiration · Artificial intelligence · Water quality · Runoff · Sediment

P. Pandey · A. P. Gupta · J. Dutta (✉)
Department of Environmental Science, UTD, Sant Gahira Guru University, Ambikapur, Sarguja (C.G.) 497001, India
e-mail: joystu.dutta@gmail.com

P. Pandey
e-mail: pp259711@gmail.com

A. P. Gupta
e-mail: avi.gupta005@gmail.com

T. K. Thakur
Department of Environmental Science, Indira Gandhi National Tribal University, Amarkantak, M.P. 484886, India
e-mail: tarun.thakur@igntu.ac.in

E. Balaji et al. (eds.), *Emerging Technologies for Water Supply, Conservation and Management*, Springer Water, https://doi.org/10.1007/978-3-031-35279-9_4

4.1 Introduction

Artificial Intelligence—As far back as the ancient Chinese, Indian, and Greek philosophers, the concept of artificial intelligence is predicated on the premise that human intellect may be mechanically reproduced [14]. While the formalisation of reasoning, which would enable debate to be reduced to calculation, remained the main focus of ancient and mediaeval philosophers [17].

The definition of artificial intelligence is "brain-mimicry." Computer systems that "can perceive their environment, reason, learn, and act in response to what they sense and their programme objectives" are referred to as having artificial intelligence (AI). The term "artificial intelligence" speaks of the idea and creation of task-performing computers available today that would typically human intellect is necessary. Various expert systems, speech recognition, and trading systems are examples of AI-type systems. The terms "machine learning" and "computer vision" are used frequently when discussing two significant ideas in the area of AI.

Many of the features found in programming languages like MATLAB and FORTRAN are mathematical functions needed to employ these techniques. For the water variable forecasting, such as sediment, dam or lake water levels, stream flow, rainfall-runoff, evaporation and evapotranspiration, and water quality factors, artificial neural networks fuzzy-based models (ANN), and the majority of their hybrids often used AI technologies. As shown in Table 4.1, for instance, AI techniques have been used to correctly calculate the amount of sediment in the width of a river, rainfall-runoff, streamflow, evaporation and evapotranspiration, water quality variables, and modelling of Lake or dam water levels.

The review papers [2, 19, 20, 46, 48, 51, 66] discuss the use of AI approaches in water resources at various points in time.

Table 4.1 Is a summary of recent AI investigations organised by factors

S.n	Variables	Recent studies
1	Water level of a lake or dam	Üneş et al. [65], Li et al. [43]
2	Evaporation and evapotranspiration	Goyal et al. [31], Karimi et al. [32], Güçlü et al. [40]
3	Rainfall-runoff	Talei et al. [63], Darras et al. [24], Londhe et al. [45], Chithra and Thampi [21]
4	Sediment	Güner and Yumuk [33], Demirci et al. [25], Droppo and Krishnappan [28], Talebi et al. [62]
5	Streamflow	Cigizoglu [23], Huang et al. [35], Nourani et al. [51], Ashrafi et al. [5]
6	Water quality variables	Akkoyunlu et al. [3], Ay and Kisi [6–10], Ay [11], Chang et al. [18], Alizadeh and Kavianpour [4], Khan and Valeo [42]

4.2 Review Studies in the Modelling of Water Variables

The studies on the modelling of water variables at various times in time are discussed by [2, 12, 19, 20, 22, 46, 48, 49, 51, 53, 55, 64, 66].

4.3 Water Conservation

Water with the molecular formulae H_2O is represented by the chemical connection established involving two hydrogen atoms and one atom of oxygen. Water exists in all the three states as Gas, liquid, and solid [16]. In addition, regarded as one of the most vital sources of information for all recognized life forms. In order to understand the value of water, assure its sustainability, and leave a society where future generations will have access to clean water, regulating water, tracking it, and conserving it should therefore be among the top priorities of work. As a result, each nation is required to implement water resource restrictions in order to safeguard its own natural environment.

The rapidly expanding industrial sector has seriously contaminated the water environment with massive amounts of domestic and industrial sewage, putting the formerly scarce water resources to extreme strain [56]. Some current sewage treatment facilities and water environment treatment projects have disappointing results, which not only harm the economy and the environment but also pose major concealed risks to the lives and health of locals [37]. In order to increase the effectiveness of water environment administration, artificial intelligence algorithms are used to monitor ecological data connected to the water environment and suggest scientific processing procedures.

4.4 Managing Water in Megacities

A project called EQWATER is funded by the IMPRINT initiative to guarantee an equitable water supply in megacities. Dr. Yogesh Simmhan, an associate professor in the department of computational and data sciences at the Indian Institute of Science in Bengaluru, is the project's supervisor. The group of researchers are investigating networks-analytics for managing data from field devices and optimizing supply schedules. They have for controlling data from mobile devices and enhancing delivery schedules to ensure that everyone has access to a reliable as well as affordable source of healthy water. Megacities (cities with a population of at least 10 million) are particularly susceptible to this problem because many of them have access issues to potable drinking water.

Smart gadgets with online access and embedded systems like microcontrollers, sensors, and communication hardware are referred to as Internet of Things (IoT) devices. IoT is also gathering, sending, and acting on the data that devices gather from the environment. IoT gadgets share sensor information from an edge device, such as an IoT gateway, where the information is either analysed locally or transmitted to the cloud for further study. These devices communicate with one another and behave in response to the information they exchange.

4.5 Government of India Water Resource Management Policies and Programs

The National Water Mission (NWM), one of eight national missions established by the Indian government in compliance with the National Action Plan on Climate Change (NAPCC). On April 6, 2011, the comprehensive Mission Document for the National Water Mission was adopted by the Union Cabinet (NWM). The primary goal of NWM is "Conservation of water, minimizing wastage and ensuring its more equitable distribution both across and within States through integrated water resources development and management".

The NWM's five main objectives are:

- A substantial public water data database and an examination of how climate change is affecting water supplies;
- Assisting state and civilian water augmentation, conservation, and preservation programs;
- Illustrated sensitive areas, especially those that are overused;
- An increase in water use efficiency of 20%; and the development of basin-level integrated water resource management.

Government of India has further designed and implemented phase-wise strategies for realizing the goals. This includes coordinated planning for sustainable development with key stakeholder involvement. Moreover, a Mission Directorate was created by the Ministry of Water Resources, River Development, and Ganga Rejuvenation. According to the NWM Mission Statement, eight advisory groups and committees have been established. Under the Central Scheme "Implementation of National Water Mission (NWM)," the Ministry of Water Resources planned and used a total cash outlay of Rs. 196 crores (Rs. one hundred ninety-six crores only) during the XII Plan period, or from April 2016 to March 2017.

4.5.1 Components

- Directorate for the National Water Mission
- Making state-specific action plans (SSAP)
- Developing human resources and building capacity

- The establishment of the National Bureau of Water Use Efficiency (NBWUE)
- Baseline studies.

Creating detailed project reports for benchmarking and demonstration projects.

Water is an invaluable natural resource and a priceless national asset and its conservation is pivotal for sustenance of all life forms [54]. 4% of the world's water resources are found in India, and makes up about 18% of the global population. Most of the population in India experiences water shortages due to unequal water distribution in several areas. A severe water deficit in India causes over 2 lakh deaths yearly, and 600 million people there as well as 1.2 billion people lack access to safe water for drinking [61]. It is now crucial to safeguard all available water resources. To meet present and future human demand, water resources must be managed and conserved The Government of India periodically creates new measures to address the country's water crisis. They consist of, but are not restricted to.

4.5.2 National Water Policy

In September 1987, the first National Water Policy was passed. The NWP was launched by India's Ministry of Water Resources to control the development and planning of water resources as well as the most effective use of those resources. The major target of NWP is to provide clean drinking water in arid and rain deprived areas, plan resources for equitable distribution of water, optimal utilization of water, recycle waste water and minimization of water wastage as well as control the ground water exploitation.

4.5.3 Jal Shakti Abhiyan

It was a time-limited, two-phase mission-mode water conservation initiative. The project's first and second phases will operate from July 1 to September 15, and from October 1 to November 30, respectively. Jal Shakti Abhiyan's primary objectives are as follows:

(a) Rainwater harvesting and water conservation.
(b) Restoration of existing water bodies and other water features.
(c) Reusing water and recharging structures are options.
(d) Creation of watersheds.
(e) Extensive afforestation.

4.5.4 Jal Jeevan Mission

The mission was established in 1972, and in 2009 it changed its name to National Rural Drinking Water Programme (NRDWP). It aspires to guarantee that every home in India has access to piped water. Only 3.27 billion of the 17.87 billion households in the nation have access to piped water, a research claims.The mission's objective is to give individual household tap connections and safe, sufficient water to every household in rural India by 2024.

4.5.5 Swajal Scheme

GoI established the Swajal Scheme to provide a steady supply of safe Water supply in rural areas. The water that is typically delivered to these rural areas is contaminated which has caused numerous illnesses when consumed. Swajal Scheme is introduced in over 115 rural Indian districts.

4.5.6 The National Rural Drinking Water Programme (NRDWP)

This ambitious project was established in 2009–2010 with the aim of integrating diverse programs. It introduces the concept of rainwater harvesting in rural areas and encourages the recharge of groundwater table. Further, this scheme ensures prevention of groundwater contamination.

The **Amrita Water Distribution System** was created with the intention of improving water management and utilization. In April 2015, the initiative started in Kerala, and in July 2015, it all began in Odisha and Rajasthan. The project was an idea of Amrita University's Live-in-Labs® programme, which uses theoretical knowledge to address problems that rural communities in India are actually facing.

4.5.7 Nal Se Jal Scheme

Since August 2019, the Government of India and States have been working together to carry out the Jal Jeevan Mission, which aims to provide tap water to every rural household and public institutions in villages by 2024, including schools, anganwadi centres, ashramshalas (tribal residential schools), health centres, Gram Panchayat buildings, etc. The mission's anticipated budget is Rs. 3.60 billion, of which Rs. 2.08 billion would come from the central government.

In reality, While the government mentioned above initiatives have not utilized artificial intelligence (AI), India is gradually moving towards it. To provide safe drinking water at a price significantly lower than the market pricing, the union minister Dr. Jitendra Singh launched an Artificial Intelligence (AI) driven Start-Up by IIT alumni in January 2022. Furthermore, the Atal Bhujal Yojana is a government of India (GOI) program that uses AI and space technologies in the water sector. However, in India, using AI for managing natural resources is still in its infancy. It needs to be implemented urgently to assure sustainability and regain confidence in distribution channels.

4.6 Information and Communication Technology Tools in Water Sector

In many regions of the world, water is in short supply, as seen by the instances in the preceding paragraph. Users have access to, and can save, send, process, and alter information thanks to information and communication technologies (ICT). ICT is the convergence and integration of ICT. Since the past ten years, the potential influence and advantages of ICT technologies on the water sector have been recognized.

The many innovative new ICT applications and their effects on the water sector are being studied by experts in both ICT and water. Hydro-informatics systems may now study the intricate geometry of terrestrial and aquatic environments as well as the intricacies of the physical hydrodynamic process. For instance, new sensor technologies like multi-beam sonars and Light Detection and Ranging (LIDAR) have significantly altered the quantity and quality of data about hydro-environments.

It is now possible to associate and mix data-driven and physically based hydro informatic techniques [30]. The use of ICTs to achieve sustainable growth is one of the European Union's top priorities for the next two decades Better resource management across the board is required for sustainable development, and ICTs can be used effectively to achieve this [34]. According to studies, deploying information systems in different businesses boosts the organization's performance and efficiency across a variety of its operations [58].

ICT tools are essential for enhancing food production as well as for bettering land and water management. Since the start of the 20th century, ICT tools have been increasingly being employed in water supply and irrigation management. Applications of cutting-edge technologies to water resource management enable amazingly effective water utilization, particularly in regions with acute water scarcity [59].

Many regions are worried about achieving environmental sustainability since natural resource strain is rising in the majority of developing nations. For the agriculture sector to fully expand, his use of ICT as a tool is crucial. ICT tools offer methods for gathering data from a distance. If not, it is a challenging, pricey, and time-consuming process. Today, data and knowledge are shared, processed, and managed via ICT tools [47].

In addition, ICT tools have great features such as:

- Save audio, video, text, images, diagrams, and descriptions.
- Collect information, digitally store it, and affordably produce precise copies of that information.
- Rapid transmission of information and expertise across various communication networks.
- The standardization of processes for handling enormous amounts of data is improving quickly.
- Be more engaged in creating, communicating, and sharing knowledge and information that is helpful.
- Create structured information systems from unstructured data.
- Conversation and communication with others [39, 47].

In order to more precisely measure, control, model, or forecast water supply and demand, ICT can be useful, as shown in the following subsections.

4.6.1 Meters and Sensors

Nowadays, a wide range of operations in water distribution systems, including water flow and pressure, water quality, pressure drop, and water and energy consumption, are controlled by meters and sensors. The primary goals of water supply are to transport water from one location to another without wasting it, to conserve water, and to prevent damage from leaks.

You can efficiently and simply manage water loss by finding and locating leaks. Water pressure and flow inside the pipe network and pipes are largely measured and controlled for the purpose of detecting leaks. To support advanced management, sensors are utilized as fundamental tools for monitoring water flow and pressure.

4.6.2 Sensor for Pressure Management

In order to reduce actual water losses and operating expenses in water distribution networks, a reliable and affordable solution is to use pressure management sensors. Table 4.2 displays various types of pressure sensors used to measure the flow of water through pipelines in order to identify storage water levels and other operations [36].

4.6.3 The Flow Sensors

Water distribution and manufacturing systems can be controlled by flow sensors. Electromagnetism is often used in flow sensors. This is regarded as an appropriate

Table 4.2 A sample of pressure sensors

Producer	Type and code	Communication means
Siemens AG	Sitrans P DS III, IP65/IP68	Profibus, RS 485, HART
SAE IT systems	Net-line FW-5, IP20	Ethernet, RS 485
WIKA	S-10, IP65/IP67/IP68	Analog
IFM electronic	PI2793, IP 67/IP68/IP69K	Analog

Table 4.3 A sample of flow sensors

Producer	Type	Technology	Communication
Siemens	SITRANS F M	Electrodynamics	Profibus, RS485, HART, etc.
Endress & Hauser Flexim	Promag Fluxus® ADM	Electrodynamics Ultrasonic	Ethernet port HART, ModBus, Profibus, BACNET
Isoil Industria	ISOMAG Flowiz next	Magmeter	GSM, GPRS wireless
ABB	AquaMaster3	Electrodynamics	GSM
Krohne	Optiflux Waterflux 3070	Electrodynamics	Profibus, RS485, HART, GSM

technique for evaluating water flow and environmental variables. The numerous flow sensor varieties now on the market are displayed in Table 4.3.

4.6.4 Sensors for Energy Consumption

Water pumps employ energy or power consumption sensors primarily to improve their energy management. Electric motors provide a wealth of data for power or energy measuring and monitoring systems, which also require interaction with control and monitoring systems. Sensors ensure that electrical energy is properly used to generate and distribute water systems. A number of energy consumption sensors used in the business are listed in Table 4.4.

Table 4.4 A sample of energy consumption sensors

Producer	Type	Communication means
Siemens AG	Sentron Pac 3200	Profibus, Rs. 485
Grundfos	CIM and CIU	Profibus

4.6.5 *Water Consumption Meter*

Water metres employ a number of techniques to gauge and track water usage over time. Water metres aid with management, leak detection, and consumption measurement. Methods of measuring consumption can be broadly categorized into either speed or volume types. A well-designed hydraulic structure's water flow via one or more nozzles is measured using velocimeters. Various types of speedometers calculate water usage by integrating measured emissions over time. Velocimeters work with sensors based on mechanical, ultrasonic, electromagnetic, pressure, optical, or liquid vibrations. Volumetrics measure water volume over time using a mechanical sensor meter with a specific volume that directly impedes water flow [36, 38].

4.6.6 *Supervisory Control and Data Acquisition (SCADA)*

Over the past 30 years, SCADA (Supervisory Control and Data Acquisition) technology has evolved in order to monitor and control large-scale activities. SCADA includes software packages, but they are not the only ones, that may be incorporated into hardware and software systems to improve the security and efficacy of the operation of these huge activities. The majority of the time, crucial functions like gathering data from sensors, sending acquired data between numerous remote sites, displaying data via a centralized host computer, and managing data at operator interfaces or workstations are handled by SCADA systems. function [36]. According to Bentley [13], these systems typically include the following subsystems.

A remote terminal unit (RTU) or programmable logic controller (PLC) that is connected to the sensors in your process:

- A communications network that links distant terminals to a central host computer or monitoring system.
- A monitoring (computer) system that collects (captures) data about processes and sends commands (control) to processes. Also called SCADA central unit, master station, master terminal unit, or MTU.
- To provide SCADA centralized host and operator terminal applications, to support communication systems, and to monitor and control remote field data interface devices.
- An operator workstation-supporting communication system.
- Machines using common Human Machine Interface (HMI) or Human Machine Interface (MMI) software.

4.7 Communication Facilities

Protocols, industrial control systems, and established registration structures are the mainstays of conventional water management systems. As a result, it is challenging to adopt fresh communication trends right away. The ability to incorporate existing facilities into a more adaptable IP-based monitoring system is now available with water supply networks.

Collecting alarms, keeping an eye on water quality, forecasting demand, conserving energy, and finding and repairing leaks are a few examples. SCADA systems are utilised to manage regionally scattered resources because of their advantageously high degree of distribution. In this case, centralised data collection and management are essential to system performance. It is currently the approach that is most frequently utilised in sewage collecting and water distribution systems. The network for long-distance communication is managed by a system controller, who also provides centralised oversight. includes tracking the status of alarms and data processing. The integration of wireless and direct wired systems is supported by this technique [41].

General packet radio services (GPRS) and the global system for mobile communication are the most often utilized wireless technologies in cellular networks for water metering infrastructure (GSM). This is because several telecom manufacturers and operators utilise and support them extensively. A packet-data technology called GPRS enables GSM users to use wireless data services, such as email [1].

4.8 Mechanics Models

Hydrological models support the management, forecasting, and sound decision-making of available water resources by enterprises, universities, municipal, regional, federal agencies, meteorological agencies, and other water sectors. Helpful. Hydraulic prototype modelling and network optimization research has increased during the last few decades (WDNs). Simulation models using sensor networks and prediction models for the administration of water distribution systems are included in the ICe-Water decision support system components. Based on the creative application of conventional global simulation-optimization algorithms, a new simulation-optimization linked technique was created. Various businesses offer water network operators with models, simulations, and optimization products for energy and cost design, optimization, water loss reduction, and effective control strategies.

Hydrological models assist water experts, businesses, and Siemens AG's SIWA technology. It is a computer-assisted technique for forecasting the hydraulic behavior of water sources and streamlining related operations. SIWA's water management technology consists of SIWA OPTIM module for water supply operation optimization and SIWA LEAK control for leakage management, reducing operating costs through dynamic simulation and network and pipe distribution system optimization and reliability [57].

4.9 Decision-Making Aid

Systems for supporting decisions in decision-making assist in managing water distribution networks, associated techniques, and technical solutions. Systems for managing water resources and networks also aid in the optimization of energy use, water quality, and demand control. The division of water distribution networks is a critical task that greatly simplifies the management of very large and highly complex structures within the water distribution system (district metering areas). The technique is also helpful for demand management, asset management and maintenance, water quality control, energy consumption reduction, and business information and analytics. Many tools based on business intelligence have been created to analyze data pertaining to water resource management. Companies like Oracle, Peer Water eXchange, VENTIX, NETBASE, and IBM provide solutions that are somewhat integrated with utilities [36].

4.10 Design and Management of Water Supply and Irrigation

Since the early 1950s, mechanisation and information and communication technology (ICT) tools have been utilised in water systems and network facilities, and modern, high-tech water supply facilities in industrialised countries are fully automated. Water supply and demand are synchronised using a variety of ICT technologies, which are also used to regulate water withdrawal from various sources and reservoirs, coordinate pump operation to save energy, and streamline purification procedures in wastewater reclamation plants.

The major freshwater issues that humanity will encounter over the next 50 years are addressed by increasing water consumption in agriculture. This helps him achieve his three objectives, which are to ensure food security, end poverty, and preserve ecosystems [29]. ICT tools are employed in agricultural development activities to enhance irrigation system hydraulic and network design.

For the purpose of calculating the water head losses that happen while water flows through pipes, several simple software programmes have been created. Furthermore, it is now simpler than ever to optimise irrigation system pressure flow and simulate water flow in a complicated water network thanks to modern, sophisticated software programmes. Irrigation network design requires extensive software development combining GIS data, aerial photography, and topography. This facilitates the development of computer-based irrigation network systems for improved water resource management. As a result of the systems' evolution, irrigation system design and monitoring may now be a part of a comprehensive agriculture strategy [50].

Scientists have used their acumen in machine learning methods which includes:

(1) Mimicking the hydrodynamic/hydrological model's behavior in the Apure River basin (Venezuela) in order to use artificial neural networks (ANNs) for model-based optimum control of reservoirs [60],
(2) Utilising ANN to model a channel network [52],
(3) Building an intelligent controller using artificial neural networks to control water levels in a polder in real-time [44],
(4) Using ANNs to predict the rainfall-runoff process [26];
(5) Utilising ANN to anticipate the surge water level in the ship guidance problem;
(6) Utilising ANN to recreate the stage-discharge relationship [15],
(7) Predicting river discharge using M5 model trees (see example below);
(8) Predicting water flows with SVMs for flood management [27].

4.11 Advantages of AI in Water Conservation

Below is a list of benefits for various AI technologies. They serve as the main inspirations for putting the ideas into practise in the assigned work:

- Dimensionality reduction and feature extraction of the enormous characteristics;
- Utilising the advantages of parallel computing to solve a challenging task;
- Precision in predicting the target variables to the appropriate degree;
- Using several data points in some applications
- For the study and prediction of time-series data, algorithms like RNN are helpful;
- Algorithms like DNN provide faster training and prediction;
- ANN is utilised for multidimensional datasets, has a high degree of arbitrary function, and allows for speedier prediction.

4.12 Conclusion

The creation of expert systems for decision-making and problem-solving marked the beginning of artificial intelligence's (AI) significance in natural resource management. Other AI techniques relevant to natural resource management were created as a result of the use of expert systems. With the help of data analytics, regression models, and algorithms, AI can simplify the process of managing water resources. The development of effective water networks and systems is facilitated by this cutting-edge technology. AI help policy planners to make water budgets from micro level (households) to district level thus ensuring equitable and optimal water distribution. AI also helps us to get the status of water resources. AI also helps in controlling water quality thus ensuring distribution of potable drinking water beside wastewater treatment. The AI-assisted solutions can be delivered through a software programme like Emagin.

It employs machine learning to analyse wastewater treatment plant data and generate predictive recommendations, enabling facilities to meet clean water and sanitation targets at the lowest feasible operational costs. In India, employing AI for natural resources management is still in its infancy and requires urgent and immediate roll-out as a strategy to ensure sustainability and restore faith in industrial markets. The use of hybrid models, which combine models of many types and adhere to various modelling paradigms, is considered as the way of the future. It is anticipated that computational intelligence (machine learning) would be used to create hybrid models with optimal and adaptive model structures in addition to data-driven models.

References

1. 4G-Americas (2014) General packet radio service. http://www.4gamericas.org/index.cfm?fuseaction=page&sectionid=243
2. Afan HA, El-shafie A, Mohtar WHMW, Yaseen ZM (2016) Past, present and prospect of an artificial intelligence (AI) based model for sediment transport prediction. J Hydrol 541(Part B):902–913
3. Akkoyunlu A, Altun H, Cigizoglu HK (2011) Depth integrated estimation of the lake dissolved oxygen (DO). J Environ Eng 137(10):961–967
4. Alizadeh MJ, Kavianpour MR (2015) Development of wavelet-ANN models to predict water quality parameters in Hilo Bay, Pacific Ocean. Marine Poll Bull 98(1–2):171–178
5. Ashrafi M, Chua LHC, Quek C, Qin X (2017) A fully-online neuro-fuzzy model for flow forecasting in basins with limited data. J Hydrol 545:424–435
6. Ay M, Kisi O (2012) Modeling of dissolved oxygen concentration using different neural network techniques in Foundation Creek, El Paso County, Colorado, USA. J Environ Eng 138(6):654–662
7. Ay M, Kisi O (2013a) Modeling dissolved oxygen concentration using neural network and ARMA techniques. In: 6th International perspective on water resources & the environment (IPWE-2013). İzmir, Turkey
8. Ay M, Kisi O (2013b) Modelling COD concentration by using different artificial intelligence methods. J Selçuk Univ Nat Appl Sci (2,477489). http://www.josunas.org/login/index.php/josunas/article/view/181/147. ISSN: 2147-3781
9. Ay M, Kisi O (2014) Modelling of chemical oxygen demand by using ANNs, ANFIS and k-means clustering techniques. J Hydrol 511:279–289
10. Ay M, Kisi O (2017) Estimation of dissolved oxygen by using neural networks and neuro fuzzy computing techniques. KSCE J Civ Eng 21(5):1631–1639
11. Ay M (2014) Sukalitesiparametrelerininyapayzekâyöntemleriiledeğerlendirilmesi (evaluation of water quality parameters by using artificial intelligence methods). YayımlanmışDoktoraTezi. Tez no: 360613. 136 sayfa
12. Banerjee K, Bali V, Nawaz N, Bali S, Mathur S, Mishra RK, Rani S (2022) A Machine-learning approach for prediction of water contamination using latitude, longitude, and elevation. Water 14:728
13. Bentley (2004) The fundamentals of supervisory control and data acquisition. Retrieved from ftp://ftp2.bentley.com/dist/collateral/whitepaper/fundscada_whitepaper.pdf
14. Berlinski D (2000) The advent of the algorithm: the 300-year journey from an idea to the computer. San Diego, USA, Harcourt
15. Bhattacharya B, Solomatine DP (2000) Application of artificial neural network in stage-discharge relationship. In: Proceedings of the 4th international conference on hydro informatics, Iowa City, USA

16. Brady JE, Holum JR (1988) Fundamentals of chemistry, 3rd edn, 1055 pp. ISBN: 978-0471844730
17. Buchanan BG (2005) A (very) brief history of artificial intelligence. AI Mag 26:53–60
18. Chang FJ, Chung CH, Chen PA, Liu CW, Coynel A, Vachaud G (2014) Assessment of arsenic concentration in stream water using neuro fuzzy networks with factor analysis. Sci Total Environ 494–495:202–210
19. Chau KW (2006) A review on integration of artificial intelligence into water quality modelling. Mar Pollut Bull 52:726–733
20. Cherkassy V, Krasnopolsky V, Solomatine D, Valdes J (2006) Computational intelligence in earth sciences and environmental applications: issue and challenges. Neural Netw 19:113–121
21. Chithra NR, Thampi SG (2016) Downscaling future projections of monthly precipitation in a catchment with varying physiography. ISH J Hydraul Eng. https://doi.org/10.1080/09715010.2016.1264895
22. Chowdury MSU, Emran TB, Ghosh S, Pathak A, Alam MM, Absar N, Andersson K, Hossain MS (2019) IoT based real-time river water quality monitoring system. Procedia Comput Sci 155:161–168
23. Cigizoglu HK (2003) Estimation, forecasting and extrapolation of river flows by artificial neural networks. Hydrol Sci J 48(3):363–379
24. Darras T, Estupina VB, Kong-A-Siou L, Vayssade B, Johannet A, Pistre S (2015) Identification of spatial and temporal contributions of rainfalls to flash floods using neural network modelling: case study on the Lez basin (southern France). Hydrol Earth Syst Sci 19:4397–4410
25. Demirci M, Üneş F, Saydemir S (2015) Suspended sediment estimation using an artificial intelligence approach. In: Heininger P, Cullmann J (eds) Sediment matters. Springer International Publishing, pp 83–95
26. Dibike Y, Solomatine DP, Abbott MB (1991) On the encapsulation ofnumerical-hydraulic models in artificial neural network. J Hydraul Res 2:147–161
27. Dibike YB, Velickov S, Solomatine DP, Abbott MB (2001) Modelinduction with support vector machines: introduction and applications. J Comput Civil Eng Am Soc Civil Eng (ASCE) 15(3):208–216
28. Droppo IG, Krishnappan BG (2016) Modeling of hydrophobic cohesive sediment transport in the Ells River Alberta, Canada. J Soils Sediments 16(12):2753–2765
29. Earthscan (2007) Water for food water for life. Retrieved from http://www.fao.org/nr/water/docs/summary_synthesisbook.pdf
30. Gourbesville P (2011) ICT for water efficiency. http://www.intechopen.com/books/environmental-monitoring/ict-for-water-efficiency
31. Goyal MK, Bharti B, Quilty J, Adamowskic J, Pandey A (2014) Modeling of daily pan evaporation in subtropical climates using ANN, LS-SVR, Fuzzy Logic, and ANFIS. Expert Syst Appl 41(11):5267–5276
32. Güçlü YS, Subyani AM, Şen Z (2017) Regional fuzzy chain model for evapotranspiration estimation. J Hydrol 544:233–241
33. Güner HAA, Yumuk HA (2014) Application of a fuzzy inference system for the prediction of longshore sediment transport. Appl Ocean Res 48:162–175
34. Holz K, Hildebrandt G, Weber L (2006) Concept for a web-based information system for. Nat Hazards 38:121–140. http://rd.springer.com/article/10.1007%2Fs11069-005-8605-5#page-1
35. Huang W, Xu B, Hilton AC (2004) Forecasting flows in Apalachicola River using neural networks. Hydrol Process 18(13):2545–2564
36. ICeWater (2012) ICeWater—Deliverable. http://icewater-project.eu/index.php?id=050000&spid=en&site=login&filecat=1
37. Lee J-H, Oh J-E (2016) A comprehensive survey on the occurrence and fate of nitrosamines in sewage treatment plants and water environment. Sci Total Environ 556:330–337
38. Kamstrup (2014) Experience a new dimesion of water meters. http://kamstrup.com/16849/new-water-meter-not-only-measures-water-consumption
39. Kappor T (2006) Role of information and communication technology in adaptive integrated water resource management. World Environ Water Resour Congr 1–10. http://ascelibrary.org/doi/abs/https://doi.org/10.1061/40856%28200%291

40. Karimi S, Kisi O, Kim S, Nazemi AH, Shiri J (2016) Modelling daily reference evapotranspiration in humid locations of South Korea using local and cross-station data management scenarios. Int J Climatol
41. Keith S, Joe F, Karen K (2006) Guide to supervisory control and data acquisition (SCADA) and industrial control systems security. In: NIST, National institute of Standards and Technology. U.S Department of Commerce. http://www.dhs.gov/sites/default/files/publications/csd-nistguidetosupervisoryanddataccquisition-scadaandindustrialcontrolsystemssecurity2007.pdf
42. Khan UT, Valeo C (2015) Dissolved oxygen prediction using a possibility-theory based fuzzy neural network. Hydrol Earth Syst Sci Discuss 12:12311–12376
43. Li B, Yang G, Wan R, Dai X, Zhang Y (2016) Comparison of random forests and other statistical methods for the prediction of lake water level: a case study of the Poyang Lake in China. Hydrol Res (in press)
44. Lobbrecht AH, Solomatine DP (1999) Control of water levels in polder areas using neural networks and fuzzy adaptive systems. In: Savic D, Walters G (eds) Water industry systems: modelling and optimization applications. Research Studies Press Ltd., pp 509–518
45. Londhe S, Dixit P, Shah S, Narkhede S (2015) Infilling of missing daily rainfall records using artificial neural network. ISH J Hydraul Eng 21(3):255–264
46. Maier HR, Jain A, Dandy GC, Sudheer KP (2010) Methods used for development of neural networks for the prediction of water resource variables in river systems: current status and future directions. Environ Model Softw 25:891–909
47. Meena M, Singh K (2012) Information and communication technology for sustainable natural resource management. http://mpra.ub.uni-muenchen.de/45818/1/MPRA_paper_45818.pdf
48. Mi X, Sivakumar M, Hagare D (2004) A general review of applications of artificial neural network to water industry. In: Mowlaei M, Rose A, Lamborn J (eds) Environmental sustainability through multidisciplinary integration. Environmental Engineering Research Event, Australia, pp. 234–243
49. Mishra KN, Chakraborty C (2020) A novel approach toward enhancing the quality of life in smart cities using clouds and IoT-based technologies. In: Digital twin technologies and smart cities. Springer, Berlin/Heidelberg, Germany, pp 19–35
50. Moshe S (2005) ICT in water supply and irrigation management. http://departments.agri.huji.ac.il/economics/gelb-sne.pdf
51. Nourani V, Baghana AH, Adamowski J, Kisi O (2014) Applications of hybrid wavelet—Artificial Intelligence models in hydrology: a review. J Hydrol 514:358–377
52. Price RK, Samedov J, Solomatine DP (1998) Network modelling using artificial neural networks. In: Proceedings international conference on hydro informatics, Balkema, Rotterdam
53. Radhakrishnan V, Wu W (2018) IoT technology for smart water system. In: 2018 IEEE 20th international conference on high performance computing and communications; IEEE 16th international conference on smart city. Proceedings of the IEEE 4th international conference on data science and systems (HPCC/SmartCity/DSS), Exeter, UK, pp 1491–1496
54. Raina NY, Alam P (2014) Waste water treatment and management in rural areas—A case study of Rajouri District, Jammu & Kashmir, India, pp 1–2. www.journalijdr.com
55. Ray P, Kaluri R, Reddy T, Lakshmanna K (2021) Contemporary developments and technologies in deep learning—Based IoT. In: Deep learning for internet of things infrastructure. CRC Press: Boca Raton, FL, USA, pp 61–82
56. Tsai SB, Chien MF, Xue Y, Li L et al (2015) Using the fuzzy dematel to determine environmental performance: a case of printed circuit board industry in Taiwan. Plos One 10(6), Article ID e0129153
57. Siemens (2012) Increasing efficiency with SIWA pipeline management system. Nuremberg, Germany: Simens AG. http://www.automation.siemens.com/mcms/water-industry/de/Documents/E20001-A110-T122-X-7600_WS_SIWA%20PMS_EN.pdf
58. Silver M, Markus ML, Mathis BC (1995). The information technology interaction model: a foundation for the MBA core course, 361–390. http://www.jstor.org/discover/https://doi.org/10.2307/249600?uid=17241608&uid=3737760&uid=2129&uid=2&uid=70&uid=3&uid=17241584&uid=67&uid=62&sid=21103241689267

59. Sne M (2005) ICT in water supply and irrigation management. http://departments.agri.huji.ac.il/economics/gelb-sne.pdf
60. Solomatine DP, Torres LA (1996) Neural network approximation of a hydro dynamic model in optimizing reservoir operation. In: Proceedings of the2nd international conference on hydro informatics, Zurich, pp 201–206
61. Solomon HG (2014) The role of information and communication technology (ICI) in water resource management, p 124. https://www.recearchgate.net/publication/279852273
62. Talebi A, Mahjoobi J, Dastorani MT, Moosavi V (2016) Estimation of suspended sediment load using regression trees and model trees approaches. Hyderabad drainage basin in Iran, Case study
63. Talei A, Chua LHC, Quek C, Jansson P-E (2013) Runoff forecasting using a Takagi-Sugeno neuro-fuzzy model with online learning. J Hydrol 488:17–32
64. Tayfur G (2017) modern optimization methods in water resources planning, engineering and management. Water Resour Manag
65. Üneş F, Demirci M, Kişi Ö (2015) Prediction of millers ferry dam reservoir level in USA using artificial neural network. Periodica Polytech Civil Eng 59(3):309–318
66. Yaseen ZM, El-shafie A, Jaafar O, Afan HA, Sayl KN (2015) Artificial intelligence-based models for stream-flow forecasting: 2000–2015 (Review Paper). J Hydrol 530:829–844

Chapter 5
Remote Sensing and GIS Based Techniques for Monitoring and Conserving Water on Newly Developed Farmlands

Abdul Rehman Zahoor, Shahbaz Nasir Khan, Arfan Arshad, and Rana Ammar Aslam

Abstract Global governments face rapid urbanization. South Asia is urbanizing at 3% per year, causing this issue. Urbanization permanently destroys agricultural land through land acquisition and non-productive rural use. Unfortunately, this urbanization is taking place on agricultural land which not only affects farming activities, energy losses but also disturbs the canal water which is lost, either it is used illegally or thrown in sea or ocean. Thus, urbanization is shifting the command area from where it is happening now to the newest places where no canal network exists. We know 3% of the water on Earth is freshwater from which only around 1.2% of that can be used; the remainder is trapped in glaciers, permafrost, and ice caps, or is buried far below. According to a NASA-led study, global water demand is expected to rise by 55% between 2000 and 2050. We need better water supply, conservation, and management methods. Remote sensing has been a significant instrument in studying the Earth's surface and thus in providing valuable information essential for hydrologic analysis since the arrival of remote sensing earth-observing satellites. Remote sensing techniques have positively impacted water resource management and assessment procedures because of their capacity for capturing spatial variations in hydro-meteorological variables and frequent temporal resolution sufficient

A. R. Zahoor · S. N. Khan (✉) · R. A. Aslam
Department of Structures and Environmental Engineering, University of Agriculture, Faisalabad 38040, Pakistan
e-mail: shahbaz.nasir.khan@uaf.edu.pk

A. R. Zahoor
e-mail: abdulrehmnazahoor456@gmail.com

R. A. Aslam
e-mail: ammar.aslam@uaf.edu.pk

A. Arshad
Department of Biosystems and Agricultural Engineering, Oklahoma State University, Stillwater, OK 74078, USA
e-mail: aarshad@okstate.edu

E. Balaji et al. (eds.), *Emerging Technologies for Water Supply, Conservation and Management*, Springer Water, https://doi.org/10.1007/978-3-031-35279-9_5

to describe the hydrologic processes. For this we can use satellite extracted images, process and classify it. The ArcGIS and ERDAS Image softwares can be used for spectral analysis to classify images for the identification of terrestrial characteristics. The multispectral data can be used to categories terrestrial objects, vegetation, water bodies, canals, and tree shadows for the classification of the satellite extracted image. We can use normalized difference vegetation index (NDVI) and normalized difference water index (NDWI) for the reconstruction/diversion of our canal system for the better supply, conservation, and water management.

Keywords Urbanization · Canal water system · Command area · Supply · Conservation and management of water · Remote sensing · Earth-observing satellites · ArcGIS and ERDAS image software · NDVI and NDWI

5.1 Introduction

Urbanization describes the movement of people from rural to urban regions and "the continuous increase in the number of people living in urban areas," as well as how different societies accommodate this shift. It's the change from a more rural to more urban lifestyle that defines urbanization [20].

By 2050, it is predicted that 64% of Africa and 86% of Asia would be urbanized [26]. It is noteworthy that the United Nations recently predicted that cities will absorb almost all of the global population growth from 2017 to 2030, absorbing about 1.1 billion more city dwellers during the following 13 years [8].

The lack of a universal criterion by which to classify cityscapes presents the greatest obstacle to research on urbanization. Almost everyone in the world makes a clear distinction between urban and rural dwellers, but what constitutes an urban area is defined differently depending on location and, in some cases, even historical context.

To put it simply, urbanization processes have always been a problem for farmers who are located close to large cities. The loss of these areas to urbanization has had serious effects on the environment, society, and the economy [18].

Wherever there is urbanization, local irrigation systems are often disrupted, especially in peril-urban areas where urban and agricultural land uses overlap. The places that have witnessed rapid urbanization in the last few decades are particularly affected by this [10].

In 2021, Biswajit Das and his team used Digital Elevation Model data and LANDSAT—8 (30 m) satellite imagery to perform an analysis of the morphometric parameters and to generate land-use–land-cover data for the basin using supervised image classification technique over Gomti River. This was done over the region that is centered on the Gomti River [9]. In order to manage, monitor, and save ground water, we will also use various remote sensing and geographic information system

(GIS) techniques. Some methods are described below in the chapter. The objectives of this study are following:

- To know about urbanization, how it affects are farmlands
- To know the importance to conserve water
- Importance of remote sensing and GIS in this era
- The literature is written very simple and general way so a person with no prior or little knowledge can understand easily
- To know the whole process from sources, way of extraction to the actual work
- To get to know about methods for monitoring and conservation with simple examples.

5.1.1 Urbanization Trends

The growth of the world's urban population is outpacing that of the rural population by a significant margin. Between 1950 and 2003, the world's urban population increased by a factor of four, whereas the world's rural population increased by a factor of less than twice as much, from 1.8 billion to 3.2 throughout the same period [7]. According to the report World's Urbanization Prospects 2004 published by the United Nations, the world's urban population is projected to rise by nearly two billion over the next 30 years. On the other hand, the world's rural population is predicted to decrease slightly, falling from 3.3 billion in 2003 to 3.2 billion in 2030. As a direct consequence of this, it is anticipated that, for the foreseeable future at least, the majority of new people moving into the world will be settling in metropolitan areas. According to the most recent research [28], the percentage of urban residents living in poverty is increasing at a rate that is higher than the total rate of growth in urban population in a significant number of the world's most impoverished nations. One estimate places the number of people living in shantytowns in Africa at 72% of the continent's total urban population. The percentage for the Asia-Pacific Region is 43%, while the percentage for Latin America is 32%, and the percentage for Northern Africa and the Middle East is 30% [11]. Cities are continuously trying to attract new migrants, which results in the growth of their populations and contributes to the worsening of issues such as unemployment, poverty, transit, and housing, to name a few. The ability of most towns to provide sufficient basic amenities for their residents has been completely outpaced as a result of rapid urban growth and the obstacles that it has caused all throughout the developing world. This is because rapid urban growth has caused an increase in the population of cities, which in turn has caused an increase in the demand for basic amenities [7].

5.1.2 Urban Characteristics

In any metropolitan center, buildings are primarily created for the following functions, each with a corresponding scale: Roads are 18.0%, Residential 60.0%, Industrial 4.0%, Commercial 2.0%, Administration 4.0%, Recreational 10.0%, Others 2.0%. The total is 100.0%.

In any urban settlement, the residential sector takes up the largest portion of the land use. Residential land use sectors see heavy crisscross movements of people and vehicles throughout the working days of the week since they are areas of a concentrated population. Any community's vitality can be found in its economic, administrative, and industrial sectors. These are a wonderful employment center Extremes of class coexist in metropolitan areas, including the richest and poorest citizens. In a metropolis, the wealthy's opulent bungalows coexist with the poor's slums and the middle class's apartment. The majority of jobs in cities are found in the industrial, managerial, and professional sectors. Vocational specialization and division of labor are more common in urban centers. Love weddings and inter-caste marriages are more common in metropolitan communities. There are also more divorces to be found. In terms of the urban community, the individual is given more weight than the family. In urban regions, nuclear families are more prevalent. Urban areas have a higher population density than rural areas. Urbanization and density have a beneficial relationship. An urban community's size is substantially larger than that of a rural community in the same nation and period. As a result, there is a positive correlation between urbanization and size such as those for the provision of water, electricity, telephone service, and the disposal of solid waste, are widespread in metropolitan areas. An effective network of roads and a transportation system improve movement efficiency for both people and vehicles. Streets that are narrow or uneven cause turmoil and congestion. The room for enough lanes and the construction of infrastructure is provided by wide road reservations with sufficient buffers. A system of communication linkages binds urban area's structures together.

Cities have a wide variety of people living in them, while rural places are more uniform. Cities are home to individuals of many different backgrounds and cultures. There is a large diversity among urban residents concerning their dietary preferences, fashion sense, housing situations, religious beliefs, cultural activities, and traditional behaviors. Social isolation is a natural consequence of anonymity and diversity. Most people's interactions with one another in a city are superficial and compartmentalized.

Interest groups form the foundation of urban societies' social structures. In comparison to the countryside, the social circles in cities are larger. City life is incredibly intricate and diverse. because there is a larger interaction system per individual and per aggregate.

In urban areas, it is easier to move up the social ladder. Ability, intelligence, and persistence are what determine a man's social status in a city. As a result, there is a positive correlation between urbanization and accessibility [19].

5.1.3 *Influences that Drive the Urbanization Process*

Three competing forces drive urbanization. According to studies, there are primarily three pathways that might lead to urbanization.

(a) Western Liberal
(b) Marxist Capitalist
(c) Ecological.

5.1.3.1 Western Liberal

It is argued that urbanization is a natural consequence of progress. It argues that the presence of employment opportunities draws people from rural areas to metropolitan centers. Both the "country push" and "urban pull" ideas lend credence to this idea. People leave the countryside for the city in search of better opportunities, so the modernization hypothesis goes. Major setbacks and inefficient economic growth were faced by countries in Africa and Asia as a result of economic growth, non-agricultural profession focus, uneven welfare, and fast migration from rural to urban cities [24].

5.1.3.2 Capitalist Marxist

According to this viewpoint, urbanization is a result of capitalism. In order to maximize their wealth, capitalists made decisions that primarily benefited them. They control the economy and ensure the migration of people to cities in order to equip their multinational corporations, local, national, and regional businesses. The worst part is that these capitalists are concentrating their efforts in Africa and Asia mainly [24].

5.1.3.3 Ecological

In the 1920s, urban ecological theory was proposed by scientists Ernest Burgess and Robert Park [5]. These social scientists argue that metropolitan areas serve as technological infrastructure for powerful groups that are constantly at odds with one another. Two presumptions underpin urban ecology. The first is that different parts of the city cater to specific needs, such as high-end housing or heavy industry. Second, as cities grew denser and more competitive, the most lucrative areas would inevitably be carved out, even if the most firmly established areas followed the concentric zone that spread outward from a central business district [24].

5.2 Urbanization Over Agricultural Land

Subsistence agriculture supported the great majority of people from the beginning of urbanization in Mesopotamia and Egypt until the 18th century, while tiny towns thrived on market trade and handicraft production. Since agricultural productivity was relatively undeveloped and stable during this time period, the rural-to-urban population ratio stayed the same. Furthermore, the rise of urbanization may be traced back to Mughal India, where, throughout the 16th and 17th centuries, a bigger percentage of the population resided in urban areas than in Europe as a whole [12, 17].

Rapid urbanization swept across the Western hemisphere, but it has only lately taken root in the developing economies of Asia and Africa. At the turn of the twentieth century, only 15% of the world's population was urban. The United Nations [27] and [22] used urbanization data published by Yale University in June 2016 to create a film portraying the growth of cities around the world from 3700 BC to 2000 AD.

Therefore, urbanization processes have always been a challenge for farmers who live close to major cities. For Europe, this has meant that most regions now have either a high degree of urbanization and land that is exceptionally suitable for agriculture, or a low degree of urbanization and land that is less suitable for agriculture. Over the past few decades, western European landscapes have changed mostly due to the expansion of the housing unit at the expense of profitable agricultural land and natural regions. As shown in (Fig. 5.1).

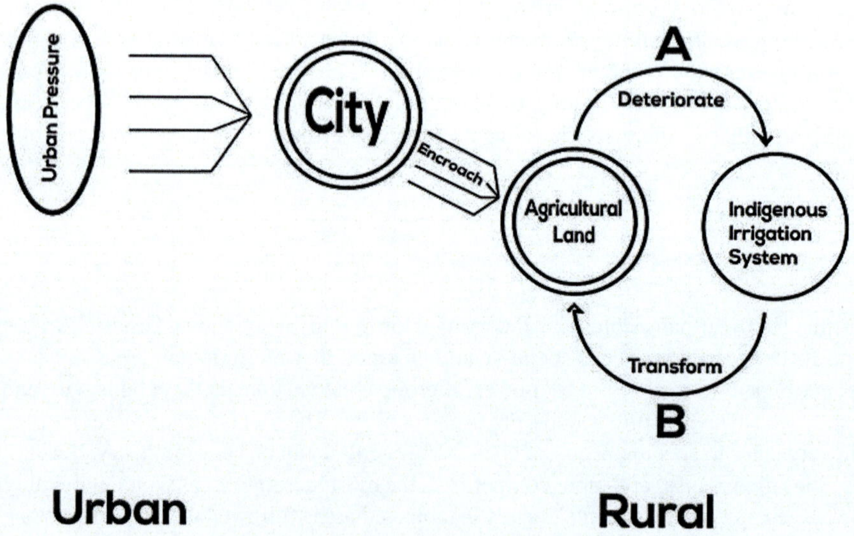

Fig. 5.1 Processes of urbanization, the deterioration of irrigation systems causes, **a** urbanization on farmland, while **b** urbanization on farmland causes irrigation system deterioration

Many of these cities grew throughout time as a result of a rising population, with many of the largest ones being established and expanded on the most fertile grounds, and this urban expansion has frequently persisted up until the present day [30].

Significant environmental and social and economic ramifications have resulted from the loss of these lands due to urban growth. As previously discussed [6], predict that urbanization will persist and even accelerate in the years ahead, with urbanization rising from 2.06% of ice-free area in 2000 to a predicted 4.71% by 2040 [30].

According to FAO [13], both urban and peril-urban agriculture take place within and around the borders of metropolitan centers, and the world at a time when densification and its effects are greatest, limiting access to land for farming and frequently generating food insecurity. New rules on agricultural land-use and management techniques that aim to promote sound land management are needed, as a result of this trend. Unchecked urbanization has contributed to the loss of urban agricultural land in Africa.

The effects of land loss in Asia were described by [23]. He predicted that by the year 2050, 57% of all farmlands in the United States would vanish. The research indicates that Southeast Asia, South Asia, Central Asia, and East Asia might all lose up to 7% of their arable land, while Western Asia and North Africa could lose close to 10%. It was found that road building and industrialization were responsible for the loss of over 1 million and 400,000 hectares of farmland in China and the United States, respectively. India's total agricultural land area has shrunk by 16.31% as a result of urbanization in the periphery and core areas.

Some farmlands will be lost to urban sprawl, and changes in land values and land markets in and around cities sometimes lead to land being abandoned while its owners wait for a better price before selling or putting it to another use [25].

Urbanization is an extreme example of how human activity modifies the usage of land to suit their requirements. The natural land surface is replaced by an area that has been built up and has little soil moisture, making the developed region a dry region [3].

These numbers show the tremendous demand for urban sprawl and the consequent impact on farmland. Concern over the poorly managed conversion of land on the edges of cities, especially from agricultural to residential usage, has been around for quite some time.

Reduced biodiversity and altered species ranges and interactions may result from urbanization's usual repercussions, such as deforestation, habitat degradation, and the withdrawal of freshwater from the environment. Pollutants in the environment, which can be harmful to human and animal health, have increased due to human activities in metropolitan areas, such as the burning of fossil fuels and the disposal of industrial waste.

The cost of repurposing land is significant. Converting farmland and forests to make room for cities limits the quantity of land accessible for agricultural and forestry purposes. Land resource quality and future agricultural output are diminished as a result of soil degradations brought on by intensive agriculture and deforestation, such as erosion, salinization, water problems, desertification, and so on [16].

5.3 Urbanization Leads to Water Lose

Local irrigation systems are often impacted wherever there is urbanization, particularly in peril-urban areas where urban and agricultural land uses coexist. This is especially true in areas that have experienced rapid urbanization in recent decades. It is not uncommon to make predictions regarding the occurrence of certain types of issues, such as the pollution of water, the degeneration of canals, and even the complete destruction of irrigation systems.

The transformation of peril-urban land into agricultural use has been significantly aided by the use of indigenous irrigation techniques. There are two processes that seem to happen, affect each other, and then repeat themselves in a cycle. These processes are interrelated and appear to happen.

The degradation of canals is one of the significant problems that are made worse by urbanization, which occurs when agricultural lands are replaced by new developments. The deterioration of irrigation systems leads to the cessation of farming, which in turn leads to an increase in urbanization. Farmers in the area have voiced their concerns regarding the many sorts of canal deterioration, most notably "canal filling," which refers to the weakening of the indigenous irrigation system that is caused by canal ownership. Even though canal water is commonly understood to be a resource that is accessible to all.

The conveyor is protected by both public and private property rights at the same time. Canals that are situated on land that is held by the public sector are referred to as "public canals," whereas canals that are situated on land that is owned by private entities or individuals are known as "private canals." The legal owners of the properties on which canals are situated are the ones who are granted management rights for those waterways.

Similarly, real estate developers who bought the lands upon which urbanization was taking place are eligible to make claims to canal rights within their holdings. They have the authority to carry out any activity that is permitted on their lands, including the filling of canals, provided that the activity does not violate any laws. The impact of canal ownership on canal deterioration is ignored by administrators and policymakers, despite the significance of the relationship between the two; furthermore, it has never been subjected to a comprehensive research effort. This is a matter that requires additional investigation and consideration.

5.4 Importance to Conserve Water

The availability of the resource water is never an issue for us. If you are fortunate enough to live in an area that does not experience repeated droughts or have water supply systems that are inadequate, it is easy to lose sight of how essential water truly is. After that point, it is easy to forget how much water you waste because you are doing normal tasks that are relatively straightforward [29].

While the total amount of freshwater on Earth has, for the most part, remained relatively constant over time—being continually recycled via the atmosphere and back into our cups—the number of people living on the planet has risen. This indicates that each year there is increasing competition for a bountiful supply of clean water that can be utilized for drinking, cooking, bathing, and generally keeping life going.

However, as there are more people and the climate is changing, there is less water available. Because of this, we all have a responsibility to do what we can to save water whenever and wherever we can.

Although some consider water scarcity to be merely a theoretical issue, for certain individuals it is a stark reality. It was caused by a complex interplay of many different political, economic, social, and environmental forces.

Freshwater makes up a negligible fraction of the total water on the planet. Despite the fact that water makes up around 70% of the planet, only 2.5% of the world's total water supply is considered to be freshwater. The rest is marine and salty in character like ocean water. Even so, only 1% of our freshwater is easily accessible, as most of it is locked up in glaciers and snowfields. To put it another way, only 0.007% of the water on the world is usable to provide for the survival and well-being of its 6.8 billion inhabitants. As the pie chart shows in (Fig. 5.2).

Some regions give the impression of having an abundance of water due to geography, climate, engineering, regulation, and resource competition; on the other hand, other regions suffer from drought and crippling pollution. Access to clean water is

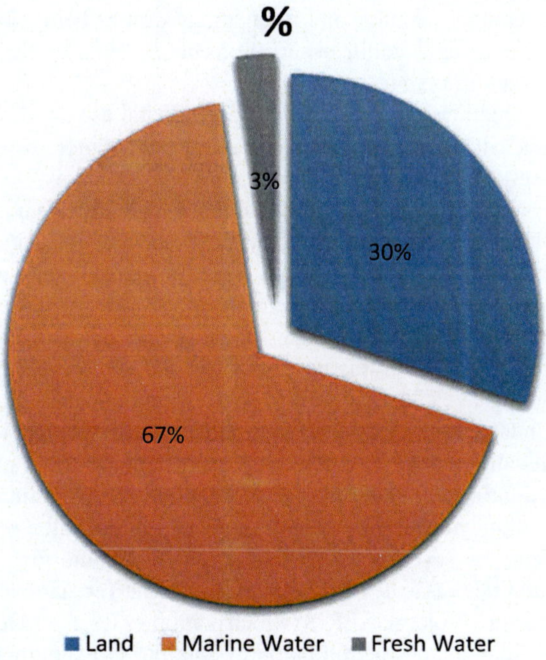

Fig. 5.2 Water %: total water on earth of which only 2.5–3% is fresh water

challenging, either because it is expensive or because it requires a lot of manual labor across a significant section of the developing globe.

No matter where they are, people will always need water to live. In addition to making up sixty percent of the human body, water is necessary for the manufacture of food, clothing, and computers; the transportation of trash; the preservation of human and environmental health; and the operation of computer systems.

Humans, it seems, are poor water managers. About 630 gallons (2,400 L) of water are used in the manufacture of one standard hamburger. Cotton is only one example of a crop that requires a lot of water that is typically grown in arid regions.

According to the United Nations, water consumption has surged at a rate that is more than twice as rapid as population growth over the past century. By 2025, due to population growth, consumption, and climate change, over 1.8 billion people, or one-third of the world's population, would live in water-stressed regions. The challenge we have going forward is figuring out how to save, manage, and distribute our water resources effectively.

5.5 Advantages of Using Earth-Observing Satellites

Data from Earth observation satellites are critical for research and practical uses at the local, national, and global levels. Satellite observations have the advantage of being systematic and frequent, as well as being able to be made uniformly over wider areas. Satellite observations help to monitor the world climate and ecosystem, as well as map resources.

Satellite observations have been a vital aspect of many operations in the last ten years, including weather forecasting, marine monitoring, monitoring of wildfires and degradation, thematic mapping, and polar studies.

Space not only benefits humanity but also provides a fresh viewpoint on our planet. Satellite Earth observations are useful in many societal contexts, such as environmental management, resource management, agricultural and food security, transportation, air quality and health, risk assessment, and safety [15].

To persuade policymakers to make greater use of Earth observation data and products from space, it is necessary to conduct comprehensive benefit and cost evaluations that evaluate Remotely sensed solutions in relation to preexisting frameworks. Understanding the value chain and all the players in it is crucial. There is no one-size-fits-all approach to calculating the benefits of Earth observation from space; rather, a multidisciplinary approach is required for each situation.

Policymakers need to be convinced of the value of space-based Remote sensing data, but several challenges must be overcome first. Data from Earth observation satellites has to be freely available. ForM@Ter (a French Solid Earth Research Infrastructure Project), AERIS (which combines Ether and ICARE), and Aviso (Archiving, Validation and Interpretation of Satellite Oceanographic data) are just a few of the scholarly data hubs that are getting easier to access. The Copernicus Sentinel data and processing infrastructure is made available via the Platform d'Exploitation des

Produits Sentinel (PEPS). Data from SPOT's (Satellite for observation of Earth) World Heritage program is now available through Theia.

The French Ministry of Environment (MEDDE) and Ministry of Overseas Territories (DGOM) have strengthened their relationship with CNES to consider facilitate the integration of satellite Earth observation data and their decision-making systems after becoming convinced that space Earth observation could be an effective tool for the implementation and control of policy. Important factors to are dependability, performance, and cost.

Remote sensing is a methodology for detecting an object's size, form, and properties from a distance. As a result, it is extremely beneficial in cases where it is not possible to touch the thing or visit the area to view it. Remote sensing from the sky, has the benefit of being able to acquire information over a large area at once, making it valuable for monitoring catastrophe situations, forecasting weather, and analyzing the global environment. The proposed platform on which the sensor is installed will change according to the type of sensor and how high we wish to observe, as will the information collected from it, such as the observing target, monitoring range, resolution, and observation frequency [14].

Drones, aircraft, the International Space Station (ISS), and satellites are common platforms for mounting sensors while undertaking remote sensing from the sky. In general, the larger the region that can be observed, the higher the observing altitude.

As a result, Earth-observing satellites have the benefit of being able to see the entire Earth equally, albeit at a lesser resolution than drones and airplanes that observe from close to the ground surface. Furthermore, once a satellite is deployed into space, it will continuously make observations until the end of its life, considerably contributing to the understanding of the worldwide energy and climate change, both of which require long-term and ongoing monitoring. On the other side, the resolution deteriorates over time. The use of satellite data in ecological economics studies is becoming more popular.

1. This is because satellite data is accessible for every region of the world,
2. provides frequent data throughout time,
3. is becoming more affordable, and is becoming simpler to handle.

5.6 ERDAS IMAGINE

Professionals in geographic imaging are required to analyze massive volumes of geospatial data every day, and they frequently rely on software that was developed for different purposes or add-on apps that create almost as many problems as they solve. Utilizing ERDAS IMAGINE can help you save time and money, make better use of the data you already have invested in, and increase your ability to analyze images. The integration of remote sensing, photogrammetry, LiDAR (light detection and ranging) analysis, standard vector analysis, and radar processing into a single product is one of the many ways in which ERDAS IMAGINE demonstrates its superior value.

The raster-based, user-friendly program known as ERDAS IMAGINE was developed expressly for the purpose of gleaning information from digital photographs. Easy to pick up and play, making it ideal for both novices and seasoned pros. Regardless of how much experience you have or don't have with geographic imaging, ERDAS IMAGINE gives you the ability to process images like a seasoned professional. ERDAS stands for Earth Resources Data Analysis System.

The first version, ERDAS 4, was released in 1978. Furthermore, Cromemco microcomputers were used, and users had access to 80 MB hard drives and big digitizing tablets. A new version of ERDAS, version 400, was released in 1980. Back then, only government agencies could afford to employ computers, which included NASA, the US Forest Service, and the US Environmental Protection Agency. The release of ERDAS 7 in 1982 prompted a partnership between ERDAS and ARC/INFO developed by ESRI. Users were allowed to combine the strengths of remote sensing and GIS mapping. At long last, the original version of what is now known as the ERDAS Imagine software package was released. Furthermore, the graphical user interface's power to facilitate data visualization, map making, and image processing is a crucial feature.

Earth Resources Data Analysis System offers a different tool as described below

- The Intelligent Viewer
- LiDAR Tools
- Terrain Tools
- Spectral Tools
- Radar Tools
- Spatial Model Editor.

Only spectral tools will be dealt with during the water conservation work.

5.6.1 Spectral Tools

Every part of our planet is made up of a different set of elements. It reflects visible light while absorbing infrared, ultraviolet, and other wavelengths. In other words, the spectral signature of each feature is unique. Consider this concept, scientists are compiling spectral libraries to record the varied chemical and physical properties of minerals, plants, and other natural materials. Similarly, to the USGS Spectral Characteristics Viewer, ERDAS Imagine provides its spectral profile library so you can delve even deeper into this topic.

More spectral bands (like those found in hyperspectral data) means more potential for accurate feature classification. However, ERDAS Imagine does have wizards for hyperspectral analysis, so don't let that put you off.

Both supervised and unsupervised categorization tools are available. There are powerful image classification tools, but unlike Trimble ECogntion Definiens Developer, you can't use them to analyze images for objects. In remote sensing, there is no shortage of indices (NDVI, SAVI, GNDVI, RVI, MSAVI, DVI, etc.). The same holds true for a wide variety of other substances, including plants, sulphates, water, snow, and ice. You can get this information out of a multispectral image if you have the right bands.

5.7 USGS

U.S. mapping had previously relied heavily on military expeditions and a number of separate government surveys prior to the establishment of the USGS. The United States Geological Survey (USGS) was formed to improve surveying practices and categories of public lands based on their geological composition, mineral wealth, and economic potential. The government's attitude toward surveying shifted because of this scientific evaluation of land potential and mineral resources, which in turn prompted conservation, economic growth, and more efficient development across the country.

Understanding Earth better is one of USGS's top priorities. Because of this, the Survey is universally recognized as the authority on matters pertaining to the natural sciences. The United States Geological Survey (USGS) provides accurate scientific data about Earth to the public. The scientific efforts of the USGS aid in the measurement and study of water, biological, energy, and mineral resources, and contribute to the reduction of human and material losses caused by natural disasters. Ecosystems, Energy and Minerals, Environmental Health, Environmental Hazards, and Water are only few of the primary mission areas that guide USGS's research and analysis. Maps developed with modern digital geographic information systems and scanned historical topographic maps from 1884 to 2006 are both available through the USGS National Geospatial Program.

The U.S. Geological Survey keeps tabs on one of the nation's most valuable assets (water). In all likelihood, the water supply you are using has been subjected to some sort of USGS testing, monitoring, or research. The United States Geological Survey (USGS) gathers vital information, such as water quality and stream trends, to aid in the prediction and prevention of local losses of life and property. The US Geological Survey (USGS) maintains a real-time streamflow map that anybody can use to keep tabs on the water levels of rivers throughout the country.

So basically, the United States Geological Survey (USGS) is a division of the U.S. Department of the Interior that focuses on scientific research The United States Geological Survey (USGS) collects and disseminates scientific data on topics including natural hazards that threaten human life and livelihoods; water, energy, minerals, and other natural resources; ecosystem and environmental health; and the consequences of climate and land-use change. The United States Geological Survey (USGS) offers scientific information on natural hazards that endanger human life and livelihoods; water, energy, minerals, and other natural resources upon which we depend; ecosystem and environmental health; and the effects of climate and land-use change. Use the USGS Earth Explorer for a simple, streamlined process to access high-resolution satellite and aerial photos at no cost. This instrument provides a great deal of flexibility. Time range, area covered, and type of image can all be customized.

5.8 Landsat

Since 1972, Landsat satellites have continuously captured high-resolution images of Earth's terrestrial surface, coastal shallows, and coral reefs. NASA and the United States Geological Survey (USGS) have joined forces to create the Landsat Project. Earth's land surface, coastal shallows, and coral reefs have been constantly imaged from space by Landsat satellites since 1972. The Landsat Project is a cooperative venture between the National Aeronautics and Space Administration (NASA) and the US Government. The National Aeronautics and Space Administration (NASA) was created to regularly collect satellite pictures of the earth's land. Using remote sensing technology, NASA launched and validated instruments and spacecraft how well the instruments and satellites functioned. The mid-1960s saw the United States make significant strides in planetary exploration with unmanned remote-sensing satellites, prompting the Interior Department, NASA, and the Department of Defense to invest in the construction of a human space program. An extensive program to modernize and expand agriculture was initiated. Send into orbit the first civilian Earth-observing spacecraft. It's in their best interest to successfully launched Landsat 1 on July 23, 1972. This satellite was formerly known as the Earth Resources Technology Satellite, or ERTS. Since then, Landsat satellites have supplied research institutions all around the world with high-quality, as well as those involved in resource management by providing a repository for Land remotely sensed data gathered from space: a useful tool for Workers in fields such as agriculture, geology, forestry, education, research on global shifts, regional development, and cartography [32].

Images captured by Landsat satellites span 185 km (115 miles) swath of Earth's surface as the satellite descends through its orbit (north to south) over the face of the planet that is facing the sun. Both Landsat 7 and Landsat 8 follow a 705-km-high orbit around the planet. A height of 267 km (or 438 miles). During that time, they each complete one orbit around the minutes, make 14 full orbits every day, and pass through every apex of the Earth's orbit once every 16 days. While every satellite does have a 16-day Earth-spanning orbital cycle, their asymmetrical paths make it possible for them to always be in the right place at the right time. Landsat data can be re-collected every 8 days from any location on Earth. Both Landsat 4 and 5 used this same orbit. Landsats 1–3 moved into a 920 km (572 km) orbit, making repeated trips around orbits the Earth once every 103 min, resulting in daily repeats of 18 days.

Government, commercial, industrial, civilian, military, and educational communities use Landsat data globally. The data support global change research, agriculture, forestry, geology, resource management, geography, mapping, water quality, and coastal studies.

Even though the SLC failed in May of 2003, Landsat 7 has continued to produce vital Earth observation data ever since. More than 3.7 million Landsat scenes had been collected by USGS ground receiving stations as of March 2013. Full coverage of Earth's landmasses is provided by this network and the ability of Landsat 7 and 8 to record and downlink data to foreign sites; however, the satellite's orbit does not include a direct pass over either the North or the South Pole. Consistent Landsat data allows for side-by-side comparison of recent site photographs with those taken months, years, or decades ago. By comparing the two, we can see if the land cover is shifting gradually or rapidly. Users can utilize time series data over wide geographical areas, thanks to the repository's comprehensive archive and free data policy, to detect long-term patterns and monitor land surface change. Landsat images, both before and after a crisis, are extremely helpful for relief efforts. Disaster response teams can get satellite imagery from the USGS EROS Center in Sioux Falls, South Dakota, within hours after data capture the interplay of density, altitude, and other elements.

EarthExplorer makes all of the Landsat data that is stored in the USGS archives accessible for free download, and there are no limits placed on who can use the data.

5.8.1 Extraction from Landsat

Google Maps is the base map for the USGS Earth Explorer interface. Just as in Google Maps, you may use the scroll wheel to zoom in and out. Additionally, Google street view is turned on so that you can place a marker and see the area as it is.

To begin, you must register for a USGS account. To sign up, use the Register tab in the upper right. The steps to activate your account will be sent to you, and they're easy enough that you won't mind. Here are the four steps you need to do to get data from USGS Earth Explorer:

- Set your search criteria
- Sclcct your data to download
- Filter out your data
- Check your results and download.

5.8.1.1 Step 1: In the "Search Criteria" Section, Specify Your Desired Subject Matter

Simply by double-clicking the browser's window, users can designate specific areas for further exploration. To restrict the scope of the search for data, a "region of interest" (ROI) must be defined. To construct a zone of interest in another fashion, you can utilize one of the following alternatives:

- Locating a location through its address
- Bringing in a KML or compressed shapefile
- You can also just double-click the map to get your ROI.

USGS Earth Explorer also allows you to schedule of satellite and aerial imagery. You may now quickly identify the right date without having to scroll through a big list of purchases. You may save time and energy by using this effective search tool.

5.8.1.2 Step 2: In the "Data Sets" Tab, Choose the Data You Want to Download

Click the Datasets Tab to see what kinds of satellite or aerial imagery are available. Through USGS Earth Explorer, you can access remote sensing datasets like aerial images, commercial images, digital elevation models, Landsat data, LiDAR data, MODIS data, Radar data, and many more.

Downloadable Landsat scenes are time- and date-specific. Landsat > Landsat Collection 1—Level 1 contains the most up-to-date Landsat imagery, which is L8 OLI/TIRS and L7 ETM+. The collections vary in terms of data quality and depth of processing, respectively. The US Geological Survey has established a hierarchy of image quality and processing depth.

5.8.1.3 Step 3: Use the "Additional Criteria" Tab to Narrow Down Your Dat

Most of the users click the tab labelled "Additional Criteria" to quickly eliminate cloudy images. You can no longer use a filter to achieve that desired clear sky background. But if you want, you can reduce cloudiness to less than 10%, which is what most people prefer. In most cases, fewer filters will suffice for the typical user. Without worrying about the clouds getting in the way, you can get on with downloading your satellite imagery.

5.8.1.4 Step 4: Go to the "Results" Tab and Click on "Free Landsat Images"

After specifying a time frame, data type, and other parameters, the search results tab will be updated with datasets that meet the specified conditions. Images can be downloaded individually by clicking on them in the "Results" tab. It's important to inspect the scene's footprint to get a sense of its precise location. It's possible to have a look at a sample of the information first, which can help pinpoint where in the picture the clouds are. When you're ready, use the "Download" button to get your hands on the info.

5.9 Monitoring and Conservation Through ERDAS

Spectrum analysis is used by the ERDAS Picture software to do the categorization of an image identify terrestrial features. The multispectral data is utilized in the process of classifying the SSC image. This includes the categorization of terrestrial objects, as well as vegetation and the shadows cast by trees. Supervised classification and unsupervised classification are the two methods that can be used to divide pixels into several groups. The identification of terrestrial objects in the Study Image was facilitated thanks to the classification of data carried out using ERDAS Image (SSC). The numerical basis for categorization is determined by using the spectral pattern that is contained within the data for every pixel. In the first stage of the analysis of the Image SSC, a generic classification system consisting of four categories is used (Grass, Trees, Man-Made and Unknown). The outcome of the unsupervised picture analysis is used as a starting point for the supervised image classification analysis that follows. The capacity of the supervised image to decipher related images, such as the roofs of the buildings and the shadows of the trees, is the primary distinction between the two pictures. Image stacking is used to build a fully categorized image so that shadow, grass, man-made objects, and trees can be distinguished from one another. On this poster, the processes of viewing and measuring images will be broken down into two categories: computer-guided (unsupervised) and user-guided (Supervised) Detailed instructions on how to stack images in order to view each classification one at a time and compile the results into a whole Classified Image are provided.

5.9.1 Supervised Image Classification

Human intervention is required for the majority of the classification proccss in supervised image analysis. Image analysts made of humans play an essential part. They detail the values of each land cover type and land use in terms of the multispectral reflection emittance. In a nutshell, the analysts are going to be in charge of supervising the process of pixel categorization through its three stages: training, allocation, and testing.

During training, the analysts are allowed to determine the membership of a known class based on a sample of pixels obtained from referenced data. Aerial pictures and existing maps are two examples of such types of data. To generate various data pertaining to the various land cover classes, the training pixels are utilized. During the stage of allocation, the photos are first classified and then assigned to the classes to which they will be most appropriately suited depending on the results of the statistics. In the final step, known as the testing stage, a sample of testing pixels is chosen, and then the various class identities are compared to one another. The comparison is carried out using the reference data as well as the spectral parameters of each pixel contained inside the image. The findings are determined via an error matrix that considers the percentage of test samples that agree and disagree with

Table 5.1 Generally, these colors show specific classes (these can be changed according to your study)

Color	Specific class
Green	Vegetation
Red	Urban
Yellow	Grasslands
Blue	Waterbodies
White	Barren lands
Grey	Roads/railways

each hypothesis. After all, three processes have been completed, an analyst will be able to evaluate the picture categorization for each land cover type. Table 5.1 Shows the color of specific classes which can be changed according to the study.

5.9.1.1 Method

- Launch ERDAS IMAGINE program, then select Preferences by clicking the File icon. When the preferences pane appears, click the IMAGINE Preferences button, followed by the User Interface button, and then click the User Interface & Session button.
- Next, choose File > Save As > and File > Save Output to their defaults.
- To do supervised classification, add a layer or some data.
- Click the Drawing tab, then the Insert Geometry group, then Grow, and then Lock. Find a specific land feature that you can use to make an eye estimate. Just click on it and it will appear.
- The most crucial step is to access Growing Properties and select a location to mask the Feature Pixels value. You can also modify this manually to obtain more precise pixels.
- To get all the pixels you need for a value, go to the Drawing Tab, expand Select, and choose Select by Box from the drop-down menu. Just drag it over the whole Satellite image, and you'll see that the value of every pixel is chosen.
- Open the Signature Editor as the next step. Choose Raster from the menu, then click Supervised and then Signature Editor.
- Click Create New Signature in the Signature Editor window that pops up. Now, on the Signature Editor, you can see a list of all the Pixels values that have been chosen. All of the signatures are listed, along with all of the pixels.
- Choose all of the signatures or the whole class, and then click Marge Selected Signatures. If you look at the table of signatures, a new class called "class 15" will be made.
- Remove all of the selected signatures or classes (1–14) except the Marge Class (class 15). Right-click on the classes you want to get rid of, and then click "Delete Selection." Now, change the class's name and color.

- For the other types of Land use features, follow the same steps. This image from space shows water bodies, farmland, plants, buildings, and marshy land. Change the values in both the Value column and the Order column.
- Finally, do the Supervised Classification. Choose Supervised in the Signature Editor's Classify menu and spend money. In the next window, select Output Destination and Name. You can also make error data by checking the box next to the Output Distance file. This error information lets us figure out how accurate the supervised image is.
- Choose each Decision Rule one at a time.

  (i) Non-parametric Rule—Parallelepiped
  (ii) Overlap Rule—Parametric Rule
  (iii) Unclassified Rule—Parametric Rule
  (iv) Parametric Rule—Maximum Likelihood.

- When you've finished with everything, all that's left to do is click the OK button. In Eras Imagine you have access to a viewer with two windows: Satellite picture known as a Raster Image, as well as a Supervised Classification Image.

5.9.2 Unsupervised Image Classification

Unsupervised classification occurs when a piece of software analyses an image and groups pixels that share similar characteristics without the need for the user to first build up training fields specific to each land cover category. This is all accomplished without making use of any training data or knowledge that was previously in existence. It is the responsibility of the picture analyst to determine how the spectral classes that are generated by the algorithm correspond with one another. There are two main steps to follow in unsupervised classification. Among these are making clusters and giving classes to them. Using the software for remote sensing, an analyst will first make clusters and decide how many groups to make [1].

5.9.2.1 Method

Add a picture of your study area to the ERDAS IMAGINE Window

- Visit the Unsupervised Classification Tool page.
- Name the input, output, and signature output features and include a true color scheme that is appropriate.
- Add the classified image carefully, making sure to check the box for "clear display."
- When you click the Cursor Button, you'll be shown the Class of the selected Feature in the original and classed image you linked.

- Change the color and class name in the image's Attribute Table as you know for sure that Class 8 is Water.
- Fill in the names of the classes (how many will depend on how many you have). And you are done!

The next step is to calculate the area of each class.

5.9.3 Area

Simply right-click the output raster you just made, select Properties, go to the Symbology tab, and take note of the Count for the displayed pixels to pinpoint the location. You may see the cell size and linear units by going to the Source tab. It's as easy as multiplying the cell size by the total number of cells to obtain the area in square units of the original linear units If you know the area of a single raster cell in projection, which should be 30 m by 30 m (which is 900 m^2 or 0.09 hectares), then you can multiply the cell area by the cell count to determine the area of each class's cells. For instance, if a class has 14,478 cells and the cell size is 0.090 ha, then the area is 1,303.8 ha.

Or to be familiar with the surface area of the features What we can do is to go to the options table that is given under the raster and choose "Add Area". This will cause a new window to open, and we can now choose between the units of hectares, acres, or square miles depending on what kind of analysis you are doing. After that, we can press ok, and the area column will add up against the classes.

The classification will be done for two satellite images, one for the most recent and one for the old; for the water perspective, we can take a difference of decade, for example, an image from 2022 and the other one from 2012; however, we can take a difference of even more than a decade. As we must plane for long term. Now that the classification is complete, we can manage, monitor, and conserve water because we know where the features are located and what is the exact area of each class.

Because we are aware of the areas of vegetation (agricultural land), barren ground, and present urbanization that is taking place on formerly agricultural land, we can control and monitor water easily. The conservation of deficient water can be done by planning a new agricultural irrigation system. ArcGIS must be used to design this system utilizing data from Eras, and the water must be diverted to newly formed agricultural land for sufficient use, saving unlawful water consumption, and resolving the canal filling issue frequently encountered by urbanization.

5.9.4 Estimated Canal Water Deficit Through RS

The global population is growing at an unprecedented rate of 100 million per year, creating competition for water supply amongst diverse industries. Agriculture is under a lot of pressure due to the rising demand for water from other sectors like urbanization, industrialization, and the energy industry. More than 70% of all freshwater is used in agriculture, and in the Asia-Pacific region, that number rises to more than 90%, all for the sake of producing food and fiber. Precision in irrigation application is essential given the current trend of decreasing water availability and the intended increase in crop yields per unit of water used from poorly managed irrigation system. In order to guarantee worldwide food safety, it is crucial to use reliable methods of measuring the amount of water crops needs. Accurate and up-to-date information on land use and land cover (LULC) in the canal command area could help manage irrigation water based on how much water crops need. It can be done to map the LULC of the irrigated command area so that the canal water deficit can be found (CWD).

We are going to download clear, high-quality photographs of LANDSAT-7 TM from the USGS website and then categorize them. This process will take place from the beginning of the season until its end. The unsupervised categorization is going to take place. To differentiate between bare soil and vegetation, the ISODATA clustering that consists of six classes will be utilized. To categorize the various types of vegetation, a division will be made using the ISODATA findings that have a greater number of clusters.

5.9.4.1 Irrigation Water Demand

Any nearby weather station will provide the essential meteorological data for determining the ETp for the research area. The Penman-Monteith equation will be used to determine ETp. Using FAO Penman-Monteith definition for ETo, crop coefficients can be found at sites by comparing the crop evapotranspiration (ETc) measurements with the ETo calculations. ET_o, i.e., $K_c = ET_c/ET_o$. [2]

According to [28], all crops in the canal command area had their water needs evaluated using Kc values for 10 days [26].

$$ET_a = K_c ET_p \tag{5.1}$$

where ETc is the crop's actual water needs in meters per growing season, Kc is the crop coefficient, and ETp is the crop's reference evapotranspiration in meters per growing season.

5.9.4.2 Irrigation Water Requirement Calculation

From the seedling stage through harvest, the crop's total water needs are estimated. After factoring in the losses due to conveyance and seepage, the overall irrigation water need (IWR) for a given crop and the total irrigation water demand for the whole command areas of distributaries are determined [26].

$$IWR = \sum_{i=1}^{i=n} K_{ci} ET_p CA \tag{5.2}$$

where IWR is the irrigation water requirement in ha-m (Hectare meters), CA (cropping area) of the corresponding crop, in ha. Kc is crop coefficient and ETp is reference evapotranspiration in m/growing season. The total amount of irrigation water needed each month is found by multiplying the crop area in square meters from satellite data by the amount of irrigation water needed each day. The total amount of water needed for irrigation was estimated by adding up the amount of water needed for 10 days of irrigation during the season. The amount of water that goes into canals every day is added up to figure out the 10-day and seasonal water supplies. The Irrigation Department's data on irrigation is used to make this estimate.

By multiplying the efficiency factor, you can figure out how much water can be utilized for irrigation.

5.9.4.3 Estimated Canal Water Deficit

Meteorological data with 10-day Kc values were used to estimate how much water crops need to grow during the season. With a regional-level study, we figured out how much water is used for irrigation in the distributaries. After taking into account how efficient the irrigation system is, the amount of water needed for irrigation is estimated and compared to how much water is available. Overall, the Pakistan canal irrigation system is only 45% effective, which shows that there are very high conveyance losses [4]. There is a significant disparity between irrigation water demand and canal water supply in the latter days of the season, as seen by the comparison.

The canal water deficit was calculated by subtracting the amount of water available in the canals during the Rabi season from the amount of water needed for irrigation. The average CWD in Khurrianwala, Killianwala, and Mungi distributary was 64% in 2009–2010 and 72% in 2010–2011 during the Rabi seasons [31].

5.9.5 Monitoring Using NDVI and NDWI

Natural wetlands and the benefits they provided to biodiversity have been altered and lost as a result of various forms of competition for water and intensified agricultural techniques, as well as the expansion of urban areas. It is essential to document and keep an eye on wetlands as well as the land characteristics that surround them to preserve and guard the wetland resources. When measuring the dynamic of wetlands, particularly across broad areas, the application of spatial science techniques like remote sensing has proven to be quite beneficial.

Free acquisition of Landsat images can be obtained from the USGS database. The image is going to be classified into four distinct categories, which are going to be vegetation, sparse vegetation, water body, quagmire, and marshland respectively. The created classed images are then examined to see whether there has been a shift in any of the land feature categories by estimating these features in a Geographic Information system ArcGIS tool.

In addition to this, NDVI (Normalized Difference Vegetation Index) and NDWI (Normalized Difference Water Index) will be assessed using satellite data [21].

$$NDVI = \frac{NIR - RED}{NIR + RED} \tag{5.3}$$

where RED are the measurements of spectral reflectance taken in the red (visible) and near-infrared (NIR) regions, respectively. So, the NDVI ranges from − 1.0 to + 1.0 NDVI is functionally but not linearly, equivalent to the simple infrared/ red ratio (NIR/VIS). The formula for NDWI is given below [21];

$$NDWI = \frac{X_{nir} - X_{swir}}{X_{nir} + X_{swir}} \tag{5.4}$$

whereas NIR refers to wavelengths in the near infrared, SWIR refers to wavelengths in the short-wave infrared, The NDVI and NDWI are used to evaluate the state of the water and the vegetation pattern at present. According to findings from earlier research, the regions with low NDWI values are more likely to experience wetland shrinkage, whereas the regions with high NDWI values are predicted to experience little to no wetland shrinkage at all [21]. The mapping of wetland dynamics in the research area was accomplished with the use of four satellite photos spanning the years 1987, 1997, 2007 (Landsat 5 Thematic Mapper), and 2017 (Landsat 8 Operational Land Imager). The conclusion revealed that the natural landscapes in the area have witnessed changes in the last three decades. Between 1987 and 2017, the areas covered by dense vegetation, sparse vegetation, and water bodies all expanded by around 14% (5976.495 km^2), 23% (10,349.631 km^2), and 1% (324.621 km^2), respectively. During the same period, wetland features such as marshland and quag endured dry conditions. A significant reduction with an area coverage of approximately 16,651.07 km^2 (38%) The results of this investigation showed that there was a change in the vegetation and in recent years, the Isimangaliso Wetland Park has experienced a dramatic deterioration, which can be directly attributed to the negative effects of water body extents [21]. Figure 5.3 shows the procedure of monitoring and conservation of water through RS.

5.10 Conclusion

We are all aware of the ways in which urbanization has impacted our lands and waterbodies, and the rate at which urbanization is occurring is accelerating at an alarming rate. As a result, it is imperative that they must be managed, monitored,

Fig. 5.3 Procedure of monitoring and conservation of water through RS

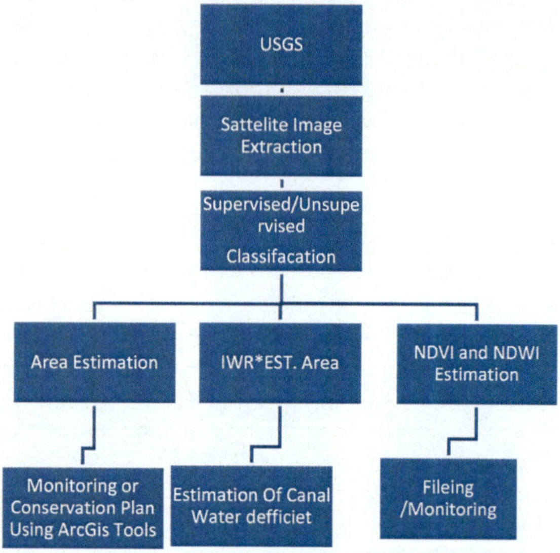

and documented in order to ensure their conservation for the benefit of our future. It would be very tough to do it with old and conventional means, thus in order to cope up with it, we need to utilize satellite images. Free and open access to this data has encouraged the use of satellite data to answer a vast number of scientific questions, to better resource management, and to inform actions related to reporting. The strategies that were presented earlier in this chapter have the potential to be a straightforward and efficient solution for managing and conserving our groundwater.

References

1. Ali A (2017) Unsupervised image classification in ERDAS imagine
2. Allen RG, Pereira LS, Raes D, Smith M (1998) Crop evapotranspiration-guidelines for computing crop water requirements-FAO irrigation and drainage paper 56. Fao Rome 300(9):D05109
3. Arshad A, Mirchi A, Samimi M, Ahmad B (2022) Combining downscaled-GRACE data with SWAT to improve the estimation of groundwater storage and depletion variations in the irrigated Indus basin. Sci Total Environ 156044
4. Bhatti AM, Suttinon P, Nasu S (2009) Agriculture water demand management in Pakistan: a review and perspective. Soc Soc Manag Syst 9(172):1–7
5. Burgess EW (2012) The growth of the city: an introduction to a research project. In: Park R et al (eds) The city (1925). In The urban sociology reader. Routledge, pp 91–99
6. Chen J (2007) Rapid urbanization in China: a real challenge to soil protection and food security. Catena 69(1):1–15
7. Cohen B (2006) Urbanization in developing countries: current trends, future projections, and key challenges for sustainability. Technol Soc 28(1–2):63–80
8. Cohen B (2015) Urbanization, City growth, and the New United Nations development agenda

9. Das B, Singh S, Jain SK, Thakur PK (2021) Prioritization of sub-basins of Gomti river for soil and water conservation through morphometric and LULC analysis using remote sensing and GIS. J Indian Soc Remote Sens 49:2503–2522
10. Deng Z, Zhao Q, Bao HX (2020) The impact of urbanization on farmland productivity: implications for China's requisition–compensation balance of farmland policy. Land 9(9):311
11. Dinye RD, Acheampong EO (2013) Challenges of slum dwellers in Ghana: the case study of Ayigya, Kumasi. Mod Soc Sci J 2:228–255
12. Eraly A (2007) The Mughal world: life in India's last golden age. Penguin Books, p 5
13. FAO (2001) The state of food insecurity in the world. Rome
14. Hitomi N, Bang H, Selva D (2017) Extracting and applying knowledge with adaptive knowledge-driven optimization to architect an earth observing satellite system. In: AIAA Information Systems-AIAA Infotech@ Aerospace, p 0794
15. Kelly AC, Loverro A, Case WF, Quéruel N, Maréchal C, Barroso T (2010) Small earth observing satellites flying with large satellites in the a-train. In: Small satellite missions for earth observation. Springer, Berlin, Heidelberg, pp 19–28
16. Lubowski RN, Bucholtz S, Claassen R, Roberts MJ, Cooper JC, Gueorguieva A, Johansson RC (2006) Environmental effects of agricultural land-use change: the role of economics and policy (No. 1477-2016-121102)
17. Malanima P (2009) Pre-modern european economy: one thousand years (10th–19th centuries). Brill Publishers, p 244
18. Marzluff JM (2001) Worldwide urbanization and its effects on birds. Avian ecology and conservation in an urbanizing world, pp 19–47
19. Mr. Abasilim (2018) Lecture notes on PAD324: government and administration of the
20. National Library of Medicine (NLM) (2014) Urbanization
21. Orimoloye IR, Mazinyo SP, Kalumba AM, Nel W, Adigun AI, Ololade OO (2019) Wetland shift monitoring using remote sensing and GIS techniques: landscape dynamics and its implications on Isimangaliso Wetland Park, South Africa. Earth Sci Inform 12(4):553–563
22. Reba M, Reitsma F, Seto KC (2016) Spatializing 6,000 years of global urbanization from 3700 BC to AD 2000. Sci Data 3(1):1–16
23. Rimal B, Sharma R, Kunwar R, Keshtkar H, Stork NE, Rijal S, Rahman SA, Baral H (2019) Effects of land use and land cover change on ecosystem services in the Koshi River Basin, Eastern Nepal. Ecosyst Serv 38:100963
24. Sanyaolu P, Sanyaolu C (2018) Urbanization 1. https://doi.org/10.13140/RG.2.2.23495.96161
25. Satterthwaite D, McGranahan G, Tacoli C (2010) Urbanization and its implications for food and farming. Philos Trans R Soc B: Biol Sci 365(1554):2809–2820
26. Ullah MK, Habib Z, Muhammad S (2001) Spatial distribution of reference and potential evapotranspiration across the Indus Basin Irrigation Systems, vol 24. IWMI
27. United Nations (2015) Department of Economic and Social Affairs, Population Division, World Urbanization Prospects: The 2014 Revision (ST/ESA/SER.A/366)
28. United Nations Human Settlements Programme (2004) The State of the World's Cities 2004/2005: globalization and urban culture, vol 2. UN-HABITAT
29. Untaru EN, Ispas A, Candrea AN, Luca M, Epuran G (2016) Predictors of individuals' intention to conserve water in a lodging context: the application of an extended theory of reasoned action. Int J Hosp Manag 59:50–59
30. van Vliet J, Eitelberg DA, Verburg PH (2017) A global analysis of land take in cropland areas and production displacement from urbanization. Glob Environ Chang 43:107–115
31. Waqas MM, Awan UK, Cheema MJM, Ahmad I, Ahmad M, Ali S, Shah SHH, Bakhsh A, Iqbal M (2019) Estimation of canal water deficit using satellite remote sensing and GIS: A case study in lower chenab canal system. J Indian Soc Remote Sens 47(7):1153–1162
32. Williams DL, Goward S, Arvidson T (2006) Landsat. Photogramm Eng Remote Sens 72(10):1171–1178

Chapter 6
A Comprehensive Review on Mapping of Groundwater Potential Zones: Past, Present and Future Recommendations

Sourav Choudhary, Jagriti Jain, Santosh Murlidhar Pingale, and Deepak Khare

Abstract The over exploitation of groundwater resources is a highly thought-provoking issue, which hinders the goal of sustainable water management worldwide. Hence, it is utmost necessary to identify the groundwater reserves in terms of potential areas/zones, average yield, and seasonal recharge. Within last decades a substantial progress has been observed in the delineation of Groundwater Potential Zones (GPZ) and have successfully applied bi-variate model, Multi-criteria Decision Making (MCDM) models, state of the art Machine Learning (ML) model, Ensemble model and metaheuristics models in the development of GPZ. However, still a research gap exists in the demarcation of GPZ both in terms of groundwater potential model development and groundwater conditioning factor selection which is very significant from the scientific and policy maker's point of view. Thus, the present review article aspires to render a more vivid understanding of future aspects of groundwater potential model development and the milestone achieved in the past. This review article covers all types of models applied in the demarcation of GPZs, selection of different groundwater conditioning factors, and type of data used (remote sensing and ground truth). The present article also comes up with all possible criteria and statistical methods for the evaluation of the model's performance and accuracy. Furthermore, recommendation for potential future research direction to enhance the model prediction accuracy is also outlined in the present article which will be highly effective for the groundwater agencies and organisation.

S. Choudhary (✉) · J. Jain · D. Khare
Department of Water Resources Development and Management, Indian Institute of Technology Roorkee, Roorkee 247667, India
e-mail: schoudhary@wr.iitr.ac.in

J. Jain
e-mail: jjain@wr.iitr.ac.in

D. Khare
e-mail: deepak.khare@wr.iitr.ac.in

S. M. Pingale
Hydrological Investigations Division, National Institute of Hydrology, Roorkee 247667, India
e-mail: pingalesm.nihr@gov.in

E. Balaji et al. (eds.), *Emerging Technologies for Water Supply, Conservation and Management*, Springer Water, https://doi.org/10.1007/978-3-031-35279-9_6

Keywords Groundwater potential zone (GPZ) · Bivariate and MCDM models · Machine learning (ML) models · Ensemble models and metaheuristics models

6.1 Introduction

Groundwater is one the most important resource on the planet, which can provide a consistent supply of good quantity and quality of water in surface water scare regions. The scarcity of surface water has necessitated research on the use of groundwater resources in recent decades [21, 20, 23, 42, 109]. The impeding stress on this valuable resource is increasing as the quality and quantity of surface water resources deteriorate [55]. Excessive groundwater extraction and unsuitable aquifer recharge are the primary causes of groundwater stress in many areas. Socioeconomic water stress (such as population and urbanisation) is also inextricably linked to groundwater resources [106]. A majority of population is experiencing drinking water crisis due to uneven spatial distribution of surface water resources and limited available groundwater resource [12]. Even industrial and agricultural sectors depend heavily on this valuable resource to meet their needs. The agricultural sector's reliance on groundwater has risen to 89% in many areas, and this figure is expected to rise further as the urban population grows [16] resource on the planet, inextricably linked to human development and national infrastructures.

Hence, a sustainable use of water is necessary and there is an urgent need to consumptively use the groundwater resources. The efficient use of groundwater resources requires identification of groundwater potential zones (GPZ) and study of groundwater recharge and discharge from all sources. The GPZ maps are becoming highly significant in many scientific and engineering fields [18]. GPZ distinguishes proper sites for groundwater well exploration and management of surface and ground water resources. Numerous techniques (such as well drilling, electrical resistivity based geophysical exploration) has already been used to map GPZs. Such methods are appropriate for identifying the hydrological characteristics of groundwater and consumes a lot of time and money. Hence, this leads to the dependence on the Remote sensing (RS) and Geographic Information System (GIS) techniques for the delineation of GPZs.

In recent decades many methods for delineating GPZs have been developed. It includes RS and GIS methods, statistical and probabilistic methods and Machine Learning (ML) algorithms based modelling techniques [23, 38, 57]. The Analytical Hierarchy Process (AHP) is a weight based most fundamental method for GPZ delineation [25]. AHP have been successfully applied in a lot of studies such as Avalanche mapping, GPZ and flood inundation mapping [36, 44]. Weight of Evidence (WoE) [52], certainty factor, Evidential Belief Function (EBF) [63, 82, 87], Index of Entropy (IOE), Frequency Ratio (FR) and Shannon's entropy are some statistical models [31]. A comprehensive review of the literature based on GPZ mapping using different techniques is presented in Table 6.1.

Table 6.1 Literature review of the analysis of GPZs

S. N	Author	Input parameters	Model	Result
1	[102]	ELV, Cur, TRI, Asp, Slp, TWI	EML, RF, FL, NBT	NBT best model has highest ROC of 0.892
2	[58]	ELV, SLP, DD, LNMTSOIL, LITH, RF, LULC	FAHP, FR, BWOE	FR (AUC = 0.964, BWOE (AUC = 0.838) and FAHP (AUC = 0.829)
3	[50]	LITH, SOIL, GEOL, YIELD DD, RF	AHP	(R2) of 0.59
4	[40]	SLP, DD, TWI, DFR, LNMT, NDVI, SOIL, RF, GEOMOR, LULC, GWP	AHP, RF, NB,	AUC of FR, AHP, NB and RF are 0.81,0.780.855 and 0.853
5	[24]	SLP, AP, ALT, PL, CUR, TWI, NDBI, RF, DD, LULC, SOIL, GWP	WOE, TOPSIS, FFR, XGB, ET, ENSEMBLE	AUC of WoE_TOPSIS = 0.94 WoE_F-FR = 0.93 TOPSIS_F-FR = 0.94 WoE_TOPSIS_F-FR = 0.95
6	[15]	ELE, SLP, ASP, LS, CUR, LULC, LNMT, TWI	CFR, MFR, DT, RF	AUC CFR = 74%, MFR = 71.4%, DT 79.3%, RF = 95%
7	[108]	ELE, SLP, ASP, LS, CUR, LULC, LNMT, TWI	FR, RBF, IOE, EBF, FAM, ENSEMBLE	(AUC = 88.9%, FR-FAM (AUC = 86.9%, EBF-FAM (AUC = 86.4%, EBF-RBF (AUC = 85.4%, FR-IOE (AUC = 83.6%, and EBF-IOE AUC = 83.3%
8	[105]	LITH, SLP, LULC, RF, LNMT, SOIL, DD	EIGHTED OVERLAY	2.22, 26.93, 56.74, 13.84, and 0.26% Very good to very poor
9	[67]	SLP, GEOMOR, RF, LITH	AHP, FR, AHP-FR	RF, AHP, HM-A and HM-B show the AUC value of 0.72, 0.605, 0.719 and 0.498
10	[19]	RF, SOIL, GEOL, LULC, RF, DD	AHP, HYBRID FUZZY SET THEORY	Accuracy of 81.81 and 75.75%
11	[93]	TWI, TRI, SPI, TPI, MVBF SLP, ASP, SLPL, LULC, GEOL	MABLAR, MPL, LR, SVM,	MABLR is more effective at reducing bias
12	[69]	LITH, SLP, LULC, RF, LNMT, SOIL, DD, TRMM	AHP	0.66, 38, 58.5, 1.8, 0.004% Best to poor GPZ
13	[57]	RF, SOIL, STRM, GEOL, WJM, SLP, LULC, NDVI, NDWI, LINM	AHP, M-AHP, PCA	Delineation of GPZ
14	[83]	RF, SOIL, STRM, GEOL, WJM, SLP, LULC, NDVI, NDWI, LINM, FLTD	LR, MARS	AUC MARS and LR models was 0.867 and 0.838

(continued)

Table 6.1 (continued)

S. N	Author	Input parameters	Model	Result
15	[64]	SLP, ASP, ALT, STPI, TWI, CUR, LULC, DD, SOIL	FR	AUC = 75.99%

ELV = Elevation, SLP = slope, ALT = Altitude, TWI = Topographic Wetness Index, CUR = Curvature, LULC = Landuse and land cover, DD = Drainage Density, STPI = Sediment Transport Potential Index, GEOL = Geology, RF = Rainfall, LITH = Lithology, LINM = Limnology, NDBI = Normalised Difference Built up Index, TRMM = Tropical Rainfall Measuring Mission

The GPZ studies incorporates various factors for prioritization, which includes hydrological, hydrometeorological, topographical and groundwater influence. The most important groundwater conditioning factors includes aspect, lithology, land use/land cover (LULC), rainfall, slope, curvature, distance to roads. The prominent parameters in the mountain were geology and lineament, while in the plain region they were altitude, rainfall and slope [14, 26, 51, 57, 88, 103]. Hence, proper selection of groundwater conditioning factors is necessary to eliminate the unnecessary parameters for GPZ suitability [80]. Recent GPZ studies have shown that ML and ensembled models outperform multivariate statistical and bi-variate models [72] due to its more robust. Hence, the efficiency of stand-alone GPZ models is improved by integrating statistical and ML models [25, 62]. The ensembled ML model have outperformed bivariate statistical models in many studies related to GPZ, landslide suitability modelling, flood potential inundation, and crop zone monitoring [101]. Studies have shown increased accuracy of ensembled meta heuristics and ML models over conventional GPZ models [49]. According to [104], a high predictive performance is observed when particle swarm optimization (PSO) is combined with extreme learning machine (ELM) techniques. Similarly, Razavi-Termeh et al. [92] studied the ensembled performance of ANFIS-DE, ANFIS-ACO and ANFIS-PSO models and observed AUC of 81.6%, 75.8% and 80.9% respectively. Khosravi et al. [47] studied the performance of four ensembled meta heuristics models and observed the higher AUC for ANFIS-DE (87.5%) followed by ANFIS-IWO model (87.4%), then ANFIS-FA model (87.3%), ANFIS-PSO model (86.5%) and ANFIS-BA model (83.9%).

Thus, the current paper focuses on the following topics: (i) Definition of GPZ (ii) Different parameters required for GPZ studies, (iii) Modelling techniques for groundwater potential studies, (iv) Validation and accuracy assessment of GPZ (v) Present scenario and future challenges for GPZ research.

6.2 Definition of GPZ

Groundwater potential of any region is generally defined as entire volume of water in the aquifers which is stored and can be extracted for an extended period. Likewise, it can also be defined as the amount of groundwater that can be extracted from an aquifer without affecting the aquifer's yield and quality. Surprisingly, the past literature search yielded no universal definition of groundwater potential, and a closer examination of the literature reveals that different authors define groundwater potential mapping differently. While majority of researchers believe that GPZ maps fluctuation in aquifer storage, other represents it as location of groundwater presence and maximum groundwater yield. A minority focuses on identifying optimal locations for boreholes at the local scale. As a result, despite methodological agreement, GPZ mapping are mostly developed based on a series of indirect indicators.

6.3 Different Parameters for Groundwater Potential Studies

6.3.1 Topographical Parameters

Various topographic parameters are used for the analysis of GPZ such as Topographic Wetness Index (TWI), altitude, slope, curvature (plan and profile), flow direction and flow index. Furthermore, most important commonly used topographic parameters include slope, slope length, TWI, aspect and altitude. Choudhary et al. [24] considered all the important factors such as aspect, altitude, slope, TWI, and experienced gentle slopes contributes more towards the groundwater filtration which enhances the groundwater potential of any region whereas steep slopes have huge runoff and less residence times for groundwater infiltration. Similarly, Naghibi and Pourghasemi [72] showed the influence of slope and aspect on the hydrological process by determining the snow melting, rainfall direction and plant growth. Tahmassebipoor et al. [99] concluded that at high elevation groundwater is frequently scares than low elevation. Rahmati et al. [89] used TWI to calculate the amount of runoff accumulation in any location within a basin. The TWI index and groundwater yield have a strong inverse correlation, which is explained in part by GWP. Lee et al. [53] also prompted that the curvature of the area has the greatest influence on acceleration and deacceleration, as well as flow convergence and divergence. Jaafarzadeh et al. [43] observed that elevation has a significant impact on vegetation and climate, which are linked to the recharge distribution area. The topographical parameters are extracted from the Digital Elevation Model (DEM) which is prepared from the SRTM or OLS data. Different affecting factors of GPZ analysis is mentioned in Table 6.2.

Table 6.2 Various factors used in the literature for the analysis of GPZs

GPZ FACTORS	[24]	[60]	[8]	[73]	[58]	[50]	[40]	[15]	[105]	[67]	[25]	[69]	[85]	[5]	[87]	[1]	[38]
Aspect	*	*	*	*				*				*		*			
Altitude	*	*	*	*	*	*		*		*				*	*		*
Slope	*	*	*	*	*	*	*	*	*	*	*	*	*	*	*	*	*
Curvature	*	*	*	*				*									
TWI	*			*			*	*							*		*
NDBI	*												*				
NDVI	*						*						*	*		*	*
TRI			*	*													
Lineament D					*		*	*	*	*		*		*		*	
Lithology					*	*	*	*				*	*				
Drainage Density	*				*	*	*	*	*	*		*		*	*	*	*
Dist to streams	*			*			*	*			*		*	*			*
Dist to roads	*																
Geology				*				*			*			*	*	*	*
Geomorphology	*					*	*		*	*	*	*	*	*	*	*	*
Rainfall	*	*	*	*	*	*	*			*		*	*	*		*	*
LULC	*	*	*	*	*	*	*	*	*	*	*	*	*	*	*	*	*
SPI	*			*											*		
STI											*						
Soil	*	*			*	*	*		*	*	*	*	*	*	*	*	*
Gw well logs	*		*	*	*		*	*	*		*	*			*		
Gw yield						*					*			*			

6.3.2 Hydrological and Meteorological Parameters

Hydrological parameters are the key variables for the delineation of GPZ. Some of the key variables includes precipitation, river density, distance to streams, surface runoff, drainage, Drainage Density (DD), evaporation, evapotranspiration, and net recharge. Precipitation, river density are the most common parameters for the groundwater potential of any region. Studies have confirmed precipitation as the key factor for GPZ as it provides as the major source for the groundwater recharge. Precipitation contributes significantly to aquifers via subsurface infiltration phenomenon. The level of groundwater rises as rainfall increases. The rainy season has a higher groundwater recharge potential than the dry season, which raises groundwater levels. Similarly, DD interprets groundwater residence times [25], with a high DD implying a huge water loss capability and vice versa for low DD. The presence of flows such as rivers and streams, expresses the concentration of DD. The number of rivers is often more concentrated in delta regions than in hill regions [68]. The higher value of DD provides more probability of groundwater potential [75]. Al-Hurban et al. [4] defines DD as the ratio of total length of rivers outflows to the total surface area of the locality. The length factor of river and flow systems has a direct impact on a region's DD. Other factors influencing DD, in addition to flow system length, include runoff, vegetation cover, runoff and infiltration. As a result, DD is an essential component in many groundwater studies.

6.3.3 Land Cover Indices Parameters

The various landuse indices parameters include Land Use/Land Cover (LULC), geology, geomorphology, and soil health. LULC and other indices covers the human and anthropogenic influence over different period. These indices NDVI, NDBI etc. depicts the influence of urbanisation and shrinkage of surface water resources scenarios. These indices affect the infiltration capacity of the region and affects the permeability of the strata. Permeability is lower in built-up areas than in vegetation areas. Based on soil erosion, evapotranspiration, and runoff, each land use section has a different impact on groundwater retention. Similarly, the soil texture is a vital parameter for the recharge determination of any region [24, 80]. Mollinedo et al. [66] also stated that, the water holding capacity of a region is determined by the type, texture, and depth of soil. According to Díaz-Alcaide and Martínez-Santos [30], sandy and gravel soils have high percolation rates, whereas clayey and silty soils have the lowest infiltration rates. Meanwhile, fine sand and loamy soils are associated with moderate infiltration.

6.3.4 Parameters Related to Aquifers

Basic aquifer parameters include aquifer thickness, groundwater yield, resistivity of aquifer, artesian pressure of aquifer, groundwater quality, groundwater depth and geological factors. The depth and volume of aquifer determines the GPZ of any basin or geographical area. Aquifer areas with thick weathering have more groundwater volume and yield than those with thin weathering. In some studies, aquifer resistivity is used in addition to aquifer thickness to calculate GWP. Confined aquifers emerge in wide and thick permeable zones with limited artesian pressure, yielding little groundwater, whereas unconfined aquifer systems occur in thin permeable formations with high pressure, yielding significant groundwater. Zones having high water table depicts probability of high groundwater potential and high average groundwater yield [2, 80]. Hence, the accuracy of GPZ studies can be increased by properly selecting the data from valid sources. Table 6.3 depicts some of the important sources of thematic data layers.

Table 6.3 Various thematic layers employed in the recent studies

Thematic layers	Source and resolution
Aspect	DEM data, 30 m
Altitude	DEM data, 30 m
Slope	DEM data, 30 m
Plan curvature	DEM data, 30 m
Profile curvature	DEM data, 30 m
TWI NDVI	Landsat 8 images, 30 m Landsat 8 images, 30 m
NDBI	Landsat 8 images, 30 m
Drainage density	DEM data, 30 m
LULC	Sentinel-2 datasets at 10 m,
Distance to Roads	SRTM, equation based
Distance to Streams Geology STI Limnology STPI	SRTM, equation based From Bhukosh and state agencies DEM Data, 30 m From Bhukosh and state agencies DEM Data, 30 m
Rainfall Lithology	IMD, TRMM, CHIRPS From Bhukosh and state agencies
SPI STI	SRTM, equation based [64] SRTM datasets, 30 m
Soil Geomorphology	FAO soil map, state agencies Bhuvan data centre
GW well points	Data from CGWB

6.4 Modelling Techniques for Groundwater Potential Studies

6.4.1 Bi Variate Models

There are various models for the delineation of GPZ and one of the important statistical models are covered under the bi variate models criteria. The various bi variate models include Wight of Evidence (WOE), Functional ratio (FR), Fuzzy Functional ratio (FFR), Set theory-based fuzzy logic method, Index of Entropy (IOP) and Evidential Belief Function (EBF).

6.4.2 Weight of Evidence (WOE) Models

It works on the Bayes' rule principle which integrates different groundwater conditioning datasets by applying prior and posterior probability concept. The weight of different conditioning factors are calculated based on their influence (absence or presence) in the study area [13]. This method assists in transforming a continuous independent set of variable based on number of similar events or non-events [100]. As a result, determining the exact number of input set of variables is a requirement for WOE type studies, which aids in obtaining the more precise weight of input variables in the model. Rane and Jayaraj [90] employed the WOE model in the basaltic aquifer systems and found it to be more effective than frequency ratio technique for GPZ. According to Lee et al. [52] for the regional groundwater productivity potential mapping, WOE model was well validated with the groundwater yield results and showed high effectiveness.

6.4.3 Functional Ratio (FR) and Fuzzified Functional Ratio (FFR) Models

Functional ratio model interprets the occurrence probability of a certain event [13]. It observes the direct correlation between different conditioning factors and groundwater wells. This method is very helpful in determining the groundwater conditioning occurrence rate of probability and the occurrence of the relationship characteristics based on their spatial correlation and model quantification [91]. FR method has been used by many research for the development of GPZ in last decade. One such study by Ozdemir [79] discovered the efficacy of FR model over WOE and Logistic regression (LR) model in mapping GWP in Turkey's Sultan Mountains. Oh et al. [77] created the GWP map in Pohang City, Korea, using the FR model. The model training cases were used to predict the availability of groundwater resource based on its thematic layers such as geology, soil, topography data and lineament. Their research findings revealed a strong influence of soil texture on GWP, while ground elevation had the least. However, in F-FR type of modelling the alteration is done for depicting the GPZ which ranges between 0 and 1. The high value depicts strong presence and low value depicts absence of GPZ of any region.

6.4.4 *Evidential Belief Functional (EBF) Models*

The EBF model's framework is based on the Dempster-Shafer theory of evidence [27]. The EBF model includes [1, 1] degrees of belief (Bel), degrees of disbelief (Dis), degrees of uncertainty (Unc), and degrees of plausibility (Pls) [17]. The lower probability of EBF method implies belief and its higher probability implies plausibility in the Dempster-Shafer theory model. Plausibility (Pls) is thus greater than or equal to belief (Bel). Uncertainty (Unc) represents ignorance or doubt about a condition that whether the evidence supports a proposition and in turn distinguishes Pls from Bel. It is also an effective model for the mapping of potential zones, as documented in many studies which includes mineral potential mapping [7], landslide susceptibility [52, 65, 86], and aquifer vulnerability mapping [54].

6.4.5 *Union and Intersection Models*

Fuzzy sets, production and membership rules are the three main components of a fuzzy system [11, 19, 45, 56]. This set contains the objects and their membership based on their grade. In this approach, fuzzy set operators such as union (AND), intersection (Or), product, and so on can be used to group two membership functions. The GPZ are investigated using the fuzzy overlay method and the integration of thematic layers. The fuzzy logic membership value ranges from 0 to 1. The return values of Boolean logic are between 0 and 1, whereas the return values of fuzzy logic are between 0 and 1. The fuzzy logic method is said to be very close to human observation because it assigns a value based on the magnitude of truth. Palacios et al. [81] has successfully applied this method for GPZ and found to be among one of the effective methods for GPZ determination.

6.5 Multi Criteria Decision Making (MCDM) Models

Researchers have recently discovered multi-criteria decision making (MCDM) as an efficient tool for groundwater management by adding auditability, structure, rigor and transparency to decisions. According to Hajkowicz et al. [39], while selecting an MCDM technique is important for groundwater resource management, more emphasis is needed on the initial structuring of a decision problem, which includes selection criteria and decision options. Saaty's Analytic Hierarchy Process (AHP) method is a popular MCDM technique quite often used in water resources engineering.

6.5.1 *The Technique for Order Preferences to Ideal Solution (TOPSIS) Model*

It is a multi-criteria decision-making process that presumes the best solution will have the shortest Euclidean distance for the ideal positive solution and the longest distance for the ideal negative solution. TOPSIS has several advantages: it considers a large number of criteria, it is simple to implement, the method for determining weights is optional, prioritisation is based on the ideal solution, and for the best and worst criteria, a scalar value is calculated simultaneously [48].

6.5.2 *Analytical Hierarchy Process (AHP)*

Saaty [94, 95] proposed the AHP MCDM method. It aids decision-makers in resolving spatially complex environmental issues [110]. To determine a potential zone for groundwater, multiple affecting parameters are considered. The weights of all the parameter are assigned based on their relative importance in the process of decision-making for locating groundwater regions. AHP's important features include hierarchical formulation, cost-effectiveness and less time-consumption and precise results, and so on [25]. In such cases, the AHP method is used successfully for water resource management all over the world. AHP's qualitative and quantitative and approach is a significant strength [33].

6.5.3 *Shannon Entropy (SE)*

Shannon Entropy quantifies a system's changeability, abnormality, unstable behaviour, degree of disorder, and uncertainty in relation to its likely initial state [3, 32, 61, 71]. The abnormality of outcome and causes is generally depicted by SE. The SE is a measure of the difference between the average shares of individual groups in a confined overall system [6]. The SE model benefits from an extremely close relationship of entropy and the quantity known as Boltzmann, which is used to depict a system's thermodynamic conditions.

6.5.4 *Criteria Importance Through Inter-Criteria Correlation (CRITIC)*

This method was proposed by [28] for the evaluation of the different criteria weights. The obtained weight accounts both for contradictory nature and intensity of the evaluation criteria. The key advantage of this method is the application of weights in the absence of decision maker. The overall judgement regarding the importance of any criteria and minimal computation makes this algorithm a worthy selection for MCDM modelling. The weights obtained from CRITIC method are often selected as inputs for the TOPSIS model. CRITIC model has several applications in the field of industry analyst, textile, electronics, and hydrology.

6.6 Machine Learning (ML) and Ensembled Methods

ML is generally defined as a collection of statistical approaches which aims at disclosing hidden patterns in big datasets. ML assists in determining the closeness of variable associated to groundwater based on field evidence for groundwater potential studies. The algorithm examines the spatial relationships between all explanatory variables (aspect, slope, altitude, LULC, NDVI etc.) to calculate the most significant variable based upon ground truth data (successful drilling spots, borehole yield, groundwater level) [71, 79]. The detailed methodology of any GPZ analysis is presented in Fig. 6.1.

ML techniques can be either supervised or unsupervised based on objective to be achieved. Supervised models create an algorithm that connects a set of explanatory independent variables to a known output (dependent variable). However, unsupervised ML are generally employed for classification type of problems and doesn't have a defined output. Hence, the correlation among the independent variable in case of unsupervised variable is generally done for clustering or grouping. Both ML algorithm have different outcome and based on the problem involved a regression or a classification option is selected. Support vector machines (SVR), statistical learners (SL), and classification trees (CT) are examples of supervised approaches, whereas clustering and dimensionality reduction are examples of unsupervised learning [97].

6.6.1 *Support Vector Regression (SVR)*

Smola and Schölkopf [98] developed the SVR algorithm, which is specifically based on the Support vector machine (SVM) algorithm-based method, to solve regression problems. Based on structural risk minimization, this method is classified under supervised model and establishes a direct relationship between the input

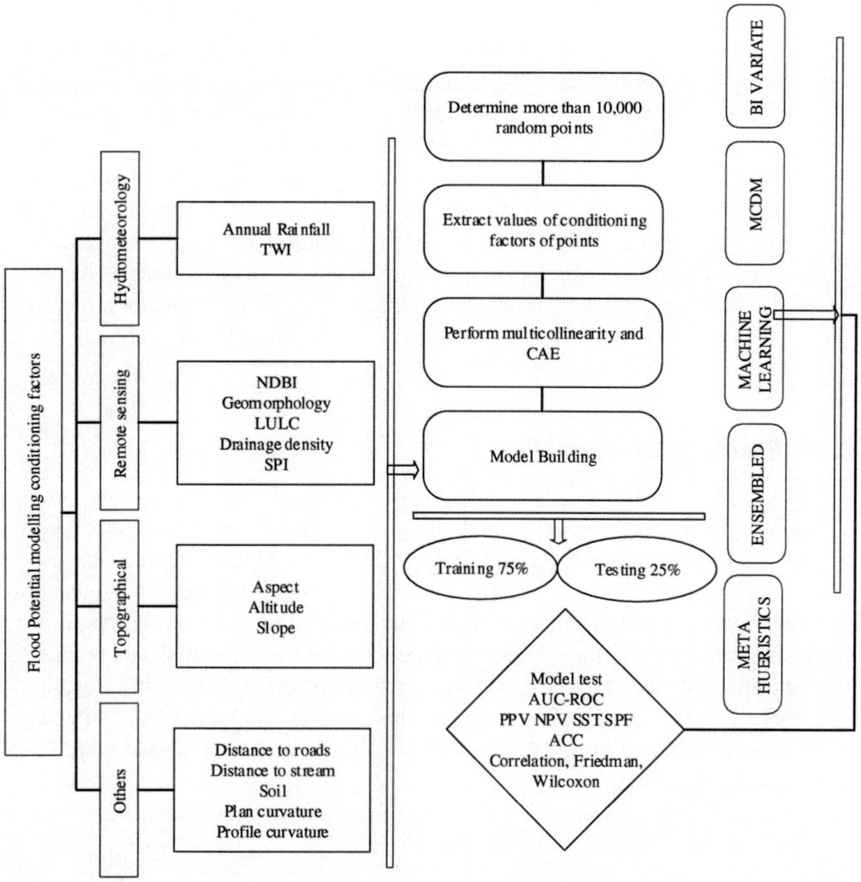

Fig. 6.1 The methodology for all the models for the delineation of GPZ

data (groundwater conditioning factors) and the target variable (groundwater occurrence). SVR separates the input vectors in a multidimensional space by a hyperplane with a greatest possible distance [46]. For regression problems, the SVR method is commonly used. Because of its powerful structure, it is also widely used in a variety of fields [59].

6.6.2 Logistic Regression (LR)

The LR model is mostly used in water resources prediction based on past scenarios. It is recently been applied to other groundwater-related issues such as ground subsidence [77] and potential groundwater springs [74]. The LR is a regression based model that calculates the coefficient by determining the relationship between dependent and independent parameters [76]. The LR yields only a 0 or a 1. (Binary model). In GWP studies, the LR can be used to predict groundwater presence or absence based on trained ground truth data. Once the regression model is trained, the generated relation is used to predict the groundwater presence in another region using same hyperparameters. The main advantage of this regression model is its simplicity in training the linear datasets which is very reliable in nature. The disadvantage of the LR is common overfitting which usually occurs when the total number of features exceeds the number of observations. Furthermore, the LR necessitates a discrete number set as the dependent variable and cannot process non-linear problems.

6.6.3 Random Forest (RF)

It is an important data mining models mainly used for the classification type of problems as it is an advanced form of decision trees (DT). The RF model generates many decision trees by interchanging the variables influencing the target. The algorithm then integrates all DT nodes in a predictive manner and fits the correlation among explanatory variables. During the training process, each tree's original data is chosen at random. RF model involves the development of different DT nodes and association of parameters upon the following nodes for progressing the hierarchy. The power of the RF model forecast improves by increasing the durability of autonomous trees and decreasing their correlation.

6.6.4 Extreme Gradient Boosting (XGB)

Chen and Guestrin [22] innovated XGB as one of the most effective ML algorithms which is best suitable for regression or classification predictive modelling. These are also developed from DT but has an accurate prediction than conventional ML models and are ensembled form of DT. It is a form of gradient boosting algorithm based model mainly employed for ensemble prediction using parallel processing technique [10]. The whole processing is done by allocating weights in a sequential form. These weights are then fed to all the independent variables which are processed further to the tree nodes for the prediction of results. If the weights predicted in the subsequent steps is not justified, it is done assigned to another node for recertification and correction. These shuffling and precision prediction of weight make its more accurate over other ML models. Hence, this strong learning in a repetitive manner makes it more useful for the delineation of GPZ [34]. It is also effective in determining the missing data and generally eliminates the processing time through parallel processing.

6.6.5 *Extremely Randomised Trees and Ensembled (ET) Models*

It is also an ensembled DT based algorithm which is based on bootstrap or bagging aggregation and RF principle. The algorithm progresses by developing large unpruned DT for the groundwater training datasets. The prediction for classification is done using majority voting classifier and for regression is done by averaging the DT prediction. The ET algorithm differs from RF and DT by fitting DT whole datasets rather than creation of each dataset. By hyperparameter tuning ET models, eliminates the randomness of DT and the variance generated in subsequent steps is also minimised [96]. An ensembled model of different ML models are done by voting classifier and are basically of two types, hard and soft. Hard voting involves the final prediction of result by properly selecting the class prediction which appears repeatedly among all the base models. In the soft voting method, however, a predict probability method is used [29].

6.7 Meta Heuristics Algorithm Methods

These algorithms have been inspired by most real-world optimisation processes taking place in nature. It includes law of chemistry and physics, biological process, human genetics, birds, wildlife, and evolutionary processes. These algorithms are basically clustered under five categories chiefly, human process based, physics based, game-based swarm based and evolutionary based. However, some of the famous meta heuristics algorithm in the field of GPZ are Harris hawk optimisation (HHO) and Bat algorithm.

6.7.1 *Harris Hawk Algorithm*

Heidari et al. [41] developed a method which is population-based to investigate the HHS behaviour. The rationality behind the selection of this bird was to provide an optimized algorithm by interpreting their extremely intelligent hunting strategy [41]. The algorithm consists of three main steps to achieve the optimisation objective. The first step of the algorithm consists of the exploration phase, during which falcons monitor the area for prey. Thereafter, the energy and spatial dynamics of the prey is used for mathematical modelling and at last final step involves surprise attack on prey. Paryani et al. [84] successfully implicated HHO algorithm and its ensemble with other model improved the efficiency to a greater extent.

6.7.2 *Bat Algorithm*

The BAT algorithm was proposed by [107] which stimulates the collective intelligence of wild bats. This algorithm attracted many researchers due to its solution to complex problems by incorporating sonar technique. The hunting strategy of bats is quite unique and follows echolocation procedure. Due to this technique bat can position themselves for maximus hunting probability. The overall method is helpful in predicting the velocity and position of BATs for optimized result.

6.8 Validation and Accuracy Assessment of GPZ

Validation is an important and final step in the evaluation of GWP models. In general, validation testing verifies the accuracy of the applied models and their ability to forecast for intermittent grids. The evolution of GIS, RS, and ML technologies has aided different model development which requires verification after validation with ground truth data. To all the thematic layers of the groundwater conditioning factors multicollinearity analysis has been performed which examines the similarity between the thematic layers. The upper threshold of Variable Inflation Factor (VIF) to satisfy no multicollinearity is 5 and its lowest threshold is 0.2 [24]. If VIF of any added thematic layer falls beyond the mentioned threshold values, then it is excluded from the modelling due to its negligible effect on the overall output. After the multi collinearity analysis the modelling is advanced and is finally validated by applying Receiver operating Area under the Curve (ROC-AUC) method, statistical measures, Root mean square error index (RMSE) and kappa index. The ROC is the most common validation technique for classification-based models in GWP studies. The value of AUC ranges between 0 and 100% and more accurate models approaches towards 100% value. Similarly for regression-based evaluations the RMSE technique is used, and it ranges between 0 and 1 with more accurate models approaching value 1. By determining a relation between the groundwater affecting factors with the help of GPZ models, areas having high to low potential can be calculated.

6.9 Present Scenario and Future Challenges of GPZ

A GPZ map delineates areas within a given geographical setting that may be more conducive to groundwater development.

The development of reliable GWP map necessitates advanced hydrogeology, and satellite science concepts, the results are typically presented in a visually appealing manner that anyone can interpret [9]. The GPZ incorporates data from various sources but hydrological, geological and satellite data remains as priority in the list. The modernisation in the field of data science using satellite data has pronounced considerable result in the field of prioritisation. This has resulted in creation of groundwater maps in a fraction of time using various processing methods [30]. Hence, there is need to use and understand in depth about these datasets for data scare region and validate the same with the validation technique as discussed [78]. Likewise, with the advent of technology boom many high qualities satellite images are available which are costly and needs more system resources to analyse the product and hence increases the system maintenance for high level processing. However, if the system configuration is met to process these datasets, the reliance on ground datasets can minimised to great extent. Of course, the accuracy needs to be assessed for few ground data points. From a huge chunk of data available to process GPZ analysis, the perfect data sets must be selected properly as the entire datasets may not have influence on the result. From the past studies it has been noted that the most common factor for GPZ study includes rainfall, soil texture, slope, LULC, TWI, geomorphology, altitude. Majority of the previous studies has incorporated these parameters for the delineation of GPZ.

The GPZ mapping is mainly classified into three types: statistical, ML, and hybrid or ensemble techniques. FR, EBF, MCDA-AHP, and WOE models are examples of common statistical techniques. These methods involve the calculation of weights of the groundwater affecting factors by identifying the presence or absence of groundwater points in each thematic layer. These weights are then used to delineate the GPZ of any region [10, 35, 37, 70]. However, the models were a bit time consuming and less efficient which led to the development and selection of advance ML algorithm for GPZ study. RF, LR, BRT, and SVM are some common examples ML models. These models are more efficient than statistical models can be used to train large amount datasets. Thereafter, to enhance the accuracy even more and decrease the computation time and complexity ensembled and meta heuristics models were developed for the GPZ analysis. However, these ML and ensembled learning models can also overfit and produce error in result when training huge datasets for large period. To avoid this accuracy loss some validation techniques are adopted such as ROC-AUC curve, RMSE and kappa coefficient index. These are the absolute indices to check the effectiveness of any kind of regression and classification models. The accuracy of statistical models generally ranges between 65 and 85%. While ML models produces accuracy range of 70.0% and 94%. Similarly, for ensembled prediction the range is (71.0% 95.0%) and for meta heuristics the efficiency shoots up to 97%. In one such study by [84] showed the model accuracy of SWARA-SVR-HHO, SWARA-SVR-BA, BWMSVR-HHO and BWM-SVR-BA models to be 93.7%, 92.6%, 96.4% and 95.9% respectively. Hence, it can be predicted that the metaheuristic and ensembled models will be more precise than statistical and MCDM models for the delineation of GPZ.

6.10 Conclusion

The present article provides a broad view of work done in the field of GPZ. It covers all the important methodologies and variables needed for the delineation of GPZ study. The analysis of GPZ has a great significance as it maps the location for the maximum probability of groundwater yield. The articles discuss all the possible groundwater affecting factors for GPZ study. It also points out the most common factor based on past literature. Thereafter, it discusses the different approach to delineate the GPZ. The major advancement in this field is also discussed and the future challenges are also taken care of. Hence, it can be suggested that to achieve high efficiency, statistical algorithms and ML techniques be integrated, and conditioning factors must be increased in the mapping of GWP maps. Overall, this article highlights the major work done in this field and discusses about the prospects for groundwater enthusiasts.

Acknowledgements I would like to express my sincere gratitude to all those who have contributed to the development of this book chapter. First and foremost, I am thankful to the editor of this book, for giving me the opportunity to contribute to this important publication. I am also grateful to my supervisor, who provided invaluable guidance and support throughout the writing process. Their feedback and encouragement helped me refine my ideas and improve the quality of my work. I would also like to acknowledge the support of my colleagues, whose insightful discussions and constructive criticism helped shape my thinking on this topic.

References

1. Abijith D, Saravanan S, Singh L, Jennifer JJ, Saranya T, Parthasarathy KSS (2020) GIS-based multi-criteria analysis for identification of potential groundwater recharge zones—A case study from Ponnaniyaru watershed, Tamil Nadu, India. HydroResearch 3:1–14
2. Agarwal E, Agarwal R, Garg RD, Garg PK (2013) Delineation of groundwater potential zone: an AHP/ANP approach. J Earth Syst Sci 122(3):887–898
3. Al-Abadi AM, Al-Temmeme AA, Al-Ghanimy MA (2016) A GIS-based combining of frequency ratio and index of entropy approaches for mapping groundwater availability zones at Badra–Al Al-Gharbi–Teeb areas, Iraq. Sustain Water Resour Manag 2(3):265–283
4. Al-Hurban A, Al-Ruwaih F, Al-Dughairi A (2021) Quantitative geomorphological and hydro-morphometric analysis of drainage Basins of as Sabriyah (Kuwait) using GIS techniques. J Geogr Inf Syst 13(02):166–193
5. Allafta H, Opp C, Patra S (2021) Identification of groundwater potential zones using remote sensing and GIS techniques: a case study of the Shatt Al-Arab Basin. Remote Sens 13(1):112 (Multidisciplinary Digital Publishing Institute)
6. Amiri V, Rezaei M, Sohrabi N (2014) Groundwater quality assessment using entropy weighted water quality index (EWQI) in Lenjanat, Iran. Environ Earth Sci 72(9):3479–3490
7. An P, Moon WM, Bonham-Carter GF (1994) Uncertainty management in integration of exploration data using the belief function. Nonrenew Resour 3(1):60–71
8. Arabameri A, Rezaei K, Cerda A, Lombardo L, Rodrigo-Comino J (2019) GIS-based groundwater potential mapping in Shahroud plain, Iran. A comparison among statistical (bivariate and multivariate), data mining and MCDM approaches. Sci Total Environ 658:160–177

9. Bagyaraj M, Ramkumar T, Venkatramanan S, Gurugnanam B (2013) Application of remote sensing and GIS analysis for identifying groundwater potential zone in parts of Kodaikanal Taluk, South India. Front Earth Sci 7(1):65–75
10. Bandara A, Hettiarachchi Y Hettiarachchi K, Munasinghe S, Wijesinghe I Thayasivam U (2020) A generalized ensemble machine learning approach for landslide susceptibility modelling. In: Sharma N, Chakrabarti A, Balas VE (eds) Data management, analytics and innovation. Advances in intelligent systems and computing, vol 1016. Springer Singapore, Singapore, pp 71–93. http://link.springer.com/https://doi.org/10.1007/978-981-13-9364-8_6
11. Barpi F (2004) Fuzzy modelling of powder snow avalanches. Cold Reg Sci Technol 40(3):213–227
12. Berhanu B, Seleshi Y, Melesse AM (2014) Surface water and groundwater resources of ethiopia: potentials and challenges of water resources development. In: Melesse AM, Abtew W, Setegn SG (eds) Nile River Basin: ecohydrological challenges, climate change and hydropolitics. Springer International Publishing, Cham, pp 97–117. https://doi.org/10.1007/978-3-319-02720-3_6
13. Bonham-Carter GF (1994) Geographic information systems for geoscientists-modeling with GIS. Comput Methods Geosci 13:398
14. Boori M, Choudhary K, Kupriyanov A (2019) Identification and mapping of groundwater potential zone through remote sensing and GIS technology in Kalmykia, Russia. Int J Geoinform 15(1)
15. Braham M, Boufekane A, Bourenane H, Nait Amara B, Bensalem R, Oubaiche EH, Bouhadad Y (2022) Identification of groundwater potential zones using remote sensing, GIS, machine learning and electrical resistivity tomography techniques in Guelma basin, northeastern Algeria. Geocarto Int 0(ja):1–24 (Taylor & Francis)
16. CGWB (2012) Manual on aquifer mapping, Government of India, Ministry of water resources. Government of India, Ministry of Water Resources, Central ground Water Board, p 72
17. Carranza EJM, Hale M (2003) Evidential belief functions for data-driven geologically constrained mapping of gold potential, Baguio district, Philippines. Ore Geol Rev 22(1):117–132
18. Chaminé HI, Carvalho JM, Teixeira J, Freitas L (2015) Role of hydrogeological mapping in groundwater practice: back to basics. Eur Geol J 40:34–42
19. Chaudhry AK, Kumar K, Alam MA (2021) Mapping of groundwater potential zones using the fuzzy analytic hierarchy process and geospatial technique. Geocarto Int 36(20):2323–2344 (Taylor & Francis)
20. Chen W, Pradhan B, Li S, Shahabi H, Rizeei HM, Hou E, Wang S (2019) Novel hybrid integration approach of bagging-based fisher's linear discriminant function for groundwater potential analysis. Nat Resour Res 28(4):1239–1258
21. Chen H, Zhang W, Nie N, Guo Y (2019) Long-term groundwater storage variations estimated in the Songhua River Basin by using GRACE products, land surface models, and in-situ observations. In: Science of the total environment, vol 649. Elsevier B.V., pp 372–387
22. Chen T, Guestrin C (2016) XGBoost: a scalable tree boosting system. In: Proceedings of the 22nd ACM SIGKDD International Conference On Knowledge Discovery And Data Mining. ACM, San Francisco California USA, pp 785–794. https://dl.acm.org/doi/https://doi.org/10.1145/2939672.2939785
23. Chen Y, Shi F, Kirby JT, Wu G, Liang B (2020) A computationally efficient subgrid model for coupled surface and groundwater flows. Coast Eng 157(December 2019):103665. (Elsevier B.V.)
24. Choudhary S, Pingale SM, Khare D (2022) Delineation of groundwater potential zones of upper Godavari sub-basin of India using bi-variate, MCDM and advanced machine learning algorithms. Geocarto Int 1–32
25. Das S (2019) Geospatial mapping of flood susceptibility and hydro-geomorphic response to the floods in Ulhas basin, India. Remote Sens Appl: Soc Environ 14:60–74
26. Das B, Pal SC (2020) Assessment of groundwater recharge and its potential zone identification in groundwater-stressed Goghat-I block of Hugli District, West Bengal, India. Environ Dev Sustain 22(6):5905–5923

27. Dempster AP (2008) Upper and lower probabilities induced by a multivalued mapping. In: Yager RR, Liu L (eds) Classic works of the dempster-shafer theory of belief functions, studies in fuzziness and soft computing. Springer, Berlin, Heidelberg, pp 57–72. https://doi.org/10.1007/978-3-540-44792-4_3
28. Diakoulaki D, Mavrotas G, Papayannakis L (1995) Determining objective weights in multiple criteria problems: the critic method. Comput Oper Res 22(7):763–770
29. Dietterich TG (2000) Ensemble methods in machine learning. In: Multiple classifier systems, Lecture notes in computer science. Springer, Berlin, Heidelberg, pp 1–15
30. Díaz-Alcaide S, Martínez-Santos P (2019) Review: advances in groundwater potential mapping. Hydrogeol J 27(7):2307–2324
31. Elvis BWW, Arsène M, Théophile NM, Bruno KME, Olivier OA (2022) Integration of shannon entropy (SE), frequency ratio (FR) and analytical hierarchy process (AHP) in GIS for suitable groundwater potential zones targeting in the Yoyo river basin, Méiganga area, Adamawa Cameroon. J Hydrol: Region Stud 39:100997
32. Farzin M, Avand M, Ahmadzadeh H, Zelenakova M, Tiefenbacher JP (2021) Assessment of ensemble models for groundwater potential modeling and prediction in a Karst watershed. Water 13(18):2540 (Multidisciplinary Digital Publishing Institute)
33. Forman EH (1993) Facts and fictions about the analytic hierarchy process. Math Comput Model 17(4):19–26
34. Friedman JH (2001) Greedy function approximation: a gradient boosting machine. Ann Stat 29(5). https://projecteuclid.org/journals/annals-of-statistics/volume-29/issue-5/Greedy-function-approximation-A-gradient-boosting-machine/https://doi.org/10.1214/aos/1013203451.full
35. Gaur S, Chahar BR, Graillot D (2011) Combined use of groundwater modeling and potential zone analysis for management of groundwater. Int J Appl Earth Obs Geoinf 13(1):127–139
36. Ghinoi A, Chung C-J (2005) STARTER: a statistical GIS-based model for the prediction of snow avalanche susceptibility using terrain features—Application to Alta Val Badia, Italian Dolomites. Geomorphology 66(1–4):305–325
37. Ghosh PK, Bandyopadhyay S, Jana NC (2015) Mapping of groundwater potential zones in hard rock terrain using geoinformatics: a case of Kumari watershed in western part of West Bengal. Model Earth Syst Environ 2(1):1
38. Gueretz JS, Da Silva FA, Simionatto EL, Férard JF, Radetski CM, Somensi CA (2019) A multi-parametric study of the interaction between the Parati river and Babitonga Bay in terms of water quality. J Environ Sci Health, Part B:1–8
39. Hajkowicz SA, McDonald GT, Smith PN (2000) An evaluation of multiple objective decision support weighting techniques in natural resource management. J Environ Plan Manag 43(4):505–518 (Routledge)
40. Hasanuzzaman M, Mandal MH, Hasnine M, Shit PK (2022) Groundwater potential mapping using multi-criteria decision, bivariate statistic and machine learning algorithms: evidence from Chota Nagpur Plateau, India. Appl Water Sci 12(4):58
41. Heidari AA, Mirjalili S, Faris H, Aljarah I, Mafarja M, Chen H (2019) Harris hawks optimization: algorithm and applications. Futur Gener Comput Syst 97:849–872
42. Hou E, Wang J, Chen W (2018) A comparative study on groundwater spring potential analysis based on statistical index, index of entropy and certainty factors models. Geocarto Int 33(7):754–769
43. Jaafarzadeh MS, Tahmasebipour N, Haghizadeh A, Pourghasemi HR, Rouhani H (2021) Groundwater recharge potential zonation using an ensemble of machine learning and bivariate statistical models. Sci Rep 11(1):5587 (Nature Publishing Group)
44. Jaccard C (1990) Fuzzy factorial analysis of snow avalanches. Nat Hazards 3(4):329–340
45. Kanimozhi U, Ganapathy S, Manjula D, Kannan A (2019) An intelligent risk prediction system for breast cancer using fuzzy temporal rules. Natl Acad Sci Lett 42(3):227–232
46. Kavzoglu T, Sahin EK, Colkesen I (2014) Landslide susceptibility mapping using GIS-based multi-criteria decision analysis, support vector machines, and logistic regression. Landslides 11(3):425–439

47. Khosravi K, Panahi M, Tien Bui D (2018) Spatial prediction of groundwater spring potential mapping based on an adaptive neuro-fuzzy inference system and metaheuristic optimization. Hydrol Earth Syst Sci 22(9):4771–4792 (Copernicus GmbH)
48. Kim G, Park CS, Yoon KP (1997) Identifying investment opportunities for advanced manufacturing systems with comparative-integrated performance measurement. Int J Prod Econ 50(1):23–33
49. Kisi O, Alizamir M, Zounemat-Kermani M (2017) Modeling groundwater fluctuations by three different evolutionary neural network techniques using hydroclimatic data. Nat Hazards 87(1):367–381
50. Kumar M, Singh SK, Kundu A, Tyagi K, Menon J, Frederick A, Raj A, Lal D (2022) GIS-based multi-criteria approach to delineate groundwater prospect zone and its sensitivity analysis. Appl Water Sci 12(4):71
51. Kumar A, Krishna AP (2018) Assessment of groundwater potential zones in coal mining impacted hard-rock terrain of India by integrating geospatial and analytic hierarchy process (AHP) approach. Geocarto Int 33(2):105–129 (Taylor & Francis)
52. Lee S, Kim Y-S, Oh H-J (2012) Application of a weights-of-evidence method and GIS to regional groundwater productivity potential mapping. J Environ Manage 96(1):91–105
53. Lee S, Hyun Y, Lee S, Lee MJ (2020) Groundwater potential mapping using remote sensing and gis-based machine learning techniques. Remote Sens 12(7):1200 (Multidisciplinary Digital Publishing Institute),
54. Li L, Lan H, Guo C, Zhang Y, Li Q, Wu Y (2017) A modified frequency ratio method for landslide susceptibility assessment. Landslides 14(2):727–741
55. Li S, Li Y, Liu Z (2013) Hebei rural groundwater contamination and integrated control. J. Cangzhou Normal Univ 29(4):8–10
56. Liu H, Lang B (2019) Machine learning and deep learning methods for intrusion detection systems: A survey. Appl Sci 9(20):4396. https://doi.org/10.3390/app9204396
57. Mahato S, Pal S (2019) Groundwater potential mapping in a rural River Basin by union (OR) and intersection (AND) of four multi-criteria decision-making models. Nat Resour Res 28(2):523–545
58. Maity B, Mallick SK, Das P, Rudra S (2022) Comparative analysis of groundwater potentiality zone using fuzzy AHP, frequency ratio and Bayesian weights of evidence methods. Appl Water Sci 12(4):63
59. Malik A, Tikhamarine Y, Al-Ansari N, Shahid S, Sekhon HS, Pal RK, Rai P, Pandey K, Singh P, Elbeltagi A, Sammen SS (2021) Daily pan-evaporation estimation in different agro-climatic zones using novel hybrid support vector regression optimized by Salp swarm algorithm in conjunction with gamma test. Eng Appl Comput Fluid Mech 15(1):1075–1094 (Taylor & Francis)
60. Masroor M, Rehman S, Sajjad H, Rahaman MH, Sahana M, Ahmed R, Singh R (2021) Assessing the impact of drought conditions on groundwater potential in Godavari Middle Sub-Basin, India using analytical hierarchy process and random forest machine learning algorithm. Groundwater Sustain Dev 13:100554
61. Massey DS, Denton NA (1988) The dimensions of residential segregation. Soc Forces 67(2):281–315 (Oxford University Press)
62. Mirchooli F, Motevalli A, Pourghasemi HR, Mohammadi M, Bhattacharya P, Maghsood FF, Tiefenbacher JP (2019) How do data-mining models consider arsenic contamination in sediments and variables importance? Environ Monit Assess 191(12):777
63. Mogaji KA, Omosuyi GO, Adelusi AO, Lim HS (2016) Application of GIS-based evidential belief function model to regional groundwater recharge potential zones mapping in Hardrock geologic terrain. Environ Process 3(1):93–123
64. Moghaddam DD, Rezaei M, Pourghasemi HR, Pourtaghie ZS, Pradhan B (2015) Groundwater spring potential mapping using bivariate statistical model and GIS in the Taleghan Watershed, Iran. Arab J Geosci 8(2):913–929
65. Mohammady M, Pourghasemi HR, Pradhan B (2012) Landslide susceptibility mapping at Golestan Province, Iran: a comparison between frequency ratio, Dempster–Shafer, and

weights-of-evidence models. J Asian Earth Sci 61:221–236 (Geological Anatomy of East and South Asia)
66. Mollinedo J, Schumacher TE, Chintala R (2015) Influence of feedstocks and pyrolysis on biochar's capacity to modify soil water retention characteristics. J Anal Appl Pyrol 114:100–108
67. Muavhi N, Thamaga KH, Mutoti MI (2021) Mapping groundwater potential zones using relative frequency ratio, analytic hierarchy process and their hybrid models: case of Nzhelele-Makhado area in South Africa. Geocarto Int 0(0):1–20 (Taylor & Francis)
68. Mukherjee P, Singh CK, Mukherjee S (2012) Delineation of groundwater potential zones in arid Region of India—A remote sensing and GIS approach. Water Resour Manag 26(9):2643–2672
69. Mumtaz R, Baig S, Kazmi SSA, Ahmad F, Fatima I, Ghauri B (2019) Delineation of groundwater prospective resources by exploiting geo-spatial decision-making techniques for the Kingdom of Saudi Arabia. Neural Comput Appl 31(9):5379–5399
70. Murmu P, Kumar M, Lal D, Sonker I, Singh SK (2019) Delineation of groundwater potential zones using geospatial techniques and analytical hierarchy process in Dumka district, Jharkhand, India. Groundw Sustain Dev 9:100239
71. Naghibi SA, Hashemi H, Berndtsson R, Lee S (2020) Application of extreme gradient boosting and parallel random forest algorithms for assessing groundwater spring potential using DEM-derived factors. J Hydrol 589:125197
72. Naghibi SA, Pourghasemi HR (2015) A comparative assessment between three machine learning models and their performance comparison by bivariate and multivariate statistical methods in groundwater potential mapping. Water Resour Manag 29(14):5217–5236
73. Namous M, Hssaisoune M, Pradhan B, Lee C-W, Alamri A, Elaloui A, Edahbi M, Krimissa S, Eloudi H, Ouayah M, Elhimer H, Tagma T (2021) Spatial prediction of groundwater potentiality in large semi-arid and karstic mountainous region using machine learning models. Water 13(16):2273 (Multidisciplinary Digital Publishing Institute)
74. Nampak H, Pradhan B, Manap MA (2014) Application of GIS based data driven evidential belief function model to predict groundwater potential zonation. J Hydrol 513:283–300
75. Nasir MJ, Khan S, Zahid H, Khan A (2018) Delineation of groundwater potential zones using GIS and multi influence factor (MIF) techniques: a study of district Swat, Khyber Pakhtunkhwa, Pakistan. Environ Earth Sci 77(10):367
76. Nguyen HD (2022) Hybrid models based on deep learning neural network and optimization algorithms for the spatial prediction of tropical forest fire susceptibility in Nghe An province, Vietnam. Geocarto Int 0(0):1–25 (Taylor & Francis)
77. Oh H-J, Kim Y-S, Choi J-K, Park E, Lee S (2011) GIS mapping of regional probabilistic groundwater potential in the area of Pohang City, Korea. J Hydrol 399(3):158–172
78. Oikonomidis D, Dimogianni S, Kazakis N, Voudouris K (2015) A GIS/Remote Sensing-based methodology for groundwater potentiality assessment in Tirnavos area, Greece. J Hydrol 525:197–208
79. Ozdemir A (2011) GIS-based groundwater spring potential mapping in the Sultan Mountains (Konya, Turkey) using frequency ratio, weights of evidence and logistic regression methods and their comparison. J Hydrol 411(3–4):290–308
80. Pal S, Kundu S, Mahato S (2020) Groundwater potential zones for sustainable management plans in a river basin of India and Bangladesh. J Clean Prod 257:120311 (Elsevier Ltd.)
81. Palacios AM, Palacios JL, Sánchez L, Alcalá-Fdez J (2015) Genetic learning of the membership functions for mining fuzzy association rules from low quality data. Inf Sci 295:358–378
82. Park I, Kim Y, Lee S (2014) Groundwater productivity potential mapping using evidential belief function. Groundwater 52(S1):201–207
83. Park S, Hamm S-Y, Jeon H-T, Kim J (2017) Evaluation of logistic regression and multivariate adaptive regression spline models for groundwater potential mapping using R and GIS. Sustainability 9(7):1157 (Multidisciplinary Digital Publishing Institute)

84. Paryani S, Neshat A, Pourghasemi HR, Ntona MM, Kazakis N (2022) A novel hybrid of support vector regression and metaheuristic algorithms for groundwater spring potential mapping. Sci Total Environ 807:151055
85. Patra S, Sahoo S, Mishra P, Mahapatra SC (2018) Impacts of urbanization on land use/cover changes and its probable implications on local climate and groundwater level. J Urban Manag 7(2):70–84 (Elsevier B.V.)
86. Pourghasemi HR, Pradhan B, Gokceoglu C, Mohammadi M, Moradi HR (2013) Application of weights-of-evidence and certainty factor models and their comparison in landslide susceptibility mapping at Haraz watershed, Iran. Arab J Geosci 6(7):2351–2365
87. Pourghasemi HR, Beheshtirad M (2015) Assessment of a data-driven evidential belief function model and GIS for groundwater potential mapping in the Koohrang Watershed, Iran. Geocarto Int 30(6):662–685 (Taylor & Francis)
88. Pradhan S, Kumar S, Kumar Y, Sharma HC (2019) Assessment of groundwater utilization status and prediction of water table depth using different heuristic models in an Indian interbasin. Soft Comput 23(20):10261–10285
89. Rahmati O, Moghaddam DD, Moosavi V, Kalantari Z, Samadi M, Lee S, Bui DT (2019) An automated Python language-based tool for creating absence samples in groundwater potential mapping. Remote Sens 11(11):1–22
90. Rane NL, Jayaraj GK (2022) Comparison of multi-influence factor, weight of evidence and frequency ratio techniques to evaluate groundwater potential zones of basaltic aquifer systems. Environ Dev Sustain 24(2):2315–2344
91. Razandi Y, Pourghasemi HR, Neisani NS, Rahmati O (2015) Application of analytical hierarchy process, frequency ratio, and certainty factor models for groundwater potential mapping using GIS. Earth Sci Inf 8(4):867–883
92. Razavi-Termeh SV, Khosravi K, Sadeghi-Niaraki A, Choi S-M, Singh VP (2020) Improving groundwater potential mapping using metaheuristic approaches. Hydrol Sci J 65(16):2729–2749 (Taylor & Francis)
93. Rizeei HM, Pradhan B, Saharkhiz MA, Lee S (2019) Groundwater aquifer potential modeling using an ensemble multi-adoptive boosting logistic regression technique. J Hydrol 579:124172
94. Saaty RW (1987) The analytic hierarchy process—What it is and how it is used. Math Modell 9(3):161–176
95. Saaty TL (1990) How to make a decision: the analytic hierarchy process. Euro J Oper Res 48(1):9–26 (Desicion making by the analytic hierarchy process: Theory and applications)
96. Sagi O, Rokach L (2020) Explainable decision forest: transforming a decision forest into an interpretable tree. Inf Fusion 61:124–138
97. Shalev-Shwartz S, Ben-David S (2014) Understanding machine learning: from theory to algorithms, 1st edn. Cambridge University Press. https://www.cambridge.org/core/product/identifier/9781107298019/type/book
98. Smola AJ, Schölkopf B (2004) A tutorial on support vector regression. Stat Comput 14(3):199–222
99. Tahmassebipoor N, Rahmati O, Noormohamadi F, Lee S (2015) Spatial analysis of groundwater potential using weights-of-evidence and evidential belief function models and remote sensing. Arab J Geosci 9(1):79
100. Talukdar S, Mallick J, Sarkar SK, Roy SK, Islam ARMdT, Praveen B, Naikoo MW, Rahman A, Sobnam M (2022) Novel hybrid models to enhance the efficiency of groundwater potentiality model. Appl Water Sci 12(4):62
101. Termeh SVR, Kornejady A, Pourghasemi HR, Keesstra S (2018) Flood susceptibility mapping using novel ensembles of adaptive neuro fuzzy inference system and metaheuristic algorithms. Sci Tot Environ 615:438–451
102. Thapa R, Gupta S, Gupta A, Reddy DV, Kaur H (2018) Use of geospatial technology for delineating groundwater potential zones with an emphasis on water-table analysis in Dwarka River basin, Birbhum, India. Hydrogeol J 26(3):899–922
103. Tien Bui D, Tuan TA, Klempe H, Pradhan B, Revhaug I (2016) Spatial prediction models for shallow landslide hazards: a comparative assessment of the efficacy of support vector

machines, artificial neural networks, kernel logistic regression, and logistic model tree. Landslides 13(2):361–378
104. Tolche AD (2021) Groundwater potential mapping using geospatial techniques: a case study of Dhungeta-Ramis sub-basin, Ethiopia. Geol Ecol Landsc 5(1):65–80 (Taylor & Francis)
105. Vasin S, Carle A, Lang U, Kirchholtes HJ (2016) A groundwater management plan for Stuttgart. Sci Total Environ 563–564:704–712
106. Xu Y, Liu L, Zhang X (2019) Multilattices on typical hesitant fuzzy sets. Inf Sci 491:63–73
107. Yariyan P, Avand M, Omidvar E, Pham QB, Linh NTT, Tiefenbacher JP (2021) Optimization of statistical and machine learning hybrid models for groundwater potential mapping. Geocarto Int 1–35
108. Zehtabiyan-Rezaie N, Alvandifar N, Saffaraval F, Makkiabadi M, Rahmati N, Saffar-Avval M (2019) A solar-powered solution for water shortage problem in arid and semi-arid regions in coastal countries. Sustain Energy Technol Assess 35:1–11
109. Zischg A, Fuchs S, Keiler M, Meißl G (2005) Modelling the system behaviour of wet snow avalanches using an expert system approach for risk management on high alpine traffic roads. Nat Hazards Earth Syst Sci 5(6):821–832. (Copernicus GmbH)
110. Zolekar RB, Bhagat VS (2015) Multi-criteria land suitability analysis for agriculture in hilly zone: remote sensing and GIS approach. Comput Electron Agric 118:300–321

Chapter 7
Geographic Information System and Remote Sensing in Deciphering Groundwater Potential Zones

Nguyen Ngoc Thanh and Srilert Chotpantarat

Abstract Innovations in the techniques of Remote Sensing (RS) and Geographic Information System (GIS) have recently increased the demand for the effectiveness of predicting the potential of groundwater in the world, improving map accuracy. In this chapter, you will learn how to use GIS and RS techniques to decipher groundwater potential zones (GWPZs). We start with definitions and descriptions of conditioning factors (CFs) for groundwater potential. Then, we introduce and discuss various GIS-based techniques in detail in mapping groundwater potential zones. The GIS techniques, including Analytic Hierarchy Process (AHP), Fuzzy Logic (FL), Multi-Influencing Factor (MIF), Technique for Order of Preference by Similarity to Ideal Solution (TOPSIS), and Frequency Ratio (FR), were introduced in this chapter. Notice that the term "groundwater potential" is used in this book to indicate the ability of the occurrence of a specific groundwater yield when prerequisites exist. Nevertheless, this term also refers to the ability to occur in spring or non-spring or groundwater storage. The result shows that the high potential of groundwater yield >10 m^3/h was distributed mainly in the eastern Kanchanaburi. However, a difference was observed in terms of area extent in the output maps. The MIF, AHP, and FR models reached good results according to 78%, 71%, and 70% in validation, whereas the accuracy of the FL and TOPSIS model was 64% and 51%, respectively.

N. N. Thanh
Graduate School, Interdisciplinary Program in Environmental Science, Chulalongkorn University, Bangkok 10330, Thailand

University of Agriculture and Forestry, Hue University, 102 Phung Hung Str, Hue City, Thua Thien Hue 53000, Vietnam

N. N. Thanh
e-mail: nguyenngocthanh@hueuni.edu.vn

S. Chotpantarat (✉)
Department of Geology, Faculty of Science, Chulalongkorn University, Bangkok 10330, Thailand
e-mail: Srilert.C@chula.ac.th

Center of Excellence in Environmental Innovation and Management of Metals (EnvIMM), Environmental Research Institute, Chulalongkorn University (ERIC), Bangkok 10330, Thailand

E. Balaji et al. (eds.), *Emerging Technologies for Water Supply, Conservation and Management*, Springer Water, https://doi.org/10.1007/978-3-031-35279-9_7

Keywords Conditioning factors · GIS · Groundwater potential · Remote sensing · Thailand

7.1 Groundwater Potential

Groundwater is used as a substitute over many areas of the world when surface water resources are depleting due to climate change and population growth. However, groundwater sources are facing serious challenges in reserves, especially in the Ganges Brahmaputra River basin [6], the North China Plain [61], and the Central California Valley in the US [32], the western Thailand [53]. Therefore, we need to have appropriate solutions to protect groundwater resources. Identifying groundwater potential (GWP) can certainly help to manage groundwater resources in recent decades. A groundwater potential map can act as a management tool, meaning that authorities and local people can determine optimal zones for groundwater development. Groundwater potential mapping took various approaches, such as groundwater reserve (groundwater storage), groundwater quality, groundwater yield, and groundwater spring [56]. A typical example of groundwater potential is relevant to groundwater resources' spatial distribution [18]. This map indicates groundwater storage levels over underground space. Another instance mentioned the potential of groundwater quality being used for domestic activities. The discovery of groundwater quality potential can assist managers and local people with the selection of uses for the groundwater [49]. Groundwater potential was defined in certain cases as the ability to occur groundwater spring [8, 13, 36] or a specific groundwater yield [14, 30, 43]. Broadly speaking, these maps aim to show groundwater's ability to serve any useful purpose. In this book, we work on the possibility of the occurrence of a specific groundwater yield to map a groundwater potential with different GIS-based techniques.

7.2 Application of GIS and RS in Data Acquisition

The recent development of GIS and RS has permitted highly accurate groundwater potential mapping about CFs. CFs are input parameters for mapping GWP, and their mission is control over conditions of groundwater occurrence that many researchers want to explore. The main role of GIS in this work is to analyze and manage data. The GIS database describes non-spatial and spatial information [16]. Nowadays, GIS approaches permit researchers to optimize time and budgets for groundwater investigation, compared to traditional ones. Along with the development of GIS, RS has also recorded a rapid transformation in data acquisition, allowing researchers and scientists to acquire new data. RS data includes ground-based observation data, aerial data, and satellite data. The development of RS provides scientists with information on both space and time. Modern remote sensors, including Whisk-Broom, ETM+, Push-broom, OLI/TIRS, space-bone, enable researchers to detect things on the Earth's surface with high resolution over brief repetition durations. The input data for hydrological investigations were collected via wavelengths using these sensors [26]. Aerial photos were previously only used for surface observations [37]. Nevertheless, on certain days, unobservable parameters must be calculated based on aerial photos. One striking example is precipitation computed by interpolating clouds using Tropical Rainfall Measuring Mission and Precipitation Estimation from Remotely Sensed Information using Artificial Neural Networks Climatic Data Record (PERSIANN CDR V1R1) [17]. Another example, Landsat or Aster is usually utilized to derive land use/land cover (LULC), slope, geomorphology, drainage, and lineaments which are the CFs [20, 24, 28].

RS has been commonly combined with GIS to generate significant values and to benefit the applications of hydrological techniques. As previously studied [10], there are three different approaches in terms of this combination. After being first saved in raster format, RS data is then digitally converted into a GIS context utilizing techniques for image interpretation. Additionally, GIS datasets are considered the background to improve the processes of RS analysis. In mapping of groundwater potential, the output is created by superimposing the main theme layers. Finally, groundwater potential approaches use the RS and GIS data as input. Furthermore, GIS and RS integration holds great importance in developing the GWP maps [1, 5, 23, 31]. Indeed, groundwater mapping and modeling activities were facilitated by the RS data in combination with GIS context. Continuous updates on parameters such as rainfall and temperature are required for GIS data in hydrological research, thus enabling researchers to produce thematic maps quicker than using the data on field investigations [34]. Besides, digital data composed in GIS database makes it simpler to conduct both qualitative and quantitative analysis than when utilizing traditional data [47].

Hydrological apps compile input data from publications or satellite data that has been incorporated into the GIS environment [5]. Hydrologists have experimented with statistical approaches or machine learning algorithms to produce GWP approaches using the combination of GIS and RS [3, 15, 25, 34, 48]. Classification and Regression Trees [9], Artificial Neural Network [39, 50], K-Nearest Neighbors [45], Logistic Regression [38], Random Forest [41, 55], and Support Vector Machine [35] are common machine learning models, while GIS-based models include AHP [40], FL [42], MIF [12], FR [11], and TOPSIS [59]. Compared to machine learning models, GIS models are less strict in their input data, allowing for easy map modeling. The bulk of the statistical models (index-based models) in groundwater potential research is associated with altitude, distance to rivers, precipitation, soil type, LULC, slope, lineament, and geology [2, 10, 29].

Regarding water security and climate change-related global, national, and regional environmental issues, the groundwater potential map is essential to groundwater resource management campaigns to reach sustainable development. This book's chapter aims to (a) decipher groundwater potential using GIS-based models such as AHP, FL, MIF, TOPSIS, and FR, and (b) assess the accuracy of these five GIS-based models. This chapter is not only a reference for hydrologists in the world for mapping groundwater potential but also a theoretical basis for groundwater models. Additionally, the findings of this book chapter are very beneficial in selecting suitable models for effective groundwater management, which may guarantee a steady water supply for an area's domestic, agricultural, and industrial sectors.

7.3 Deciphering GWPZs Using GIS-Based Models

As mentioned above, most GIS models involve index-based techniques, which means that groundwater potential levels will be represented by numerical value intervals. In this section, several GIS models will be discussed to decipher groundwater potential mapping, including AHP, FL, MIF, FR, and TOPSIS. The input data (CFs) include altitude, distance to rivers, geology, lineament density, LULC, slope, soil type, and precipitation. Additionally, groundwater yield data with a threshold value of 10 m^3/h is also used for observing groundwater potential (Fig. 7.1). A total of 1601 wells are used for groundwater potential surveys, which are collected by the Groundwater Source Department of Thailand. All these CFs were saved in raster format with a spatial resolution of 30 m. In this chapter, we select Kanchanaburi province in Thailand to perform illustrative examples. Kanchanaburi is located in the west of Thailand, where mountainous terrain covers more than 60% of the province's area. The province has an area of approximately 19,380 km^2. The Kanchanaburi's hydrographic network is very complex with the Khwae Noi and Khwae Yai Rivers converging into the Mae Klong River. Kanchanaburi has average annual precipitation fluctuating from 1500 to 2200 mm, especially heavy rainfall during the summertime. Regarding geology, the area consists mainly of metamorphic, igneous, and sedimentary rocks ranging in age from the Precambrian to the Quaternary. Although there are diverse river systems, groundwater is still being quarried to serve agricultural activities and domestic purposes in many regions [54]. Therefore, it is crucial to manage consistently the groundwater resources in Kanchanaburi.

7.3.1 Conditioning Factors

Altitude: This is one of the essential factors in the distribution of groundwater because water migrates from regions of high height to regions of low height. The altitude map of Kanchanaburi province was derived from satellite images of Aster Global Digital Elevation Model V003. The altitude in Kanchanaburi varied between 2 and 1812 m above sea level and was categorized into five classes by the quantile classification method: >200 m; 150–200 m; 100–150 m, 50–100 m, 2–50 m (Fig. 7.2a).

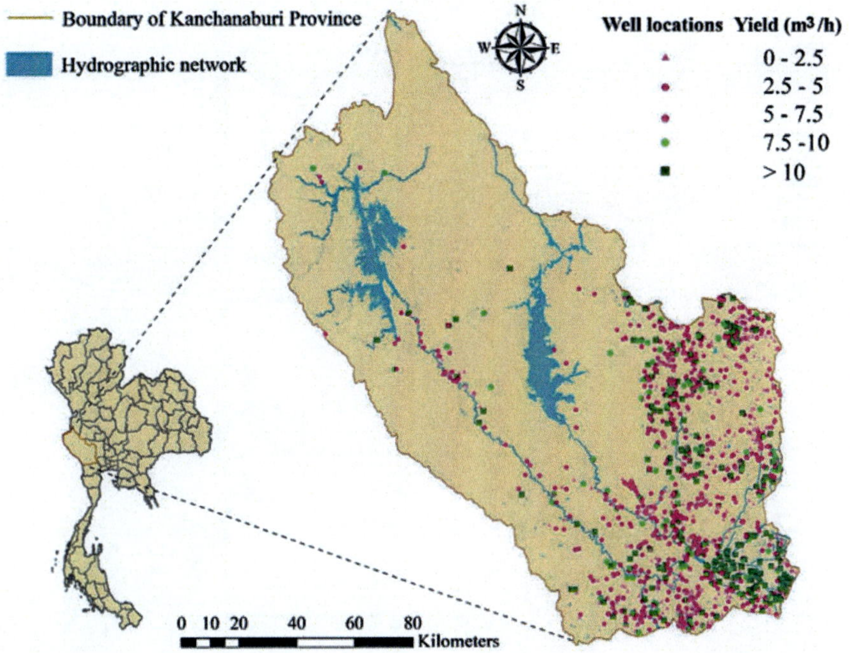

Fig. 7.1 Kanchanaburi province, Thailand

Distance to rivers: The natural penetration of groundwater and the aquifer's capacity to recharge are closely related to the rivers. The river data was first extracted from Thailand's hydrological system, and then the distance to rivers map was created by the Euclidean Distance tool in ArcGIS. Data on distance to rivers was classified into five classes by the quantile classification method: 0–2,000 m, 2,000–4,000, 4,000–6,000, 6,000–8,000, and >8,000 m (Fig. 7.2b).

Geology: Geologic structures and features decide aquifer storage capacity. Thus, geological considerations are a necessary part of deciphering groundwater potential (GWP) in any terrain. Geological data was taken from the Thailand geology map published by the Department of Mineral Resources. There were fourteen geological units in the study area. They included Cretaceous (Kgr), Triassic (Tr), Tertiary (Tmm), Silurian-Devonian-Carboniferous (SDC), Silurian-Devonian (SD), Quaternary (Q), Cambrian (E), Cambrian-Ordovician (EO), Carboniferous-Permian (CPk), Devonian (D), Jurassic (Ju), Ordovician (O), Permian (Pr), Precambrian (PE) (Fig. 7.2c).

Lineament density: Lineament (Fault) can play a role as a connection to recharge groundwater in the case of vertical subsurface flow. However, a fault zone can act as a barrier to horizontal flow. In our case, we considered the groundwater yield, so lineament serves as a barrier. Lineament data was extracted and driven from

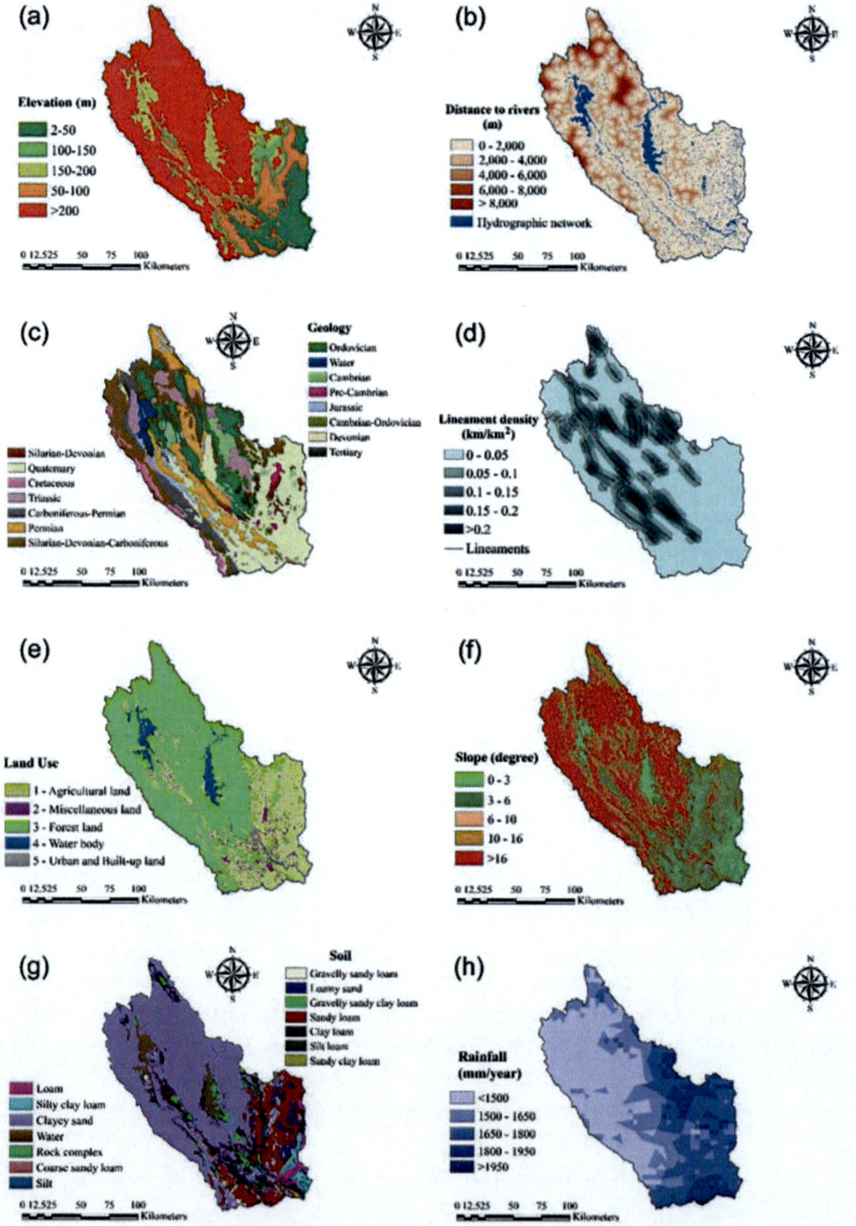

Fig. 7.2 Conditioning factors: **a** altitude; **b** distance to rivers, **c** geology, **d** lineament density, **e** land use/land cover, **f** slope, **g** soil type, **h** precipitation

the geology map combined with the Landsat eight satellite images. The lineament density (LD) map was created by the "line density" tool in ArcGIS software based on lineament data. The LD map was then divided into five categories, including 0–0.05, 0.05–0.1, 0.1–0.15, 0.15–2, >0.2 km/km^2 (Fig. 7.2d).

Land use/Land cover (LULC): LULC reflects the development and occurrence of groundwater, which relates to information on groundwater yield. The LULC map was collected by the Land Development Department of Thailand in vector format. The LULC map was then converted to raster format and classified into five classes, including agricultural land (AG), miscellaneous land (ML), forest land (FL), water bodies (WB), and urban/built-up land (UB) (Fig. 7.2e).

Slope: Slope is known as an essential factor for deciphering GWP because it reflects residence times of water surface and precipitation on the earth's surface. Terrain with low slopes facilitates the recharge of groundwater and vice versa. The Aster Global Digital Elevation Model V003 was used to extract slope digital elevation data using the "slope" tool in ArcGIS 10.8 software. The slope map was divided into five classes, including 0°–3°, 3°–6°, 6°–10°, 10°–16°, and >16° (Fig. 7.2f).

Soil type: In most groundwater studies, soil type is key for deciphering GWP because it relates to water surface infiltration capacity into aquifers. Soil type data in vector format was also required from the Thailand's Land Development Department, which was then converted to raster format. There were thirteen soil types found in Kanchanaburi, including silt, coarse sandy loam, rock complex, clayey sand, silty clay loam, loam, sandy clay loam, silt loam, clay loam, sandy loam, gravelly sandy clay loam, loamy sand, gravelly sandy loam (Fig. 7.2g).

Precipitation: The average annual precipitation contributes much to the recharge capacity potential of groundwater. The average annual precipitation map was established based on Global satellite precipitation data from 2011 to 2021. Kanchanaburi received rainfall from 1500 to 1950 mm/year. The precipitation map was categorized into five classes, including >1,950, 1,800–1,950, 1,650–1,800, 1,500–1,650, <1,500 mm/year (Fig. 7.2h).

7.3.2 Deciphering GWPZs Using AHP

The AHP concept was first proposed by Saaty [44]. It refers to decision-making procedures based on different criteria in multivariate environments to obtain various solutions for a decision, in which criteria are organized into a hierarchic structure based on the weights of criteria from the decision-maker's judgments.

We define the AHP as an approach for calculating the weight of CFs in this book based on expert opinion which provides a collocation of variables to find the most reasonable final decision in mapping GWP. A pair-wise comparison matrix is utilized to design the data structures for the AHP model. The AHP scale for CFs is consulted

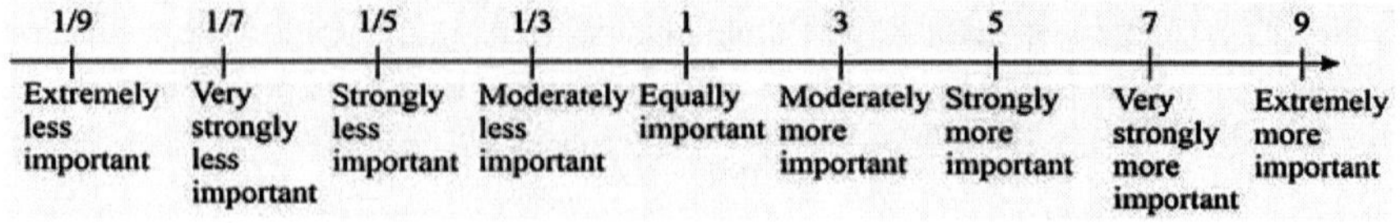

Fig. 7.3 AHP scale

based on experts' knowledge in the field of groundwater. Experts give the assessment for each CF based on the importance of each CF to the possibility of occurrence of groundwater yield >10 m^3/h (Fig. 7.3). The pair-wise comparison matrix is then established to estimate a relative weight matrix and principal eigenvalue. The result of the relative weight matrix must be validated with a consistency ratio (CR) <0.1. The mathematics of CR is expressed in Eqs. (7.1) and (7.2).

$$\mathrm{CI} = \frac{(\lambda_{\max} - n)}{(n - 1)} \tag{7.1}$$

$$\mathrm{CR} = \frac{\mathrm{CI}}{\mathrm{RI}} \tag{7.2}$$

where CI is the consistency index, $\lambda_{\max}$ is the principal eigen value, n is the number of comparisons, RI is the random consistency index (Table 7.1), and CR is the consistency ratio.

The percentage of effect of input variables is then estimated to give the final decision. Finally, all the thematic layers of CFs are integrated into the GIS environment to decipher the GWP map.

Example 7.1 Application of the AHP model to decipher GWPZs in Kanchanaburi, Thailand.

Step 1: Establishing pair-wise comparison.

To determine the evaluation of CFs, we consulted with the groundwater experts. Then, each pair-wise CF was compared together using the AHP scale to obtain the pair-wise comparison matrix. Observing the first row of matrix, altitude scored a "1" above lineament density, meaning comparisons were equally important. Altitude scored a "3" above distance to rivers and slope, meaning comparison was moderately more important. Altitude scored a "7" above geology, precipitation, and soil type, meaning comparison was very strongly more important. Altitude scored a "9" above LULC, meaning comparison was extremely more important. We did the same for

Table 7.1 Random consistency index

1	2	3	4	5	6	7	8	9	10	11	12	13	14	15
0.00	0.00	0.58	0.90	1.12	1.24	1.32	1.41	1.45	149	1.51	1.48	1.56	1.57	1.59

other pair-wise CFs to get the reciprocal matrix (pair-wise comparison matrix). The number of comparisons rely on the number of CFs calculated as Eq. (7.3).

$$N = \frac{n(n-1)}{2} \tag{7.3}$$

where N is the number of comparisons, n is the number of CFs. In this case, we had eight CFs so there are 28 comparisons. Details about the pair-wise comparison matrix (Table 7.2), the first row and first column are the names of the CFs, the diagonal elements are equal to one in the cells, and the other cells are the results of pair-wise comparisons. We now have an 8 by 8 pair-wise comparison matrix.

Step 2: Building the relative weight matrix and normalized principal eigenvalue.

In this step, we built the relative weight matrix and got the normalized principal eigenvalue by calculating priority vectors (Table 7.3). Then, element values of the pair-wise comparison matrix were divided by the sum of its columns to get the relative weight matrix. The normalized principal eigenvalue was achieved by averaging across the rows. The total normalized principal eigenvalue was one across the column.

Step 3: Calculating the CR and deciphering groundwater potential zones.

To estimate the CR, the principal eigenvalue of CFs was conducted by calculating the summation of products between the sum of columns of the pair-wise comparison matrix and each CF of the normalized principal eigenvalue. The largest principal eigenvalue was used to estimate the CR. The CR value was 0.055 using Eqs. (7.1) and (7.2). Next, classes of each CF were assigned based on heuristic approaches (knowledge-driven methods) (Table 7.4). Finally, all eight condition factor maps were integrated into ArcGIS software to map GWP zones. This process was conducted as below formula.

Table 7.2 Reciprocal matrix from pair-wise comparison

CFs	Altitude	LD	DtR	Slope	Geology	LULC	P	Soil type
Altitude	1	1	3	3	7	9	7	7
LD	1	1	3	3	7	9	7	7
DtR	1/3	1/3	1	1	5	7	5	5
Slope	1/3	1/3	1	1	5	7	5	5
Geology	1/7	1/7	1/5	1/5	1	3	1	1
LULC	1/9	1/9	1/7	1/7	1/3	1	1/3	1/3
P	1/7	1/7	1/5	1/5	1	3	1	1
Soil type	1/7	1/7	1/5	1/5	1	3	1	1
Geometric	3.21	3.21	8.74	8.74	27.33	42.00	27.33	27.33

LULC—Land use/Land cover; DtR—Distance to rivers; LD—Lineament density; P—Precipitation

Table 7.3 Relative weight matrix

CFs	AL	LD	DtR	SL	G	LULC	P	S	Sum	Weight (A_X)	Eigenvalue (λ)
AL	0.312	0.312	0.343	0.343	0.256	0.214	0.256	0.256	2.293	28.66	8.55
LD	0.312	0.312	0.343	0.343	0.256	0.214	0.256	0.256	2.293	28.66	8.55
DtR	0.104	0.104	0.114	0.114	0.183	0.167	0.183	0.183	1.152	14.40	8.43
SL	0.104	0.104	0.114	0.114	0.183	0.167	0.183	0.183	1.152	14.40	8.43
G	0.045	0.045	0.023	0.023	0.037	0.071	0.037	0.037	0.316	3.95	8.07
LULC	0.035	0.035	0.016	0.016	0.012	0.024	0.012	0.012	0.162	2.03	8.11
P	0.045	0.045	0.023	0.023	0.037	0.071	0.037	0.037	0.316	3.95	8.07
S	0.045	0.045	0.023	0.023	0.037	0.071	0.037	0.037	0.316	3.95	8.07
$\lambda_{max} = 8.55$; CI = 0.078; RI = 1.41; CR = 0.055											

AL—Altitude; LD—Lineament density; DtR—Distance to rivers; SL—Slope; G—Geology; LULC—Land use/land cover; P—Precipitation; S—Soil

$$\text{GWPI} = \sum_{i}^{n} A_X \times B_Y \tag{7.4}$$

where GWPI is the GWP index, A is the weight of the CFs, and B is the rank of the classes in CFs. The X term ($A = 1, 2, 3, \ldots, M$) represents the CF and Y term ($B = 1, 2, 3, \ldots, N$) represents the CF classes.

Based on the GWP index, the GWP map from the AHP model was deciphered into five categories, ranging from very low to very high. (Fig. 7.4). The present result reveals that approximately 5.30% of Kanchanaburi province had very poor GWP and nearly 36.22% of the area had poor GWP, focusing mainly on the west. While parts with moderate, high, and very high are distributed in the east with 26.68%, 13.48%, and 18.31% of the area, respectively.

7.3.3 Deciphering GWPZs Using FL

The Fuzzy Logic (FL) concept was first introduced by Zahdel in 1965 [27]. It has been continually developed and remains in use in many fields. Theoretically, the FL relates to complex data structures. Its result is not binary. Fuzzy theory is applied to ambiguous and imprecise information in a dataset. Each candidate object is assigned a member value which is conducted based on FL rules from the assessment of the experts in the FL model. The process of setting the FL rule is called fuzzification. Subsequently, member values are transformed into the 0–1 scale, called a fuzzy membership value. Mathematically, fuzzy theory can be explained as follows:

$$\{\text{Xz},\ \text{MF(z)}\},\ \text{z} \in \text{Z} \tag{7.5}$$

Table 7.4 Information on the AHP model

Influencing factors	Classes	Weight (Wi)	Rank (Rj)
Altitude	>200 m	28.66	1
	150–200 m		3
	100–150 m		5
	50–100 m		7
	2–50 m		9
Lineament density	>0.2 km/km^2	28.66	1
	0.15–0.2 km/km^2		3
	0.1–0.15 km/km^2		5
	0.05–0.1 km/km^2		7
	0–0.05 km/km^2		9
Distance to river	>8,000 m	14.40	1
	6,000–8,000 m		3
	4,000–6,000 m		5
	2,000–4,000 m		7
	0–2,000 m		9
Slope	>16°	14.40	1
	10°–16°		3
	6°–10°		5
	3°–6°		7
	0°–3°		9
Geology	Tmm	3.95	1
	D		1
	EO		1
	Ju		3
	PE		1
	E		7
	Water		9
	O		3
	SDC		3
	Pr		5
	CPk		3
	TR		1
	Kgr		1
	Q		9
	SD		7

(continued)

Table 7.4 (continued)

Influencing factors	Classes	Weight (Wi)	Rank (Rj)
Land use	UB	2.03	1
	WB		9
	FL		5
	ML		3
	AG		7
Precipitation	>1,950 mm/year	3.95	9
	1,800–1,950 mm/year		7
	1,650–1,800 mm/year		5
	1,500–1,650 mm/year		3
	<1500 mm/year		1
Soil type	Silt	3.95	3
	Coarse sandy loam		3
	Rock complex		3
	Water		9
	Clayey sand		3
	Silty clay loam		1
	Loam		5
	Sandy clay loam		3
	Silt loam		1
	Clay loam		3
	Sandy loam		5
	Gravelly sandy clay loam		5
	Loamy sand		9
	Gravelly sandy loam		7

where X is a fuzzy set of Z sets, MF is the membership function that reflects set X's degree of membership to Z. The FL model is introduced as a method for approximating the fuzzy membership (FM) value of CFs in this chapter based on the uncertainty and imprecision of a hydrology dataset in GIS environment, in which CFs (fuzzy membership) is assigned a discrete fuzzy value between 0 and 1, corresponding to the assessment of the occurrence ability of groundwater yield >10 m^3/h on CF's classes. Then, the FM values are integrated to give maps with FM functions using the "Fuzzy Overlay" tool in ArcGIS software, a process which is called defuzzification. The defuzzification can be processed in a variety of algorithms, including fuzzy AND, fuzzy OR, fuzzy algebraic product, fuzzy algebraic sum, and fuzzy gamma operator, in which fuzzy algebraic product is best suited to deciphering GWP due to its high sensitivity in classification [4, 27].

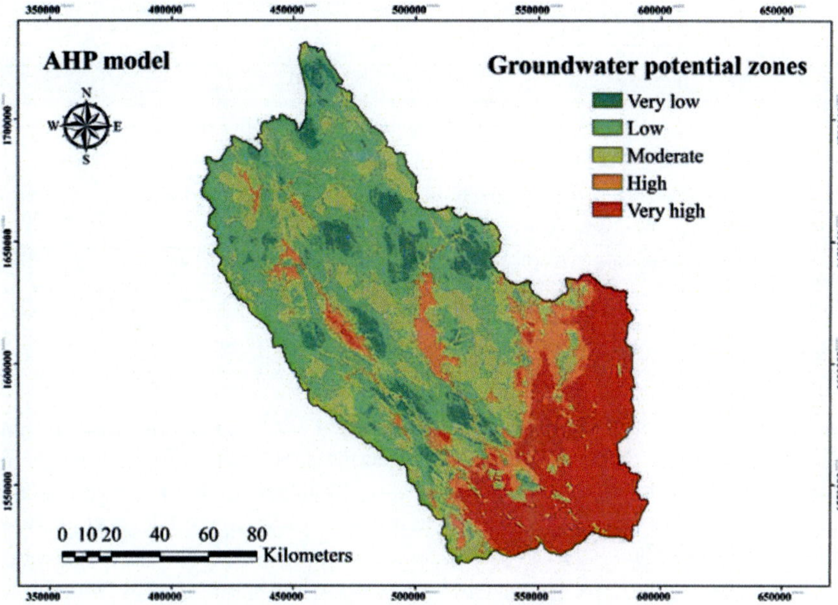

Fig. 7.4 Groundwater potential map in Kanchanaburi province using the AHP model

Example 7.2 Application of the FL model to decipher GWPZs in Kanchanaburi, Thailand.

Step 1: Set up FL rules and assign fuzzy membership values.

The FL rule was established based on the expert's knowledge and experience (Table 7.5). Table 7.6 illustrates the member values of classes from the decision-maker's judgments. Then, the member values were integrated into classes of CFs in the GIS environment corresponding to their values. This process was accomplished through the "Reclassify" function. Subsequently, classes of each CFs were computed the fuzzy membership values using the 'fuzzy membership' function in ArcGIS software. The FM values for CFs were described in Fig. 7.5. Lower fuzzy membership values imply that positions have a low potential for groundwater yield >10 m^3/h and vice versa.

Step 2: Conduct Fuzzy Overlay function.

All CFs were assigned the fuzzy membership values to build "fuzzified" layers. Then, the "fuzzy overlay" function in ArcGIS was utilized to decipher GWP zones. This integration process uses the algebraic product of the fuzzy sum and fuzzy product. The output of the FL overlay was divided into five levels: the very poor, poor, moderate, high, and very high potential of groundwater yield >10 m^3/h (Fig. 7.6). According to their means, 45.01% and 21.00% of the Kanchanaburi's area are categorized as very poor and poor GWP, respectively. Meanwhile 13.51%, 8.34% and 12.14% of the province is categorized as moderate, high, and very high GWP, respectively.

Table 7.5 Fuzzy Logic rule

No	Fuzzy logic rules	Score
1	If a class is very good GWP, it can be had very high groundwater yield	5
2	If a class is good GWP, it can be had high groundwater yield	4
3	If a class is moderate GWP, it can be had moderate groundwater yield	3
4	If a class is low GWP, it can be had low groundwater yield	2
5	If a class is very low GWP, it can be had very low groundwater yield	1

7.3.4 Deciphering GWPZs Using MIF

The MIF approach was first suggested by Shaban et al. [46] for mapping GWP within the GIS framework. This technique is based on the decision-maker's judgments to arrange criteria in a hierarchy. Additionally, it is also known as the weighted overlay method in some studies [26, 52, 58], which is used to compare CFs and analyze interrelationships. The relationship between the conditioning factor and target variable is shown by cumulative weights in the MIF modeling approach. The cumulative weight is the sum of the weights. These weights are evaluated based on minor and major effects between CFs, which helps to estimate the relative rate of changes. The minor and major variables are assigned a weight value of 0.5 and 1, respectively. To understand the MIF model, we practice the model with Example 7.3.

Example 7.3 Application of the MIF model to decipher groundwater potential zones in Kanchanaburi, Thailand.

Step 1: Determine the minor and major CFs.

In our case, we had eight CFs, including soil type, slope, lineament density, altitude, distance to rivers, geology, LULC, and precipitation. The relationship between CFs was analyzed based on the literature review [21, 51, 57, 60]. Figure 7.7 shows the interaction of CFs in the occurrence of groundwater yield >10m^3/h. To obtain a greater understanding of what this principle means, we could consider an instance from geology. Geology has major effects on soil type and has a minor effect on lineament density in the interaction. This process was also considered for other CFs.

Step 2: Calculating the cumulative weight and relative rate for each CF.

In the MIF modeling approach, we had two types of cumulative weights. The major cumulative weight of each CF was the sum of its major effects. Similarly, the minor cumulative weight was estimated from minor effects. Next, relative rates were calculated as the sum of the major and minor cumulative weights followed by the proposed weights of each CF using the below equation:

$$\text{Proposed weight} = \left[\frac{(X+Y)}{\sum(X+Y)}\right] \times 100 \tag{7.6}$$

Table 7.6 Classes of condition factors with the member values

No	Influencing factors	Classes	Member values
1	Altitude	>200 m	1
		150–200 m	2
		100–150 m	3
		50–100 m	4
		2–50 m	5
2	Lineament density	>0.2 km/km^2	1
		0.15–0.2 km/km^2	2
		0.1–0.15 km/km^2	3
		0.05–0.1 km/km^2	4
		0–0.05 km/km^2	5
3	Distance to river	>8,000 m	1
		6,000–8,000 m	2
		4,000–6,000 m	3
		2,000–4,000 m	4
		0–2,000 m	5
4	Slope	>16°	1
		10°–16°	2
		6°–10°	3
		3°–6°	4
		0°–3°	5
5	Geology	Tmm	1
		D	1
		EO	1
		Ju	2
		PE	1
		E	4
		Water	5
		O	2
		SDC	2
		Pr	3
		CPk	5
		TR	1
		Kgr	1
		Q	4
		SD	5

(continued)

Table 7.6 (continued)

No	Influencing factors	Classes	Member values
6	Land use	UB	4
		WB	5
		FL	3
		ML	2
		AG	4
7	Precipitation	>1,950 mm/year	5
		1,800–1,950 mm/year	4
		1,650–1,800 mm/year	3
		1,500–1,650 mm/year	2
		<1500 mm/year	4
8	Soil type	Silt	2
		Coarse sandy loam	2
		Rock complex	2
		Water	5
		Clayey sand	2
		Silty clay loam	1
		Loam	3
		Sandy clay loam	2
		Silt loam	1
		Clay loam	2
		Sandy loam	3
		Gravelly sandy clay loam	3
		Loamy sand	5
		Gravelly sandy loam	4

where X is the major cumulative effect of CFs and Y is the minor cumulative effect of CFs. Table 7.7 presents information on the MIF modeling approach.

Step 3: Assigning ranking for classes of CFs and deciphering GWPZs.

To decipher GWPZs, classes of CFs must assign the rank, which was done through heuristic approaches (knowledge-driven methods). Table 7.8 classifies the rank classes of each CF. Next, the weighted overlay approach was used to integrate all the CFs after giving the proposed weights and ranks to the CFs and their classes. The GWPI was estimated by Eq. (7.4) using the "Raster Calculator" tool in the GIS environment.

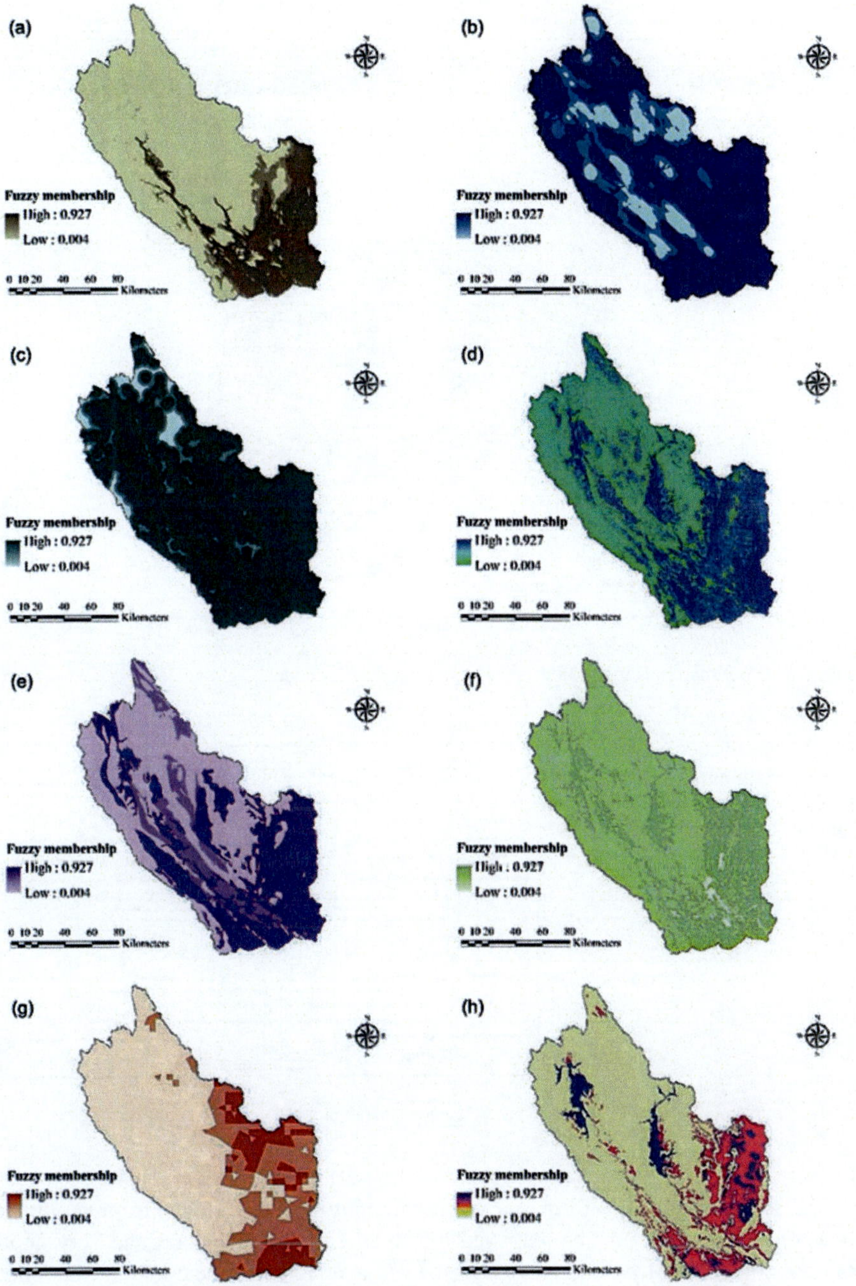

Fig. 7.5 FM value for: **a** altitude; **b** distance to rivers, **c** geology, **d** lineament density, **e** land use/ land cover, **f** slope, **g** soil type, **h** precipitation

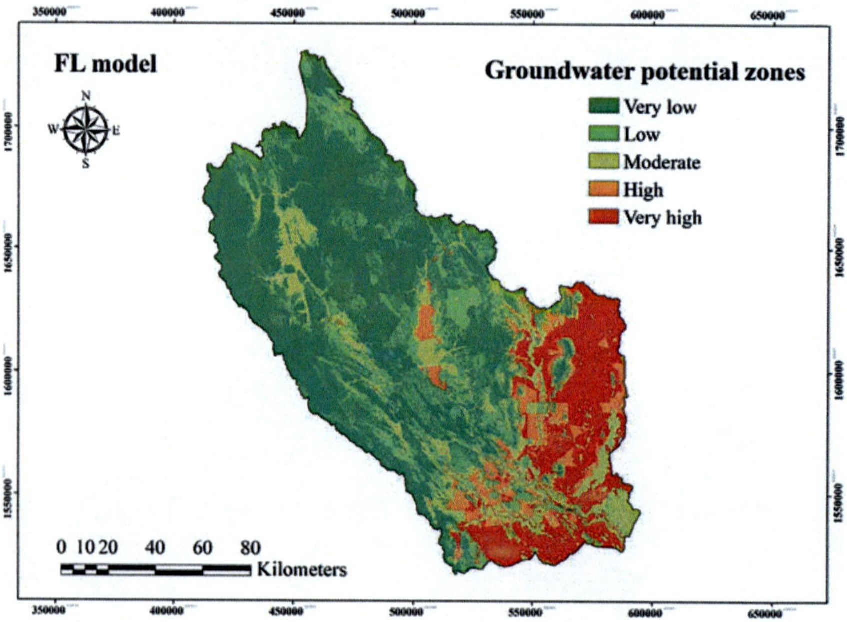

Fig. 7.6 GWP map in Kanchanaburi province using the FL model

Fig. 7.7 Interaction between CFs in determining the potential zones of groundwater yield >10 m^3/h

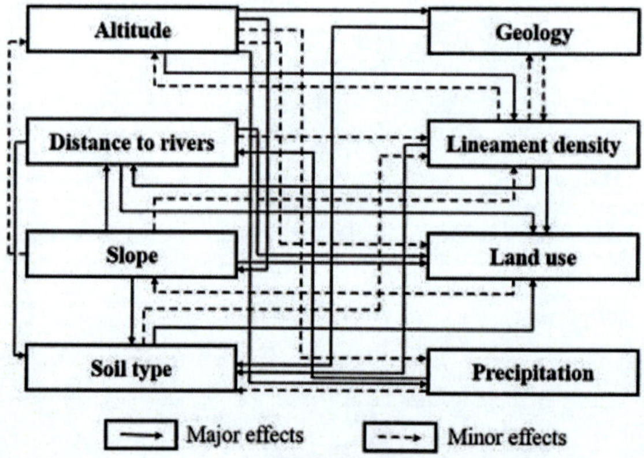

Subsequently, the GWPI map was classified into five categories, from very poor to very high (Fig. 7.8). The statistical result reveals that the good and very good GWPZ covered 16.15% and 15.99% of the whole area, while the poor and very poor potential zones consisted of 43.72% and 6.36% of the overall area, respectively. Approximately 17.79% of the remaining area of Kanchanaburi was under moderate potential.

Table 7.7 Effect of conditioning factor, relative rate, and proposed weight for each potential factor

Conditioning factors	Major cumulative weight (X)	Minor cumulative weight (Y)	Relative rate (*X* + *Y*)	Proposed weight
Geology	1	0.5	1.5	6.98
Lineament density	3	1	4	18.60
LULC	0	0.5	0.5	2.33
Precipitation	1	0.5	1.5	6.98
Altitude	4	1	5	23.26
Distance to rivers	3	0.5	3.5	16.28
Slope	3	1	4	18.60
Soil	1	0.5	1.5	6.98
Total			21.5	100.00

7.3.5 Deciphering GWPZs Using TOPSIS

The TOPSIS, a multi-criteria decision-making (MCDM) process, was first introduced by Hwang and Yoon [19]. This approach is different from other MCDM methods in that all criteria are given scores at a time as alternatives. A decision matrix is established from these scores. This means that the decision matrix helps us indicate the best alternative without having to calculate too much. It has therefore been most widely applied in the real-world involving MCDM problems. In the TOPSIS approach, the score distance between criteria is used to rank alternatives. With the ideal answer and the negative ideal solution, the smallest and largest distances are decided, respectively.

A TOPSIS model is to be started with a decision matrix (D) in the GWP mapping scope based on the decision-maker experience score for each CF Eq. (7.7).

$$\mathrm{D} = \begin{array}{c} \\ A_1 \\ A_2 \\ . \\ A_i \\ . \\ A_m \end{array} \begin{array}{c} X_1\ X_2\ .\ X_{\mathrm{j}}\ .\ X_n \\ \begin{bmatrix} x_{11} & x_{12} & . & x_{1\mathrm{j}} & . & x_{1n} \\ x_{21} & x_{22} & . & x_{2\mathrm{j}} & . & x_{2n} \\ . & . & . & . & . & . \\ x_{\mathrm{i}1} & x_{\mathrm{i}2} & . & x_{\mathrm{ij}} & . & x_{\mathrm{in}} \\ . & . & . & . & . & . \\ x_{\mathrm{m}1} & x_{\mathrm{m}2} & . & x_{\mathrm{mj}} & . & x_{\mathrm{mn}} \end{bmatrix} \end{array} \tag{7.7}$$

where D represents the decision matrix; A_i represents the alternatives with i ranging from 1 to m; X_j represents the score of the jth CF with j ranging from 1 to n. A weighted matrix (R) is then calculated based on the decision matrix. The standardized equation is explained by Eq. (7.8).

Table 7.8 Proposed weights and ranks for conditioning factors and their classes

Influencing factors	Proposed weight	Classes	Rank (Rj)
Altitude	23.26	>200 m	1
		150–200 m	3
		100–150 m	5
		50–100 m	7
		2–50 m	9
Lineament density	18.60	>0.2 km/km^2	1
		0.15–0.2 km/km^2	3
		0.1–0.15 km/km^2	5
		0.05–0.1 km/km^2	7
		0–0.05 km/km^2	9
Distance to river	16.28	>8,000 m	1
		6,000–8,000 m	3
		4,000–6,000 m	5
		2,000–4,000 m	7
		0–2,000 m	9
Slope	18.6	>16°	1
		10°–16°	3
		6°–10°	5
		3°–6°	7
		0°–3°	9
Geology	6.98	Tmm	1
		D	1
		EO	1
		Ju	3
		PE	1
		E	7
		Water	9
		O	3
		SDC	3
		Pr	5
		CPk	3
		TR	1
		Kgr	1
		Q	9
		SD	7
Land use/land cover	2.33	UB	1

(continued)

Table 7.8 (continued)

Influencing factors	Proposed weight	Classes	Rank (Rj)
		WB	9
		FL	5
		ML	3
		AG	7
Precipitation	6.98	>1,950 mm/year	9
		1,800–1,950 mm/year	7
		1,650–1,800 mm/year	5
		1,500–1,650 mm/year	3
		<1500 mm/year	1
Soil type	6.98	Silt	3
		Coarse sandy loam	3
		Rock complex	3
		Water	9
		Clayey sand	3
		Silty clay loam	1
		Loam	5
		Sandy clay loam	3
		Silt loam	1
		Clay loam	3
		Sandy loam	5
		Gravelly sandy clay loam	5
		Loamy sand	9
		Gravelly sandy loam	7

$$r_{ij} = \frac{x_{ij}}{\sqrt{\sum_i^m (x_{ij})^2}} \tag{7.8}$$

where r is the normalized score of the ijth element in the weighted matrix. The weighted matrix is defined as follows:

$$R = [r_{ij}] \tag{7.9}$$

Next, a normalized decision matrix is computed by multiplying the decision matrix with its relative weights Eq. (7.9).

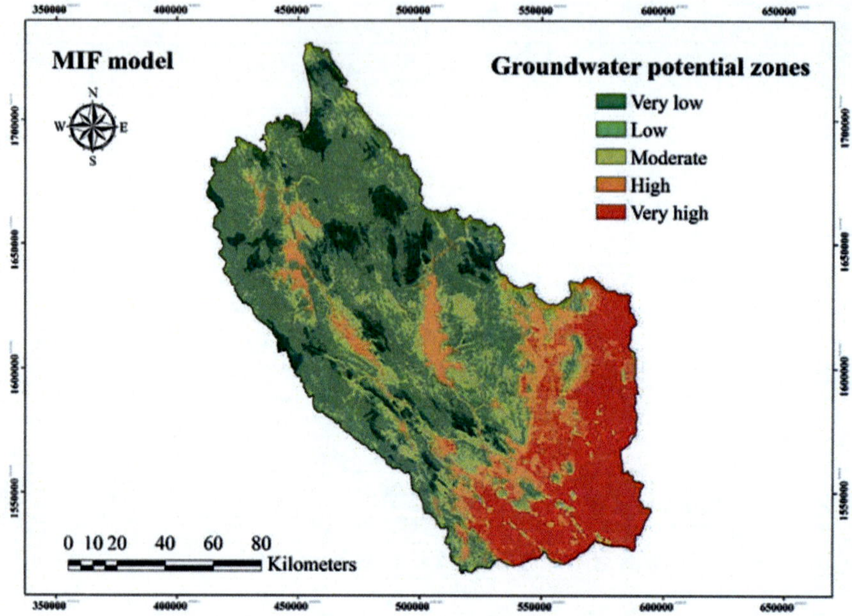

Fig. 7.8 Groundwater potential map in Kanchanaburi province using the MIF model

$$V = \begin{bmatrix} w_1 r_{11} & w_2 r_{12} & . & w_j r_{1j} & . & w_n r_{1n} \\ w_1 r_{21} & w_2 r_{22} & . & w_j r_{2j} & . & w_n r_{2n} \\ . & . & . & . & . & . \\ w_1 r_{i1} & w_2 r_{i2} & . & w_j r_{\text{ij}} & . & w_n r_{\text{in}} \\ . & . & . & . & . & . \\ w_1 r_{m1} & w_2 r_{m2} & . & w_j r_{\text{mj}} & . & w_n r_{\text{mn}} \end{bmatrix} = \begin{bmatrix} v_{11} & v_{12} & . & v_{1j} & . & v_{1n} \\ v_{21} & v_{22} & . & v_{2j} & . & v_{2n} \\ . & . & . & . & . & . \\ v_{i1} & v_{i2} & . & v_{\text{ij}} & . & v_{\text{in}} \\ . & . & . & . & . & . \\ v_{m1} & v_{m2} & . & v_{\text{mj}} & . & v_{\text{mn}} \end{bmatrix} \tag{7.10}$$

where v_{ij} is the normalized value. Then, calculating the distance value via the ideal answer (V^+) and the negative ideal solution (V^-). The solutions are estimated. Eqs. (7.11) and (7.12), followed by distance value.

$$V^+ = \left\{(\max v_{ij}|j \in J) or \left(\min v_{ij}|j \in J'\right), i = 1, 2, \ldots, m\right\} = \{v_1^+, v_2^+, \ldots, v_n^+) \tag{7.11}$$

$$V^- = \left\{(\min v_{ij}|j \in J) or \left(\max v_{ij}|j \in J'\right), i = 1, 2, \ldots, m\right\} = \{v_1^+, v_2^+, \ldots, v_n^+) \tag{7.12}$$

where $J = \left\{i = 1, 2, \ldots, n|v_{\text{ij}},\ a \text{ larger respon is desigend}\right\}$

$$J' = \left\{i = 1, 2, \ldots, n|v_{ij},\ a \text{ smaller respon is desigend}\right\}$$

$$S_i^+ = \sqrt{\sum_{i=1}^{n} (v_{ij} - v_j^+)^2} \quad (7.13)$$

$$S_i^- = \sqrt{\sum_{i=1}^{n} (v_{ij} - v_j^-)^2} \quad (7.14)$$

where S_i^+ is the distance from the ideal answer to the alternative, S_i^- is the distance from a negative ideal solution to the alternative. Consequently, the criteria are ranked based on the relative closeness (C_i) to the ideal solution. The relative closeness can be expressed in Eq. (7.15).

$$C_i = \frac{S_i^-}{S_i^+ + S_i^-} \quad (7.15)$$

where value of C_i varies from 0 to 1; if the Ci is near 1, the priority is the *i*th alternative.

Example 7.4 Application of the TOPSIS model to decipher groundwater potential zones in Kanchanaburi, Thailand.

Step 1: Create the decision matrix.

The decision matrix was built from the assessment of the attributes of criteria, which was performed by the decision-maker's judgment. The attributes for candidate locations (administrative units) were determined from the information on altitude, lineament density, distance to rivers, slope, geology, LULC, precipitation, and soil type. Thirteen candidate locations consist of Nong Prue, Dan Makham Tia, Lao Khwan, Sangkhla Buri, Tha Muang, Si Sawat, Bo Phloi, Sai Yok, Mueang Kanchanaburi, Tha Maka, Thong Pha Phum, Phanom Thuan, and Huai Krachao districts (Fig. 7.9). Table 7.9 describes the decision matrix.

Step 2: Establish the normalized decision matrix.

The weighted matrix was established from the decision matrix (Table 7.9) using Eqs. (7.8) and (7.9). Table 7.10 describes the information on the weighted matrix. Subsequently, the normalized decision matrix was calculated based on Eq. (7.10), shown in Table 7.11.

Step 3: Determine the Euclidean Distance from the ideal answer and negative ideal solution.

The ideal answer (V^+) and negative ideal solution (V^-) (Table 7.12) were computed by Eqs. (7.11) and (7.12). Then, the Eqs. (7.13), (7.14) and (7.15) were applied to calculate the distance between each candidate from V^+ and V^- and the relative closeness (C_i), respectively. Finally, candidates were ranked based on the (C_i) value (Table 7.13).

Step 4: Deciphering GWP from the ranks of administrative units.

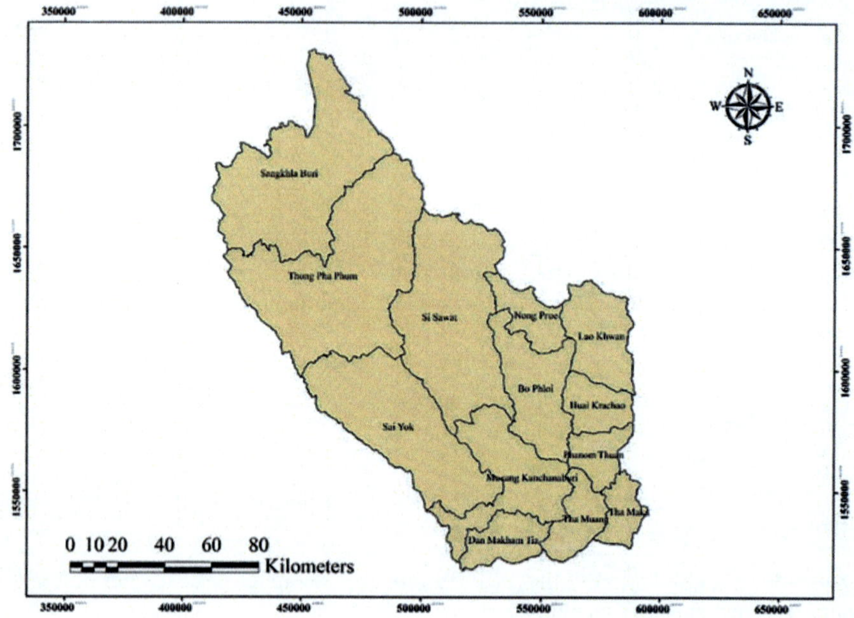

Fig. 7.9 Administrative units in Kanchanaburi province

Table 7.9 Decision matrix

Administrative units	Altitude	LD	DtR	Slope	Geology	LULC	Precipitation	Soil type
Mueang Kanchanaburi	5	5	7	3	3	1	5	3
Sai Yok	1	1	5	1	3	3	1	3
Bo Phloi	5	5	7	3	9	7	3	5
Si Sawat	1	1	5	1	7	9	3	3
Tha Maka	9	9	9	9	9	5	7	1
Tha Muang	9	9	9	9	9	5	9	5
Thong Pha Phum	1	1	7	1	3	7	1	3
Sangkhla Buri	1	1	7	1	1	7	1	3
Phanom Thuan	9	9	7	7	7	5	5	9
Lao Khwan	7	9	7	5	7	5	5	5
Dan Makham Tia	7	9	9	5	7	5	5	7
Nong Prue	3	5	5	3	3	5	7	5
Huai Krachao	7	9	7	5	7	7	5	7
Weight (%)	20	15	15	15	10	5	10	10

LULC—Land use/land cover; DtR—Distance to rivers; LD—Lineament density

Table 7.10 Weighted matrix

Administrative units	Altitude	LD	DtR	Slope	Geology	LULC	Precipitation	Soil type
Mueang Kanchanaburi	0.235	0.210	0.272	0.168	0.131	0.048	0.277	0.166
Sai Yok	0.047	0.042	0.194	0.056	0.131	0.144	0.055	0.166
Bo Phloi	0.235	0.210	0.272	0.168	0.393	0.335	0.166	0.277
Si Sawat	0.047	0.042	0.194	0.056	0.306	0.431	0.166	0.166
Tha Maka	0.423	0.379	0.350	0.505	0.393	0.239	0.388	0.055
Tha Muang	0.423	0.379	0.350	0.505	0.393	0.239	0.499	0.277
Thong Pha Phum	0.047	0.042	0.272	0.056	0.131	0.335	0.055	0.166
Sangkhla Buri	0.047	0.042	0.272	0.056	0.044	0.335	0.055	0.166
Phanom Thuan	0.423	0.379	0.272	0.393	0.306	0.239	0.277	0.499
Lao Khwan	0.329	0.379	0.272	0.281	0.306	0.239	0.277	0.277
Dan Makham Tia	0.329	0.379	0.350	0.281	0.306	0.239	0.277	0.388
Nong Prue	0.141	0.210	0.194	0.168	0.131	0.239	0.388	0.277
Huai Krachao	0.329	0.379	0.272	0.281	0.306	0.335	0.277	0.388

LULC—Land use/land cover; DtR—Distance to rivers; LD—Lineament density

Table 7.11 Normalized decision matrix

Administrative units	Altitude	LD	DtR	Slope	Geology	LULC	Precipitation	Soil type
Mueang Kanchanaburi	0.047	0.032	0.041	0.025	0.013	0.002	0.028	0.017
Sai Yok	0.009	0.006	0.029	0.008	0.013	0.007	0.006	0.017
Bo Phloi	0.047	0.032	0.041	0.025	0.039	0.017	0.017	0.028
Si Sawat	0.009	0.006	0.029	0.008	0.031	0.022	0.017	0.017
Tha Maka	0.085	0.057	0.053	0.076	0.039	0.012	0.039	0.006
Tha Muang	0.085	0.057	0.053	0.076	0.039	0.012	0.050	0.028
Thong Pha Phum	0.009	0.006	0.041	0.008	0.013	0.017	0.006	0.017
Sangkhla Buri	0.009	0.006	0.041	0.008	0.004	0.017	0.006	0.017
Phanom Thuan	0.085	0.057	0.041	0.059	0.031	0.012	0.028	0.050
Lao Khwan	0.066	0.057	0.041	0.042	0.031	0.012	0.028	0.028
Dan Makham Tia	0.066	0.057	0.053	0.042	0.031	0.012	0.028	0.039
Nong Prue	0.028	0.032	0.029	0.025	0.013	0.012	0.039	0.028
Huai Krachao	0.066	0.057	0.041	0.042	0.031	0.017	0.028	0.039

LULC—Land use/land cover; DtR—Distance to rivers; LD—Lineament density

Table 7.12 Ideal answer (V^+) and the negative ideal solution (V^-)

Criteria	V^+	V^-
Altitude	0.085	0.009
LD	0.057	0.006
DtR	0.053	0.029
Slope	0.076	0.008
Geology	0.039	0.004
LULC	0.022	0.002
Precipitation	0.050	0.006
Soil type	0.050	0.006

Table 7.13 Distance between each candidate from V^+ and V^-

Administrative units	S_i^+	S_i^-	C_i	Rank
Mueang Kanchanaburi	0.375	0.056	0.130	8
Sai Yok	0.444	0.015	0.032	13
Bo Phloi	0.364	0.067	0.156	7
Si Sawat	0.435	0.036	0.076	10
Tha Maka	0.294	0.125	0.299	3
Tha Muang	0.226	0.131	0.367	1
Thong Pha Phum	0.441	0.023	0.050	11
Sangkhla Buri	0.444	0.022	0.046	12
Phanom Thuan	0.255	0.119	0.318	2
Lao Khwan	0.308	0.094	0.233	6
Dan Makham Tia	0.296	0.099	0.250	4
Nong Prue	0.384	0.055	0.126	9
Huai Krachao	0.298	0.097	0.247	5

The GWP map was rendered from the rank of administrative units corresponding to the occurrence ability of groundwater yield >10 m^3/h. Figure 7.10 indicates that Tha Muang and Phanom Thuan districts were arranged into very high GWP. Dan Makham Tia and Tha Maka were high GWP. Boi Phloi, Lao Khwan and Huai Krachao districts were moderate GWP. Si Sawat, Nong Prue, and Mueang Kanchanaburi had low GWP. Sangkhla Buri and Thong Pha Phum had very low GWP.

7.3.6 Deciphering GWPZs Using FR

The FR, which is a bivariate probability function, is proposed by Bonham-Carter [7]. It describes the probability of an event occurring based on independent variables.

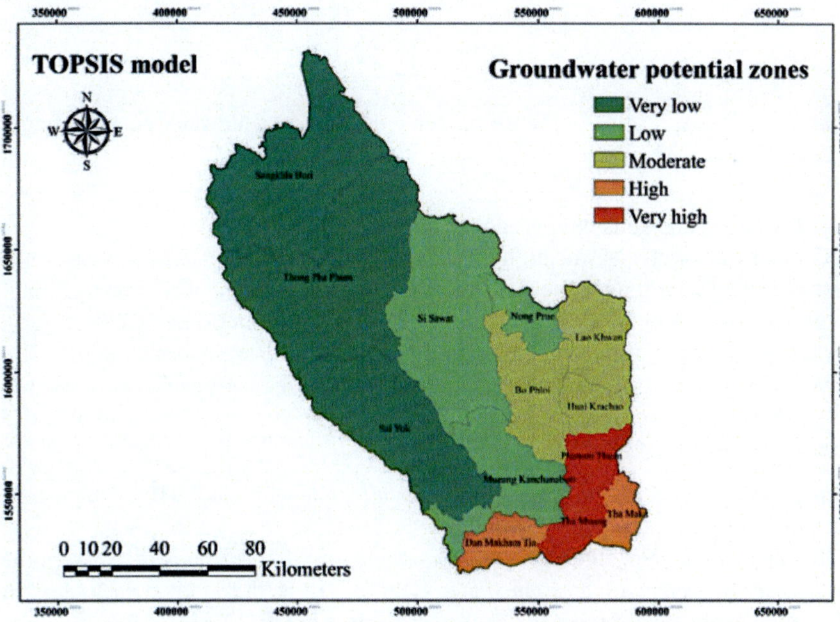

Fig. 7.10 Groundwater potential map in Kanchanaburi province using the TOPSIS model

Recently, the FR model has been so successful and versatile that it has been applied in the groundwater field [16, 22, 30, 33, 43]. In this chapter, we apply the FR model to indicate the relationship between groundwater yield >10 m^3/h and conditioning factors, followed by deducing GWP. The formulation for the FR model is explained as:

$$\mathrm{FR}_{\mathrm{ij}} = \frac{\left(\frac{A}{B}\right)}{\left(\frac{C}{D}\right)} \tag{7.16}$$

FR_{ij} is the frequency ratio of groundwater yield >10 m^3/h on each class of each conditioning factor. A is the number of wells with a groundwater yield of >10 m^3/h found in the *i*th class of the *j*th conditioning factor, and B is the number of groundwater wells in the study area. C is the number of pixels of the *i*th class of the *j*th conditioning factor. D is the number of pixels of the *j*th conditioning factor. If the FR value is lower than 1, the relationship between groundwater yield >10 m^3/h and the concerned class of conditioning factor is negative. In contrast, if the FR value is higher than 1, the relationship between groundwater yield >10 m^3/h and the concerned class of conditioning factor is positive. Finally, the GWPI at a location is calculated by the total of FRs Eq. (7.17) [16].

$$\mathrm{GWPI} = \sum \mathrm{FR}_{\mathrm{ij}} \tag{7.17}$$

Example 7.5 Application of the FR model to decipher groundwater potential zones in Kanchanaburi, Thailand.

Step 1: Classifying wells with groundwater yield.

Groundwater well classification is essential to the FR model. Groundwater wells were classified according to the yield value of 10 m^3/h. Potential wells had a groundwater yield of >10m^3/h and nonpotential wells had a groundwater yield of $\leq$ 10 m^3/h. Figure 7.11 shows the classification of groundwater wells based on the yield values of 10 m^3/h in Kanchanaburi, Thailand. A total of 1601 wells were used in this instance, in which 255 wells had a groundwater yield of more than 10 m^3/h (Fig. 7.11).

Step 2: Determining the number of groundwater wells with yield >10 m^3/h on each class of CFs.

To conduct this step, all CFs must be input into the GIS environment to determine the number of wells with a groundwater yield of >10 m^3/h found on each class of CFs, followed by determining the area of each class of CFs (Table 7.14).

Step 3: Estimating the FR value for each class of condition factors and the GWPI.

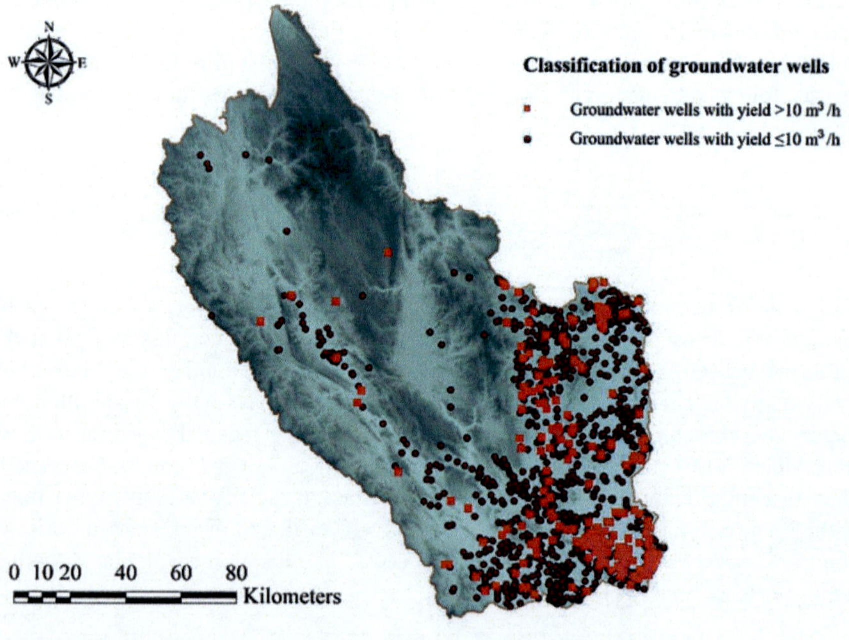

Fig. 7.11 Classification of groundwater wells

Table 7.14 Information on FR model for mapping GWPZs

Parameters	Classes	Num pixel of class	% of each classes	No. wells	% well	FR
Altitude	2–50 m	2,590,578	12.02	143	49.48	4.12
	50–100 m	2,531,187	11.75	57	19.72	1.68
	100–150 m	1,404,065	6.52	33	11.42	1.75
	150–200 m	2,389,108	11.09	14	4.84	0.44
	>200 m	12,631,163	58.62	8	2.77	0.05
Lineament density	0–0.05 km/km^2	11,753,397	5.45	233	80.62	14.78
	0.05–0.1 km/km^2	3,915,233	1.82	14	4.84	2.67
	0.1–0.15 km/km^2	3,468,278	1.61	5	1.73	1.07
	0.15–0.2 km/km^2	1,908,318	0.89	2	0.69	0.78
	>0.2 km/km^2	500,875	0.23	1	0.35	1.49
Distance to river	0–2,000 m	14,795,142	68.67	252	87.20	1.27
	2,000–4,000 m	4,396,017	20.40	2	0.69	0.03
	4,000–6,000 m	1,542,558	7.16	1	0.35	0.05
	6,000–8,000 m	543,408	2.52	0	0.00	0.00
	>8,000 m	268,976	1.25	0	0.00	0.00
Slope	0°–3°	3,671,525	17.04	85	29.41	1.73
	3°–6°	4,244,957	19.70	108	37.37	1.90
	6°–10°	3,469,374	16.10	50	17.30	1.07
	10°–16°	3,411,916	15.84	10	3.46	0.22
	>16°	6,748,329	31.32	2	0.69	0.02
Geology	SD	402,951	1.87	4	1.38	0.74
	Q	6,082,732	28.23	222	76.82	2.72
	Kgr	625,202	2.90	0	0.00	0.00
	TR	1,636,212	7.59	1	0.35	0.05
	CPk	1,412,578	6.56	2	0.69	0.11
	Pr	2,700,326	12.53	7	2.42	0.19
	SDC	3,533,408	16.40	5	1.73	0.11
	O	2,941,752	13.65	3	1.04	0.08
	Water	412,851	1.92	0	0.00	0.00
	E	672,851	3.12	6	2.08	0.66
	PE	183,734	0.85	2	0.69	0.81
	Ju	638,309	2.96	1	0.35	0.12
	EO	105,767	0.49	1	0.35	0.70
	D	12,972	0.06	0	0.00	0.00
	Tmm	175,578	0.81	1	0.35	0.42

(continued)

Table 7.14 (continued)

Parameters	Classes	Num pixel of class	% of each classes	No. wells	% well	FR
Land use	AL	5,915,278	27.46	87	30.10	1.10
	ML	504,868	2.34	9	3.11	1.33
	FL	13,582,255	63.05	5	1.73	0.03
	WB	921,919	4.28	0	0.00	0.00
	UB	619,084	2.87	154	53.29	18.54
Rainfall	<1500 mm/year	8,990,313	41.73	10	3.46	0.08
	1500–1650 mm/year	4,692,909	21.78	28	9.69	0.44
	1650–1800 mm/year	5,612,822	26.05	128	44.29	1.70
	1800–1950 mm/year	2,182,294	10.13	86	29.76	2.94
	>1950 mm/year	67,445	0.31	3	1.04	3.32
Soil	Gravelly sandy loam	349,197	1.62	4	1.38	0.85
	Loamy sand	514,108	2.39	9	3.11	1.30
	Gravelly sandy clay loam	576,966	2.68	6	2.08	0.78
	Sandy loam	3,066,194	14.23	92	31.83	2.24
	Clay loam	2,207,956	10.25	29	10.03	0.98
	Silt loam	559,393	2.60	28	9.69	3.73
	Sandy clay loam	98,007	0.45	2	0.69	1.52
	Loam	372,193	1.73	48	16.61	9.61
	Silty clay loam	366,614	1.70	31	10.73	6.30
	Clayey sand	12,433,143	57.72	4	1.38	0.02
	Water	921,425	4.28	0	0.00	0.00
	Rock complex	37,538	0.17	0	0.00	0.00
	Coarse sandy loam	2031	0.01	0	0.00	0.00
	Silt	35,696	0.17	2	0.69	4.18

The FR value for each class of condition factors was calculated based on Eq. (7.16) (Table 7.14). To decipher the GWP map, the FR values must be integrated into maps of CFs. The GWPI map was established based on Eq. (7.17) using the "Map Algebra" with Rater Calculator in ArcGIS software. Based on the GWPI, the output map was categorized into five potential classes of groundwater yield >10 m^3/h using the "Equal Interval" method, including very high, high, moderate, low, and very low, and each class is covered by 2.00%, 3.98%, 20.23%, 31.45%, and 42.34% of Kanchanaburi, respectively (Fig. 7.12).

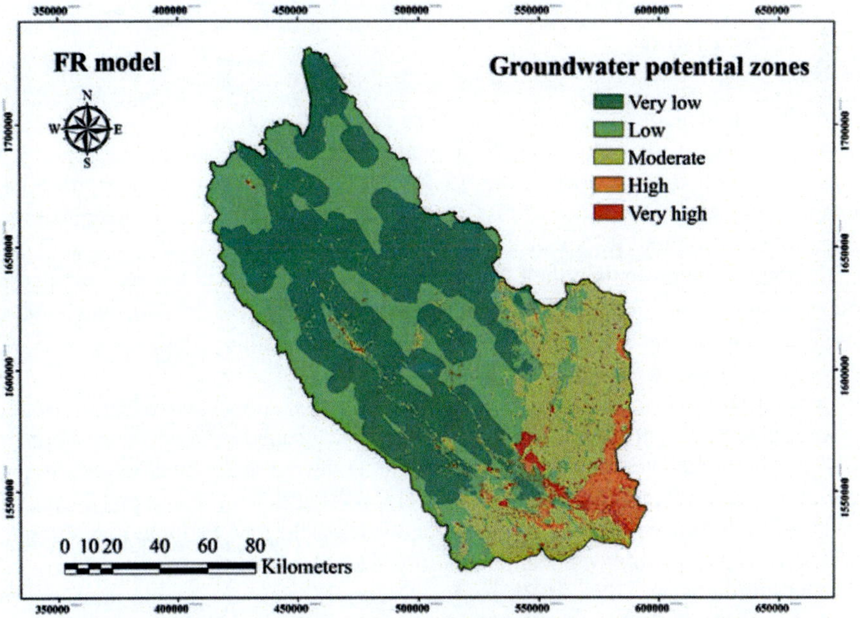

Fig. 7.12 Groundwater potential map in Kanchanaburi province using the FR model

7.3.7 Validation

Validation of maps helps to evaluate the model's performance which in turn can select an appropriate model for each specific study. In this study, 255 wells with groundwater yield >10 m^3/h were used to explore the performance of the GIS-based models. The locations of wells were converted into raster format with a pixel size of 30 × 30 m to validate the obtained groundwater potential maps. The coincidence rate between validation wells and locations with high groundwater potential on maps was applied to calculate the models' accuracy. The validation result shows that the MIF model had the highest accuracy rate at 78%, followed by the AHP model (71%), FR model (70%), and FL model (64%). The TOPSIS model achieved an accuracy rate of 51% (Table 7.15).

Table 7.15 Accuracy of the GIS-based model

Model	No. of coincidence locations	No. of validation locations	Accuracy (%)
AHP	181	255	71
FL	164	255	64
MIF	198	255	78
TOPSIS	130	255	51
FR	179	255	70

7.4 Conclusion

The development of the GWP models driven by innovation in GIS and RS has allowed scientists and authorities to efficiently reach groundwater resources, which was demonstrated in previous studies [5, 20, 23, 28, 52, 60]. The scope of this book chapter introduced the basics of GWP, the role of GIS and RS in data acquisition, and the basic GIS and RS application models for deciphering GWP.

In addition to presenting the GWP models based on GIS and RS, an example in Kanchanaburi, Thailand was also integrated to help readers get a better insight into the models. Based on the literature, we also provided the advantages and disadvantages of each technique in the summary section.

The difference involves the target output in previous studies, such as "groundwater storage potential", "groundwater quality potential", "groundwater occurrence potential" and "groundwater yield potential", indicating the research objective decides the concept of GWP. Regarding groundwater yield, GWP is defined as one that the possibility of the occurrence of a specific groundwater yield on CFs including altitude, distance to rivers, geology, lineament density, LULC, slope, and soil type.

The link between GIS and RS is strong. The RS plays a key role in access to hydrology information, especially data from the Earth's surface. The GIS is a useful tool that facilitates managing and analyzing databases, as well as developing GWP models. Generally, the integration of GIS and RS for deciphering GWP can be considered in three different approaches, including RS data utilized as input data for GIS, the base of RS analysis being GIS datasets, and GIS and RS data considered the input data for GWP models.

The technique of five common GIS models for deciphering GWP based on groundwater yield, including AHP, FL, MIF, FR, and TOPSIS, was also presented in this chapter. In which the AHP, FL, MIF, and TOPSIS models involve the multiple-criteria decision-making approach, the FR relates to the statistical approach. Although there are many differences in approach theory, the GIS models have certain features in common in deciphering GWP by the GWP index as well as in formatting the input data.

Based on the AHP scale, the AHP involves the decision-maker's judgments who can be experts in the groundwater field to assess GWP by giving scores for CFs. The CFs are compared one by one in the ability of the occurrence of a specific yield in the AHP model, which is called pair-wise comparison. The decision matrix and the weight of CFs are determined by the value of pair-wise comparison. The consistency index is used to validate the decision matrix's consistency. The AHP model's strength is its usability. It can be applied to many different data systems, especially complex data. The AHP has a simple route for deciphering GWP. Additionally, the consistency rate is used to test the decision-maker's judgment. However, the AHP model is based on the expert's experience and knowledge to assign the score for the CFs (input variables), which can lead to the result in deviations and confusion. Simultaneously, it cannot work well with a large number of CFs due to the increase of CFs leading to a rise in the number of pair-wise comparisons according to $n(n-1)/2$, which causes disappointment or irregularity in the AHP model.

The FL set fuzzy member values using ambiguous and imprecise information on a dataset. The fuzzy member value varies from 0 to 1. Each CF receives an equivalent value with a level of GWP. The FL is also based on the decision-maker's judgments. Care should be taken in assigning the fuzzy member value for CFs, with consideration of objects without information on groundwater yield, because this can lead to a variation in the resulting map. The FL model has the advantage of having a simple structure. However, the disadvantage is being unable to estimate the consistency rate of the decision-maker's judgment.

The MIF, an MCDM approach, uses the interrelationship of the different CFs to determine the conditioning factor score. The interrelationship between CFs is assessed by the decision makers' knowledge as well as experts in groundwater, which is divided into two types of interrelationships including minor and major effects. The minor effects are assigned a value of 0.5 and the major effects are assigned a value of 1. On the plus side, the MIF model allows calculating the weight of CFs in a simple process, indicating the priority of CFs on GWP, but on the minus side, the consistency rate is not applied.

The TOPSIS, an MCDM approach, is entirely based on the decision-maker's judgments or experts. In this technique, all CFs are assigned values at a time to find the best alternative, followed by the score distance assigned by the ideal answer and the negative ideal solution. The TOPSIS model has the benefits of being simple to use, ubiquitous, and consideration of distances to the ideal answer as well as to the negative ideal solution. However, its considerable drawbacks can lead to excessive subjectivity.

The FR, a bivariate probability approach, is used to decipher GWP based on the interrelationship of specific groundwater yield and CFs. Compared with the MCDM approaches, the FR can indicate the ability of the occurrence of a specific groundwater yield on CF and eliminate the subjectivity from the decision-maker's judgments. Regarding advantages, the FR has a simple process for estimating the probabilistic connection between the CFs and a specific groundwater yield. However, the technique-related elements can affect the observed outcomes.

Acknowledgements The researchers would like to thank the Interdisciplinary Program in Environmental Science, Graduate School, Chulalongkorn University and Hue University of Agriculture and Forestry, Hue University. We acknowledge partially financial support from the Thailand Science research and Innovation Fund Chulalongkorn University (CU_FRB65_dis(2)_090_23_20) and the Ratchadapisek Sompoch Endowment Fund (2022), Chulalongkorn University (765007-RES02), the ASEAN/NON-ASEAN Scholarship and the 90TH Anniversary of Chulalongkorn University Scholarship.

References

1. Abijith D, Saravanan S, Singh L, Jennifer JJ, Saranya T, Parthasarathy KSS (2020) GIS-based multi-criteria analysis for identification of potential groundwater recharge zones-a case study from Ponnaniyaru watershed, Tamil Nadu, India. HydroResearch 3:1–14. https://doi.org/10.1016/j.hydres.2020.02.002
2. Ahmed JB II, Mansor S (2018) Overview of the application of geospatial technology to groundwater potential mapping in Nigeria. Arab J Geosci 11(17):1–16. https://doi.org/10.1007/s12517-018-3852-4
3. Al-Ruzouq R, Shanableh A, Merabtene T, Siddique M, Khalil MA, Idris A, Almulla E (2019) Potential groundwater zone mapping based on geo-hydrological considerations and multi-criteria spatial analysis: North UAE. Catena 173:511–524. https://doi.org/10.1016/j.catena.2018.10.037
4. An P, Moon WM, Rencz A (1991) Application of fuzzy set theory for integration of geological, geophysical and remote sensing data. Can J Explor Geophys 27(1):1–11
5. Bagyaraj M, Mengistie AT, Gnanachandrasamy G, Gemechu B (2019) Data of remote sensing and GIS-to demarcate the potential sector of groundwater in Debre Berhan, Amhara region, Ethiopia. Data Brief 26. https://doi.org/10.1016/j.dib.2019.104542
6. Bhanja SN, Mukherjee A, Rodell M (2020) Groundwater storage change detection from in situ and GRACE-based estimates in major river basins across India. Hydrol Sci J 65(4):650–659. https://doi.org/10.1080/02626667.2020.1716238
7. Bonham-Carter GF (1994) Geographic information systems for geoscientists-modeling with GIS. Comput Methods Geosci 13:398
8. Chen W, Pradhan B, Li S, Shahabi H, Rizeei HM, Hou E, Wang S (2019) Novel hybrid integration approach of bagging-based fisher's linear discriminant function for groundwater potential analysis. Nat Resour Res 28(4):1239–1258. https://doi.org/10.1007/s11053-019-09465-w
9. Choubin B, Rahmati O, Soleimani F, Alilou H, Moradi E, Alamdari N (2019) Regional groundwater potential analysis using classification and regression trees. In: Spatial modeling in GIS and R for earth and environmental sciences. Elsevier, pp 485–498. https://doi.org/10.1016/B978-0-12-815226-3.00022-3
10. Díaz-Alcaide S, Martínez-Santos P (2019) Advances in groundwater potential mapping. Hydrogeol J 27(7):2307–2324. https://doi.org/10.1007/s10040-019-02001-3
11. Elmahdy SI, Mohamed MM (2015) Probabilistic frequency ratio model for groundwater potential mapping in Al Jaww plain, UAE. Arab J Geosci 8(4):2405–2416. https://doi.org/10.1007/s12517-014-1327-9
12. Etikala B, Golla V, Li P, Renati S (2019) Deciphering groundwater potential zones using MIF technique and GIS: a study from Tirupati area, Chittoor District, Andhra Pradesh, India. HydroResearch 1:1–7. https://doi.org/10.1016/j.hydres.2019.04.001
13. Golkarian A, Naghibi SA, Kalantar B, Pradhan B (2018) Groundwater potential mapping using C5. 0, random forest, and multivariate adaptive regression spline models in GIS. Environ Monitor Assess 190(3):1–16. https://doi.org/10.1007/s10661-018-6507-8
14. Gumma MK, Pavelic P (2013) Mapping of groundwater potential zones across Ghana using remote sensing, geographic information systems, and spatial modeling. Environ Monit Assess 185(4):3561–3579. https://doi.org/10.1007/s10661-012-2810-y
15. Gupta M, Srivastava PK (2010) Integrating GIS and remote sensing for identification of groundwater potential zones in the hilly terrain of Pavagarh, Gujarat, India. Water Int 35(2):233–245. https://doi.org/10.1080/02508061003664419
16. Guru B, Seshan K, Bera S (2017) Frequency ratio model for groundwater potential mapping and its sustainable management in cold desert, India. J King Saud Univ-Sci 29(3):333–347. https://doi.org/10.1016/j.jksus.2016.08.003
17. Harrison L, Funk C, Peterson P (2019) Identifying changing precipitation extremes in Sub-Saharan Africa with gauge and satellite products. Environ Res Lett 14(8):085007. https://doi.org/10.1088/1748-9326/ab2cae

18. Hussein AA, Govindu V, Nigusse AGM (2017) Evaluation of groundwater potential using geospatial techniques. Appl Water Sci 7(5):2447–2461. https://doi.org/10.1007/s13201-016-0433-0
19. Hwang CL, Yoon K (1981) Methods for multiple attribute decision making. In: Multiple attribute decision making. Springer, Berlin, Heidelberg, pp 58–191. https://doi.org/10.1007/978-3-642-48318-9
20. Kaewdum N, Chotpantarat S (2021) Mapping potential zones for groundwater recharge using a GIS technique in the lower Khwae Hanuman Sub-Basin Area, Prachin Buri Province, Thailand. Front Earth Sci 814. https://doi.org/10.3389/feart.2021.717313
21. Kaliraj S, Chandrasekar N, Magesh NS (2014) Identification of potential groundwater recharge zones in Vaigai upper basin, Tamil Nadu, using GIS-based analytical hierarchical process (AHP) technique. Arab J Geosci 7(4):1385–1401. https://doi.org/10.1007/s12517-013-0849-x
22. Khoshtinat S, Aminnejad B, Hassanzadeh Y, Ahmadi H (2019) Groundwater potential assessment of the Sero plain using bivariate models of the frequency ratio, Shannon entropy and evidential belief function. J Earth Syst Sci 128(6):1–16. https://doi.org/10.1007/s12040-019-1155-0
23. Konkul J, Rojborwornwittaya W, Chotpantarat S (2014) Hydrogeologic characteristics and groundwater potentiality mapping using potential surface analysis in the Huay Sai area, Phetchaburi province, Thailand. Geosci J 18(1):89–103. https://doi.org/10.1007/s12303-013-0047-6
24. Luo D, Wen X, Zhang H, Xu J, Zhang R (2020) An improved FAHP based methodology for groundwater potential zones in Longchuan River basin, Yunnan Province, China. Earth Sci Inform 13(3):847–857. https://doi.org/10.1007/s12145-020-00469-2
25. Machiwal D, Jha MK, Mal BC (2011) Assessment of groundwater potential in a semi-arid region of India using remote sensing, GIS and MCDM techniques. Water Resour Manag 25(5):1359–1386. https://doi.org/10.1007/s11269-010-9749-y
26. Magesh NS, Chandrasekar N, Soundranayagam JP (2012) Delineation of groundwater potential zones in Theni district, Tamil Nadu, using remote sensing, GIS and MIF techniques. Geosci Front 3(2):189–196. https://doi.org/10.1016/j.gsf.2011.10.007
27. Mahmoud SH, Alazba AA (2016) Integrated remote sensing and GIS-based approach for deciphering groundwater potential zones in the central region of Saudi Arabia. Environ Earth Sci 75(4):1–28. https://doi.org/10.1007/s12665-015-5156-2
28. Maity DK, Mandal S (2019) Identification of groundwater potential zones of the Kumari river basin, India: an RS & GIS based semi-quantitative approach. Environ Dev Sustain 21(2):1013–1034. https://doi.org/10.1007/s10668-017-0072-0
29. Mallick J, Khan RA, Ahmed M, Alqadhi SD, Alsubih M, Falqi I, Hasan MA (2019) Modeling groundwater potential zone in a semi-arid region of Aseer using fuzzy-AHP and geoinformation techniques. Water 11(12):2656. https://doi.org/10.3390/w11122656
30. Manap MA, Nampak H, Pradhan B, Lee S, Sulaiman WNA, Ramli MF (2014) Application of probabilistic-based frequency ratio model in groundwater potential mapping using remote sensing data and GIS. Arab J Geosci 7(2):711–724. https://doi.org/10.1007/s12517-012-0795-z
31. Martínez-Santos P, Renard P (2020) Mapping groundwater potential through an ensemble of big data methods. Groundwater 58(4):583–597. https://doi.org/10.1111/gwat.12939
32. Massoud EC, Purdy AJ, Miro ME, Famiglietti JS (2018) Projecting groundwater storage changes in California's Central Valley. Sci Rep 8(1):1–9. https://doi.org/10.1038/s41598-018-31210-1
33. Muavhi N, Thamaga KH, Mutoti MI (2021) Mapping groundwater potential zones using relative frequency ratio, analytic hierarchy process and their hybrid models: case of Nzhelele-Makhado area in South Africa. Geocarto Int 1–20. https://doi.org/10.1080/10106049.2021.1936212
34. Mukherjee P, Singh CK, Mukherjee S (2012) Delineation of groundwater potential zones in arid region of India—A remote sensing and GIS approach. Water Resour Manag 26(9):2643–2672. https://doi.org/10.1007/s11269-012-0038-9
35. Naghibi SA, Ahmadi K, Daneshi A (2017) Application of support vector machine, random forest, and genetic algorithm optimized random forest models in groundwater potential mapping. Water Resour Manag 31(9):2761–2775. https://doi.org/10.1007/s11269-017-1660-3

36. Naghibi SA, Pourghasemi HR, Dixon B (2016) GIS-based groundwater potential mapping using boosted regression tree, classification and regression tree, and random forest machine learning models in Iran. Environ Monit Assess 188(1):1–27. https://doi.org/10.1007/s10661-015-5049-6
37. Nanda S, Annadurai R, Barik KK (2017) Geospatial decipherment of groundwater potential of Kattankolathur block of Tamil Nadu using MCDM techniques. Remote Sens Appl: Soc Environ 8:240–250. https://doi.org/10.1016/j.rsase.2017.10.002
38. Nguyen PT, Ha DH, Avand M, Jaafari A, Nguyen HD, Al-Ansari N, Phong TV, Sharma R, Kumar R, Le HV, Ho LS, Prakash I, Pham BT (2020) Soft computing ensemble models based on logistic regression for groundwater potential mapping. Appl Sci 10(7):2469. https://doi.org/10.3390/app10072469
39. Nguyen PT, Ha DH, Jaafari A, Nguyen HD, Van Phong T, Al-Ansari N, Prakash I, Le HV, Pham BT (2020) Groundwater potential mapping combining artificial neural network and real AdaBoost ensemble technique: the DakNong province case-study, Vietnam. Int J Environ Res Publ Health 17(7):2473. https://doi.org/10.3390/ijerph17072473
40. Pinto D, Shrestha S, Babel MS, Ninsawat S (2017) Delineation of groundwater potential zones in the Comoro watershed, Timor Leste using GIS, remote sensing and analytic hierarchy process (AHP) technique. Appl Water Sci 7(1):503–519. https://doi.org/10.1007/s13201-015-0270-6
41. Rahmati O, Pourghasemi HR, Melesse AM (2016) Application of GIS-based data driven random forest and maximum entropy models for groundwater potential mapping: a case study at Mehran Region, Iran. Catena 137:360–372. https://doi.org/10.1016/j.catena.2015.10.010
42. Rajasekhar M, Raju GS, Sreenivasulu Y, Raju RS (2019) Delineation of groundwater potential zones in semi-arid region of Jilledubanderu river basin, Anantapur District, Andhra Pradesh, India using fuzzy logic, AHP and integrated fuzzy-AHP approaches. HydroResearch 2:97–108. https://doi.org/10.1016/j.hydres.2019.11.006
43. Razandi Y, Pourghasemi HR, Neisani NS, Rahmati O (2015) Application of analytical hierarchy process, frequency ratio, and certainty factor models for groundwater potential mapping using GIS. Earth Sci Inf 8(4):867–883. https://doi.org/10.1007/s12145-015-0220-8
44. Saaty TL (1977) A scaling method for priorities in hierarchical structures. J Math Psychol 15(3):234–281. https://doi.org/10.1016/0022-2496(77)90033-5
45. Sachdeva S, Kumar B (2020) Groundwater potential mapping using machine learning models for Northeastern Karbi Anglong district, Assam, India. In: 2020 2nd international conference on advances in computing, communication control and networking (ICACCCN). IEEE, pp 116–121. https://doi.org/10.1109/ICACCCN51052.2020.9362869
46. Shaban A, Khawlie M, Bou Kheir R, Abdallah C (2001) Assessment of road instability along a typical mountainous road using GIS and aerial photos, Lebanon–eastern Mediterranean. Bull Eng Geol Env 60(2):93–101. https://doi.org/10.1007/s100640000092
47. Shekhar S, Pandey AC (2015) Delineation of groundwater potential zone in hard rock terrain of India using remote sensing, geographical information system (GIS) and analytic hierarchy process (AHP) techniques. Geocarto Int 30(4):402–421. https://doi.org/10.1080/10106049.2014.894584
48. Singh P, Thakur JK, Kumar S (2013) Delineating groundwater potential zones in a hard-rock terrain using geospatial tool. Hydrol Sci J 58(1):213–223. https://doi.org/10.1080/02626667.2012.745644
49. Srinivasa Rao Y, Jugran DK (2003) Delineation of groundwater potential zones and zones of groundwater quality suitable for domestic purposes using remote sensing and GIS. Hydrol Sci J 48(5):821–833. https://doi.org/10.1623/hysj.48.5.821.51452
50. Sumdang N, Chotpantarat S, Cho KH, Thanh NN (2023) The risk assessment of arsenic contamination in the urbanized coastal aquifer of Rayong groundwater basin, Thailand using the machine learning approach. Ecotoxicol Environ Saf 253:114665. https://doi.org/10.1016/j.ecoenv.2023.114665
51. Taheri K, Gutiérrez F, Mohseni H, Raeisi E, Taheri M (2015) Sinkhole susceptibility mapping using the analytical hierarchy process (AHP) and magnitude–frequency relationships: a case study in Hamadan province, Iran. Geomorphology 234:64–79. https://doi.org/10.1016/j.geomorph.2015.01.005

52. Taheri K, Missimer TM, Taheri M, Moayedi H, Mohseni Pour F (2020) Critical zone assessments of an alluvial aquifer system using the multi-influencing factor (MIF) and analytical hierarchy process (AHP) models in Western Iran. Nat Resour Res 29(2):1163–1191. https://doi.org/10.1007/s11053-019-09516-2
53. Tanachaichoksirikun P, Seeboonruang U, Saraphirom P (2018) Impact of climate change on the groundwater sustainability in the lower Chao Phraya basin, Thailand. In: MATEC web of conferences, vol 192. EDP Sciences, p 02006. https://doi.org/10.1051/matecconf/201819202006
54. Thanh NN, Chotpantarat S, Ha N-T, Trung NH (2023) Determination of conditioning factors for mapping nickel contamination susceptibility in groundwater in Kanchanaburi Province, Thailand, using random forest and maximum entropy. Environ Geochem Health. https://doi.org/10.1007/s10653-023-01512-z
55. Thanh NN, Chotpantarat S, Trung NH, Ngu NH (2022) Mapping groundwater potential zones in Kanchanaburi Province, Thailand by integrating of analytic hierarchy process, frequency ratio, and random forest. Ecol Ind 145:109591. https://doi.org/10.1016/j.ecolind.2022.109591
56. Thanh NN, Thunyawatcharakul P, Ngu NH, Chotpantarat S (2022) Global review of groundwater potential models in the last decade: parameters, model techniques, and validation. J Hydrol 128501. https://doi.org/10.1016/j.jhydrol.2022.128501
57. Thapa R, Gupta S, Guin S, Kaur H (2017) Assessment of groundwater potential zones using multi-influencing factor (MIF) and GIS: a case study from Birbhum district, West Bengal. Appl Water Sci 7(7):4117–4131. https://doi.org/10.1007/s13201-017-0571-z
58. Thomas R, Duraisamy V (2018) Hydrogeological delineation of groundwater vulnerability to droughts in semi-arid areas of western Ahmednagar district. Egypt J Remote Sens Space Sci 21(2):121–137. https://doi.org/10.1016/j.ejrs.2016.11.008
59. Wang Y, Qiu M, Shi L, Xu D, Liu T, Qu X (2019) A GIS-based model of potential groundwater yield zonation for a sandstone aquifer based on the EWM and TOPSIS methods. In: IMWA mine water: technological and ecological challenges, pp 387–393
60. Zabihi M, Pourghasemi HR, Pourtaghi ZS, Behzadfar M (2016) GIS-based multivariate adaptive regression spline and random forest models for groundwater potential mapping in Iran. Environ Earth Sci 75(8):1–19. https://doi.org/10.1007/s12665-016-5424-9
61. Zhang C, Duan Q, Yeh PJF, Pan Y, Gong H, Moradkhani H, Gong W, Lei X, Liao W, Xu L, Huang Z, Zeng L, Guo X (2021) Sub-regional groundwater storage recovery in North China Plain after the South-to-North water diversion project. J Hydrol 597:126156. https://doi.org/10.1016/j.jhydrol.2021.126156

Chapter 8
Remote Sensing and GIS Based Monitoring and Management of Coastal Aquifers and Ecosystem

Somenath Ganguly and Uday Bhan

Abstract Globe has a total of 1.16 million km long continental coastline. About 40% of the global population live in the coastal cities. Therefore, coastal environmental monitoring and management is always an important task to accomplish for sustainable use of its resources like freshwater and protecting coastal ecosystem. For tectonically active or passive continental margin sedimentary basins are geologically very complicated in terms of its sedimentary environment and facies since it is an integration of different sub-environments. Considering the above-mentioned fact coastal ecosystem is also very versatile. Hydrogeologically freshwater aquifers along the coast are very vulnerable in terms of saline water intrusion. With increasing human population coastal cities are also getting populated thus freshwater demand is increasing and saline water intrusion to freshwater aquifer has become a serious concern now for coastal cities. Remote Sensing and GIS (Geographical Information System) now has become an inevitable tool for any regional scale survey and special data storage & management, therefore making high precision digital maps. Remote Sensing satellite data has been used throughout to make land-cover maps, which includes landscape pattern and anthropogenic development. In the early days of Remote Sensing, Landsat and SPOT data had limitations to capture details of coastal habitats, ecosystem, and geology since it is highly dynamic in space and time. Considering recent developments in Remote Sensing such as Light Detection and Ranging (LiDAR), Interferometric Synthetic Aperture Radar (InSAR) and Hyperspectral Remote Sensing has made significant contribution to capture and model dynamic coastal environment particularly landscape & seascape, land-cover/land-use pattern etc. Digital Elevation Model (DEM) has been successfully used to understand coastline migration, global sea level change and sometime land subsidence monitoring due to groundwater mining which is a common problem for some overpopulated coastal cities like Kolkata, India. Systematic GPS (Global Positioning System)

S. Ganguly (✉) · U. Bhan
Department of Petroleum Engineering & Earth Sciences, University of Petroleum and Energy Studies, Bidholi, Via, Prem Nagar, Dehradun, Uttarakhand 248007, India
e-mail: sganguly@ddn.upes.ac.in

U. Bhan
e-mail: ubhan@ddn.upes.ac.in

E. Balaji et al. (eds.), *Emerging Technologies for Water Supply, Conservation and Management*, Springer Water, https://doi.org/10.1007/978-3-031-35279-9_8

guided in situ measurement coupled with Remote Sensing data (airborne or spaceborne) yield high precision digital maps capturing coastal dynamics particularly vegetation mapping, drainage system mapping/analysis, flood prediction etc. Coming to groundwater management and modeling Remote Sensing and GIS is not a useful tool for that. For simulating variable density groundwater flow for coastal aquifers MODFLOW based tools like Sea Water Intrusion Package for MODFLOW-2000, SEAWAT, MOCDENS3D and MODHMS are there. However, none of the abovementioned codes can be coupled with any commercially available GIS platform. Recently advancement in ArcGIS (©ESRI) particularly ArcGIS online, MODFLOW models can be uploaded to ArcGIS online as MODFLOW layers. In this chapter an attempt has been made to give comprehensive coverage on different aspects of coastal monitoring in terms of land use pattern, drainage pattern analysis, flood prediction, ecology etc. with the help of Remote Sensing and GIS. Considering coastal groundwater management and vulnerability assessment, available variable density simulation codes and their working principle have been illustrated in detail. Lastly the chapter has been enriched by discussing two published case studies where Remote Sensing and GIS along with variable density groundwater flow model simulation have been taken into consideration for coastal groundwater management.

Keywords Coastal aquifer · Saltwater intrusion · Groundwater modeling · Coastal ecology · Remote sensing · GIS

8.1 Introduction

From aspiration to conquer space, Remote Sensing technology has made such an advancement that satellite image of Mars is not a dream anymore. Even "Mars Curiosity Rover" send by NASA (launched on 26th November 2011 and landed on Mars on 5th August 2012, 10:32 p.m. PDT) has captured views of Mars landscape and now analyzing composition of Mars atmosphere to understand if Mars ever had life sustainable atmosphere or not. However, the journey of space borne Remote Sensing started long back in 1957 when Soviet Union first launched "Sputnik-I" which is considered first artificial satellite of the world. After that many satellites were launched by several countries for different purposes [51]. Now private space agencies like SPACEX, Sierra Nevada Corporation, Boeing etc. are there to offer equipment and services toward space flight.

Narrowing down the discussion domain to application of Remote Sensing in hydrology and hydrogeology with special emphasis to coastal area which is also the focus of this chapter, there are several applications of Remote Sensing and GIS which will be discussed eventually. To begin with application of Remote Sensing in hydrology, we can have a direct measurement of watershed geometry, estimating runoff to a storm event and estimation of soil moisture from Remote Sensing data. Considering advanced water-energy balance hydrological modeling Remote Sensing contribute meteorological data, soil type & land cover data, solar radiation data, surface air humidity data etc. [1]. In the field of hydrogeology and groundwater management Remote Sensing contributes in terms of making the base map for hydrogeological modeling and qualitative interpretation of recharge and water table condition by vegetation survey with the help of NDVI (Normalized Difference Vegetation Index) [11]. For hard rock terrain, lineament mapping is done with the help of Remote Sensing to locate aquifer recharge zones [16]. Now coming to coastal ecology monitoring and management, satellite data are readily available and are used to observe and measure biophysical property/biodiversity of coastal habitat and tracking its changes through time [62]. Various GIS platforms (commercial and open source) are there now to handle and manage spatial data of any volume.

In this chapter the focus is on application of Remote Sensing and GIS in coastal environment monitoring and management with special emphasis on freshwater coastal aquifers and coastal ecosystem. For better understanding of reader, in this chapter three different subdomains have been identified for discussion that is geology of coastal system, hydrology & hydrogeology of coastal aquifers & freshwater bodies and lastly versatile coastal ecosystem. These three subdomains are somehow linked to each other and stress on any of the domains will be reflected on others through time. Remote Sensing and GIS have proven a very trusted tool to capture and monitor these spatiotemporal changes, thus helping in decision making for the policymakers for coastal environmental management.

8.2 Dynamic Coastline

The junction between ocean and continent is called coastline [14]. From simple observation the coastline is not stationary. If we just consider high and low tide, coastline migrates daily. Going back to the Geological Time Scale globe has witnessed several episodes of marine transgression and regression in continental scale. In terms of plate tectonics coastline can be classified into two categories i.e., active continental margin (when coastline is on a tectonically active plate boundary) and passive continental margin (when coastline is not on a tectonically active plate boundary). Geomorphologically passive continental margin can be farther divided into continental shelf (average slope 0.1°), continental slope (average slope 4°), continental rise (average slope 0.05°–0.6°) and abyssal plane (average slope 0.05°). According to depth, ocean floor is divided into three zones neritic (depth 200 m and above), bathyal (200–2000 m) and abyssal (below 2000 m). Coastline can also be classified as sandy coastline and rocky coastline.

8.2.1 Coastal Geology

Coastal geology is extremely dynamic. In terms of sedimentary environment there are different sub environments that can exist along a coastline like clastic shore and barrier island, associated estuaries and lagoons, deltaic environment, carbonate platforms, deep sea basins etc. All these sub environments may be laterally adjacent or may not be and it is governed by regional geology, sediment supply and ocean energy condition. In and around the passive margin continental shelf sediments may be of clastic or carbonate type.

Detail understanding of ocean dynamics is very complicated. Ocean dynamics is governed by the following three factors i.e., Earth's rotation (Coriolis force), attraction of planetary bodies (tides), and internal dynamics of Earth (mid oceanic hydrothermal plums, tsunamis etc.). Some other factors that govern the ocean dynamics are heat transfer from solar radiation, wind direction & velocity, evaporation from open ocean surface etc. [60].

Deep ocean sediments capture undisturbed records of different geological events that happened in geological past like eustatic sea level change, change in spreading rate of Mid Oceanic Ridge etc. [28]. Deep ocean sediments also keep continuous record of change in Earth's stress, thermal, magnetic and gravity field. Coastal marine sediments record evidence of relative sea level change in terms of marine transgression and regression which is a research matter of sequence stratigraphy. A detailed discussion on the application of Remote Sensing and GIS in coastal geology, hydrogeology and ecology modeling will be taken under consideration in the upcoming sections.

8.2.2 *Coastal Hydrogeology*

In terms of groundwater chemistry of coastal aquifers, it can be divided into three main categories i.e., freshwater, brackish water, and saline water. The contact between these zones is gradational and can migrate depending on subsurface groundwater conditions. Salty sea water density (1.025 gm/cm^3 at 20 °C) is little higher than freshwater density (1.000 gm/cm^3 at 20 °C). Considering this density difference and assuming salt water and freshwater are immiscible Drabbe and Ghyben [22], and Herzberg [32] derived the following relation by force balance as described by Eq. (8.1).

$$z = \frac{\rho_f}{\rho_s - \rho_f} h \tag{8.1}$$

Equation 8.1 The Ghyben-Herzberg equation.

When h = Length of freshwater head above sea level, z = Length of freshwater head below sea level and ρ_s and ρ_f are density of saline and freshwater respectively.

Considering above mentioned density of sea and freshwater Eq. (8.1) boils down to

$$z = \mathbf{40}\ \boldsymbol{h} \tag{8.2}$$

Equation 8.2 Simplified version of Ghyben-Herzberg equation.

Equation 8.2 demonstrates that considering an unconfined aquifer for every meter of freshwater head above sea level there exists a 40 m of freshwater head below sea level. Alternately unit decline in freshwater head above sea level the salt water-freshwater interface will rise 40 times.

As mentioned earlier, due to increasing population, freshwater resources of Earth are under stress. Coastal freshwater aquifers are also not out of this stress. In fact, for coastal aquifers the situation is more complicated due to saline water encroachment to freshwater aquifers due to over pumping. Not only that Ghyben-Herzberg equation (Eq. 8.1) assumes a sharp boundary between saline and freshwater front which is also oversimplification of the reality. There exists a brackish water zone in between fresh and saline waterfront. The extent of this brackish water zone may extend from few meters to few kilometers in coastal aquifers. Extent of brackish water zone is influenced by tidal fluctuations, geological structure of the coastline, sea level rise, pumping in freshwater zone and temporal variation of freshwater aquifer recharge [25, 49].

Monitoring the brackish water zone for a coastal aquifer system is one of the most important things in coastal groundwater management. Saline water intrusion in the coastal aquifer can be monitored by geophysical methods, isotope methods, geochemical methods etc. Detail description of these methods are way beyond the scope of this chapter. However, Remote Sensing data can act as a very good base map and GIS can act as a special data base repository for a coastal aquifer system which can be farther used to simulate density dependent groundwater modeling for coastal aquifers. A detailed discussion on this will be taken into consideration in the upcoming sections.

8.2.3 Coastal Ecology

Coastal ecology is one of the most versatile ecosystems of the globe. A coastal ecosystem consists of the following three sub ecosystems in terms of lateral continuity. They are of terrestrial, marine, and freshwater ecosystem. In terms of continental scale, coastal regions include a wide array of near-shore terrestrial, intertidal, benthic, and pelagic marine ecosystems [27]. The extent of marine or coastal environment may extend even 100 km inland. Coastal ecosystem is kind of an integration of sub-environments which may include wetland and estuary, marshes and mangrove flora, sandy beaches with dunes, seagrass beds and occasionally coral or oyster reefs. These sub-environments are somehow mutually connected resulting in nutrient flux and material exchange. This chain extends up to nutrient supply to deep marine fauna [20]. About 40% of the global population live along the coastline [54]. Coastal ecosystems are very important in terms of their contribution to cultural, economic, and aesthetic services and benefits to the nation. Coastal and marine ecosystems are one of the most heavily exploited ecosystems of the globe. Approximately 90% of global fisheries products directly come from marine fishing.

Due to overfishing the most upcoming problem for coastal ecosystem is deteriorating water quality like increasing occurrence of harmful algal blooms, offshore pollution and oxygen depletion which is also known as hypoxia. However, increasing human population and our interference with coastal ecosystem have made serious threat on coastal ecology. For example, due to coastal development, population growth, pollution, and other anthropogenic activities, we have already lost or degraded 50% of salt marshes, 35% of mangroves, 30% of coral reefs, and 29% of seagrasses worldwide over several decades [9, 35, 61]. Therefore, a serious thought needs to be developed and implemented globally in terms of coastal ecological monitoring and management [17, 58]. Coastal ecosystem management includes continuous monitoring and modeling of coastal phenomenon like shoreline migration, sediment input budget, vulnerability of sub-ecosystems etc. in terms of best available information. The best way to do this is making time transgressive maps of the coastal area. Remote Sensing and GIS has proven a very useful, handy, and technologically sound tool to this in terms of local or continental scale shoreline mapping [66]. A detailed discussion on this will be taken into consideration in the upcoming sections.

8.3 Recent Advancement in the Field of Remote Sensing and GIS to Monitor Coastal Dynamics

Starting from aerial photography, the science of Remote Sensing has made a significant long journey and advancement that satellite images of Mars is no longer dream to us. Transition from aerial photography to digital images, invention of digital camera made a groundbreaking contribution in the advancement of Remote Sensing. Digital images are way more accurate than aerial photograph in terms of photogrammetric accuracy/coverage and capturing multispectral data at any required ground resolution [2]. 1960 onwards space borne Remote Sensing platforms started dominating to capture and store earth's reflected radiation data in comparison to aircraft mounted sensors [15]. Gradually advancement in electronic sensors and satellite platforms, spectral range of remotely sensed data are not only restricted to visible part of the spectrum but extend up to near-infrared (NIR), thermal infrared (TIR) and microwave region of the electromagnetic spectrum [47]. Artificial satellites prove themselves very useful for monitoring changes on earth in terms of global coverage if they

Table 8.1 Classification of remote sensors in terms of application, wavelength and active or passive mode

Application	Mode	Wavelength
Imagers (Mappers)	Active	*Visible*
Photographic (Film)	*LiDAR*	*Near infrared (Reflected)*
Multispectral (Array)	*Radar*	*Thermal infrared (Emitted)*
Radar (SAR & SLAR)	*Sonar**	*Microwave*
Side-scanning sonar*		*Sound waves**
Radiometers		*Seismic waves**
Spectrometers		
Profilers (Rangers)	Passive	
Lidar	*Visible*	
Radar (Altimeter)	*Infrared*	
*Echo sounder**	*Microwave*	

* Not electromagnetic waves
Adapted from Xiaojun [65]

are polar-orbiting and continuous coverage for geostationary orbit [46]. Table 8.1 summarizes classification of remote sensors in terms of application, wavelength and active or passive mode [65]. Nowadays satellites are launched mounted with highly sensitive sensors with very fine spectral resolution designed for capturing specific features of earth. Table 8.2 summarizes some of the space missions with sensors and spectral band width to capture coastal zone related phenomenon [65].

Now focusing down the discussion to recent developments in the field of Remote Sensing for coastal area monitoring, modeling, and management, airborne light detection and ranging (LiDAR) technology has proven very effective. In this active Remote Sensing technique, the round-trip timing is measured between the sensor and target with the help of a pulse of laser energy [64]. GPS guided LiDAR remote sensing has proven itself a very useful tool for coastal zone monitoring and management in terms of its ground resolution and capturing the details in comparison to any earlier space mission for the same purpose. This technique is now widely used for coastal forest management, mapping barrier island vegetation communities, capturing detail & accurate topography (LiDAR digital elevation model) of the coastline, detecting coastal vegetation canopy height etc. [18].

In the field of active remote sensing RADAR (Radio Detection and Ranging) has a long history. Starting from SEASAT mission commissioned by NASA in 1978 there has been a significant progress to launch earth orbiting Synthetic Aperture Radar (SAR) like RADARSAT-2 in 2006 launched by Canada. In this domain the recent advancement is Interferometric Synthetic Aperture Radar (InSAR). Advantage of InSAR over SAR is that it produces an interferogram from two SAR images of the same area captured from similar vantage points at two different time. The interferogram describes range changes between ground and radar. This can be farther

Table 8.2 Space missions with sensors to capture coastal zone related phenomenon

Satellite	Sensor	Spectral band (μm)	Resolution (m)	Cycles (days)	Swath width (km)
Landsat 1, 2, 3	MSS	4 0.5–0.6 5 0.6–0.7 6 0.7–0.8 7 0.8–1.1	80	18	185
Landsat 4, 5	TM	1 0.45–0.52 2 0.52–0.60 3 0.63–0.69 4 0.76–0.90 5 1.55–1.75 6 2.08–2.35 7 10.40–12.5	30 for band 1–5 & 7 and 120 for band 6	16	185
Landsat 7	ETM+	1 0.450–0.514 2 0.525–0.605 3 0.630–0.690 4 0.750–0.900 5 1.55–1.75 6 10.40–12.50 7 2.08–2.35 8 (Pan) 0.52–0.90	30 for band 1–7 60 for band 6 15 for band 8	16	185
Landsat 8	Operational Land Imager (OLI) and Thermal Infrared Sensor (TIRS)	1 0.43–0.45 2 0.450–0.51 3 0.53–0.59 4 0.64–0.67 5 0.85–0.88 6 1.57–1.65 7 2.11–2.29 8 (PAN) 0.50–0.68 9 1.36–1.38 10 10.6–11.19 11 11.5–12.51	30 for OLI, 15 for PAN and 100 for TIRS	16	185
SPOT	HRV	1 0.50–0.59 2 0.61–0.68 3 0.79–0.89 4 0.51–0.73	20 for band 1–3 10 for band 4	26 (daily if camera tilted)	60
IKONOS		1 0.45–0.52 2 0.52–0.60 3 0.63–0.69 4 0.76–0.90 5 (Pan) 0.45–0.90	4 for band 1–4 1 for band 5	<3 days	11

(continued)

Table 8.2 (continued)

Satellite	Sensor	Spectral band (μm)	Resolution (m)	Cycles (days)	Swath width (km)
NOAA	AVHRR	1 0.58–0.68 2 0.725–1.1 3 3.55–3.93 4 10.5–11.3 5 11.5–12.5	1,000	2/day	3,000
Orbview 2	SeaWiFS	1 402–422 2 433–453 3 480–500 4 500–520 5 545–565 6 660–680 7 745–785 8 845–885	1,000	Daily	2,800

Adapted from Xiaojun [65]

processed with DEM to produce high resolution ground deformation maps of coastal areas [44]. InSAR data is helpful to detect water level monitoring at different tidal levels. InSAR images can be effectively used to study ground deformation associated with volcanism, earthquakes, land subsidence and landslides [42].

Another very important tool to monitor and model time transgressive changes in coastal areas is Digital Elevation Model (DEM). DEM can be subdivided into two broad categories. One is Digital Surface Model (DSM) and the other is Digital Terrain Model (DTM). DSM gives elevation information of vegetation; land use and other structures present in the landscape. Whereas DTM gives elevation information of bare earth without any interference of man-made structure or vegetation. DEM are useful to provide information on slope and curvature aspects of the terrain. DEM data compiled with satellite reflectance data has wide application to model coastal hydrology. Land use information obtained from Remote Sensing coupled with soil map and DEM can be very handy to classify different Hydrological Similar Units (HSU) of a drainage basin [24]. Recent advancement in this domain has been able to create three-dimensional perspective views for better visualization of the terrain variation.

In the era of computer huge volume of special data handling is done by Geographic Information System (GIS). GIS is a platform which is capable of storage, retrieval, display, and manipulating large volumes of specially referenced data. GIS has certain four basic components that is special data input & editing, storage of special database, data analysis & special modeling and data visualization & presentation. While monitoring and modeling time transgressive changes in coastal geology, hydrology, and ecology it often requires compiling several special datasets to get the desired result for decision making. GIS is fully capable of doing so. Advancement in web-based GIS like Google Earth, Google Map (©Google) has become an integral part of any reconnaissance survey before going to field since they are free of coast and very user-friendly. Not only that, but users can also upload shape files to Google Earth to make maps of desired interest. Now a days it is always possible to locate yourself in the terrain with the help of Google Map and a GPS enabled smart phone. Coming to commercial GIS platforms like ArcGIS (©ESRI) handling huge volume of special data even in cloud platform is not a problem anymore. ArcGIS has its inbuilt tools and plugins to compute complicated mathematical models for special database analysis/management and making professional quality high precision digital maps for decision making. Open-source GIS is also becoming popular now a days like QGIS, GRASS-GIS. However, they have limitations in terms of computational capability compared to other commercial GIS platforms.

8.4 Remote Sensing and GIS Based Approach to Coastal Environmental Management

As discussed earlier, the coastal environment is highly exploited, and it is tremendously vulnerable. Before going to coastal environmental management, we first need to understand the proper economic and biological evaluation of it. Considering the environmental aspect coastal geomorphology like beach, dune, shore, bays & estuaries along with associated ecosystem like mangroves, coral & oyster reefs, seagrass beds help in storm protection, erosion control of coastline, oxygen cycling, carbon storage & sequestration, nutrient cycling etc. Not only that, but diverse coastal environment also supports the near shore marine biodiversity and ecological connectivity thus nursing and protecting the deep marine aquatic habitat [5].

Considering economic appraisal of a coastal area, it considers the following points/parameters [29].

a. Original measurement units
b. Methodology for estimation
c. Estimation year
d. Currency involved for estimation
e. Ecosystem of biome characteristics
f. Ecosystem service valued.

According to Milon et al. [50] global biosphere ecosystem services worth between USD $16–54 trillion per year having average of USD $33 trillion per year. In this amount, the stack of coastal marine ecosystem is approximately USD $22.4 trillion. At the same time total global gross national product was approximately USD $18 trillion. It means that the global ecosystem and coastal marine ecosystem alone is more valuable in comparison to the direct market goods and services produced by the entire global economy. Just focusing down from global economic scale to local scale most of the fishery product of global market comes from the seagrass bed of coastal ecosystem [8]. Not only that, but the nation can also generate significant amount of revenue from coastal tourism and recreation.

From the above discussion coastal ecosystem carries a significant amount of wealth. Due to that, from the beginning of civilization this wealth has been over exploited. Now the situation has come to give serious thought on conservation of coastal ecosystem and planning for sustainable use of this wealth. Otherwise, the global ecosystem may suffer a significant damage. Remote Sensing and GIS with recent advancement has proven itself a very useful tool for regional scale monitoring and surveying of coastal region which has been discussed in the earlier section. In the upcoming section a detailed discussion has been made on how Remote Sensing and GIS can be helpful for coastal environmental management with special emphasis to coastal geology, hydrogeology, and ecology.

8.4.1 Remote Sensing Based Coastal Geological Modeling

Remote Sensing has widely been used and applied for regional scale geological modeling and reconnaissance study. Remote Sensing has wide application in geology for structural evaluation of the terrain, lineament mapping, watershed delineation, mineral exploration etc. Not only that, application of Remote Sensing as passive indicator for groundwater exploration is also widely reported. Remote Sensing has huge applications in land degradation mapping, monitoring, and management of arid, semiarid, and dry sub-humid areas [23, 26]. As discussed earlier, the coastal geology is extremely complicated and vulnerable to any change therefore coastal geology is suffering huge impact damage due to over exploitation of coastal resources particularly freshwater resource. Saline water intrusion is a continuous problem to coastal aquifer which is kind of a continuous focus of this chapter also. In this section an effort has been made to capture how Remote Sensing and GIS is applied for modeling coastal geology.

Stress on coastal geology may happen due to natural and/or anthropogenic influence. Massive coastline erosion and coastal lowland/wetland erosion is one of the challenging problems of coastal geology. This phenomenon is widely reported from Louisiana's (USA) coastal systems [10]. Consequently, the state of Louisiana is suffering from frequent hurricane disasters along with significant adverse effect on coastal ecosystem. Remote Sensing has been successfully applied to monitor spatiotemporal variation of wetlands for better understanding of environmental changes and anthropogenic influence in the process. For wetland monitoring, synthetic aperture radar (SAR) has been found quite effective particularly for cloud-prone subtropical regions because of its all-weather day & night imaging capacity. However, RADARSAT-1 data can also be successfully applied for wetland mapping and monitoring. Another important aspect of coastal geology monitoring is estimation of water storage capacity by the wetlands. This parameter is one of the governing parameters of coastal hydrological modeling as well as broad assessment of flood hazard [4]. For this purpose, InSAR data has proven itself very useful for mapping time transgressive water level change in the coastal wetlands. Not only that LiDAR and InSAR data are helpful for three-dimensional habitat mapping and morphological changes in coastal geology like shoreline migration, dune migration etc. [53]. Recent advances in three-dimensional modeling such as Shuttle Radar Topography Mission (SRTM) data can be applied for measuring ground topography, thus estimating relative sea level change.

8.4.2 Remote Sensing Based Hydrogeological Modeling

As per common practice the term hydrogeology means groundwater. However, Remote Sensing has no direct application in groundwater exploration and management. But considering hydrologic cycle there are several applications of Remote Sensing which can be informative to understand regional hydrologic cycle thus giving valuable feedback on local groundwater and surface water management [55]. In this section an effort has been made to capture how Remote Sensing is useful for gathering regional scale hydrological data thus improving our understanding on regional freshwater resource and working towards its sustainable use.

One of the important parameters for hydrologic modeling is rainfall data. In the early days of Remote Sensing rainfall data was estimated by ground-based radar [36]. With the advancement of space born Remote Sensing visible/infrared band gives good estimate of precipitation data with the help of indirect relationship between precipitation and cloud brightness temperature. Space born radar is a very good tool to measure rainfall since radar reflectivity is directly related to rainfall rate. However, there are some limitations to estimate rainfall from radar data, still it is considered as reliable tool to measure regional scale rainfall [19].

Remote Sensing has huge application in land use/land cover and river catchment area mapping [33]. For any hydrologic modeling estimation of overland flow for any storm event is very important. Overland flow corresponding to any storm event is highly dependent on land cover/ land use pattern of the catchment area. Densely vegetated catchment area with temperate climate leads to saturation overland flow whereas poorly vegetated sloping catchment area leads to direct rapid runoff [63]. For better mapping of vegetated and poorly vegetated catchment areas there are different vegetation indices, like Difference Vegetation Index (DVI), Ratio Vegetation Index (RVI), Perpendicular Vegetation Index (PVI), Normalized Difference Vegetation Index (NDVI) etc. In addition, there are other indices like Normalized Difference Water Index (NDWI) which gives a good estimate of plant water content and Modified Normalized Difference Water Index (MNDWI) which can give a good estimate of open water features. Detail description of these indices are beyond the scope of this chapter. Different image processing techniques are using these indices for better image classification and detail land cover mapping [31]. Remote Sensing has widely applied to model evapotranspiration [3] and soil moisture [13] also. Considering regional scale groundwater management Remote Sensing has applications in lineament mapping, soil data, terrain and watershed delineation, river basin management etc.

Now if we focus down the discussion domain to coastal environment, Remote Sensing is widely used for mapping coastal inland water bodies and wetlands [37]. Considering coastal areas of permafrost region like northern part of Alaska, Remote Sensing has application to monitor snow & ice cover along the coastline [21]. As discussed in the earlier section, soil erosion at coastal area is a big problem and Remote Sensing has a huge application in soil erosion mapping. Digital radiance data from satellite like Landsat, SPOT, AVHRR have been widely used to interpret coastal soil erosion [59]. In the coastal estuaries and oceans major factors that affect water quality are suspended sediments, turbidity, dissolved organic matter, nutrients, chlorophylls (algae) and chemicals like oil spill from oil tankers at the time of on or off loading the crude oil [40, 41]. All the above-mentioned parameters are measurable in the visible part of the spectrum from any Remote Sensing platform. However, change in the ocean water temperature can only be measured by emitted thermal radiation.

8.4.3 Remote Sensing Based Ecological Modeling

Remote Sensing has a huge application in coastal ecology modeling. As mentioned earlier, coastal ecology accommodates near-shore terrestrial, intertidal, benthic, and pelagic marine ecosystems [14]. Main components of coastal ecosystem that are constantly monitored for ecological modeling through Remote Sensing platforms are spaciotemporal variation of coastal marshes, mangroves, sea grass systems and planktonic flora and fauna. Sensor and band width selection to monitor above mentioned components of coastal ecosystem are case and need specific.

Another important application of Remote Sensing is monitoring estuaries. Estuaries are enclosed body of water near coastline where freshwater river meets sea. There are different micro-components of an estuary system like bays, lagoons, harbors, inlets etc. Estuaries supply nutrients to marine ecosystems as well as buffer and filter out pollutants and sediments of terrestrial runoff thus acting as lifeline for coastal birds, fish, mammals, and other wildlife. Not only that estuaries are economically important because they are potential areas for fishing, tourism and development of harbors and ports.

Most of the coastal watershed model requires land use and land cover data as input parameter in the model. With the background information of changing land use and land cover along with other input data like slope & rainfall, one can predict type of surface runoff to rivers, bays, and estuaries [34]. Landsat Thematic Mapper (TM) is one of the most reliable sources for acquiring land cover data with 30 m ground resolution. For small scale coastal wetland mapping it requires high resolution satellite data which may be obtained from IKONOS, QuickBird platforms. Airborne digital cameras of color and color infrared imagery (preferably stereopair) can also be an alternative. Airborne data integrated with GPS location can act as georeferenced layers in GIS for wide range of ecological modeling applications [45]. To map actual beach profile LiDAR data can be helpful. However, one of the limitations of LiDAR data is deterioration of optical water quality occurs due to sediment suspension.

Therefore, for the LiDAR system, attenuation coefficient should be high enough to accommodate larger depth [12]. For mapping submerged aquatic vegetation like coral reef one requires very high resolution (1–4 m) digital data [6, 56]. Biomass and vegetation indices are widely used for monitoring spaciotemporal change in coastal vegetation. Canopy thickness and density data obtained from red band and near-infrared band (Landsat/TM images) coupled with NDVI gives good estimate on wetland plant biomass quantitatively (grams dry weight per square meter). Modified Soil Adjusted Vegetation Index (MSAVI) is also another parameter that is widely used to detect changes in wetland biomass quantitatively [38]. For coastal estuaries and open ocean Remote Sensing, there are dedicated satellites like NASA's Sea-viewing Wide Field-of-view Sensor (Sea-WiFS), MODIS, MERIS and AQUA which gives important information on bio-optical properties like chlorophyll concentration of ocean water. Ocean color changes captured by satellite images are useful to estimate global ocean productivity [43] along with ocean biotic province delineation. Change in ocean water temperature can be detected by thermal IR images from NOAA's AVHRR or Sea-WiFS [57] which has also been widely used to detect temperature dependent global ocean circulation phenomenon like "El Nino". Also, Remote Sensing has been widely used to detect growth of harmful algal blooms and spread of oil spills. Therefore, from the above discussion Remote Sensing has extensive application in coastal ecology monitoring and modeling. However, capturing everything is beyond the scope of this chapter. Table 8.3 summarizes recent advances in space borne oceanic Remote Sensing techniques.

Table 8.3 Recent advances in space borne ocean sensing platforms

Sensers/platforms	Environmental parameters
Infrared radiometer	Sea surface temperature and ocean current patterns monitoring
Synthetic aperture radar	Monitoring short surface waves like swell, internal waves etc.
Altimeter	Monitoring topography and roughness of sea surface, sea level, currents, wave height etc.
Scatterometer	Measuring amplitude of short surface waves, surface wind velocity and roughness
Microwave Radiometer	By microwave brightness temperature measuring salinity, surface temperature, water vapor, soil moisture etc.
Color scanner	From ocean color measuring chlorophyll concentration, suspended sediment, attenuation coefficient etc.

Adapted from Xiaojun [65]

8.5 Available Codes to Simulate Variable Density Groundwater Flow for Coastal Aquifers

Coastal aquifers have three distinct zones like freshwater zone, brackish water zone and saline water zone. Most important aspect of coastal groundwater management is monitoring the extent of brackish waterfront since saline water intrusion to coastal aquifers is a constant problem. Hydrogeologically saline water intrusion can be modeled by "Ghyben-Herzberg equation" as discussed in Sect. 8.2.2. Also, there are other analytical solutions to address/model the saline water intrusion issue [52].

Remote Sensing has no direct application in subsurface groundwater management. However, georeferenced satellite data can be used as base map for groundwater modeling. GIS can act as a very good special data repository. Since there is density difference between freshwater and saline sea water therefore identifying the source of saline water intrusion is a tough job. There may be different reasons of saline water encroachment to a freshwater well for coastal aquifers. Identifying over pumping as the main reason, there may be several mechanisms of saltwater intrusion to the well like horizontal saltwater intrusion, salt water up coning to freshwater well or simply brackish surface water intrusion to unconfined aquifer due to relative sea level rise. Once the reason/source of saline water intrusion is identified, a conceptual model is developed. Eventually the conceptual model is simulated with available variable density groundwater modeling code for groundwater management.

In most cases two-dimensional model is built to understand the reason and mechanism of saltwater intrusion to coastal aquifers. Another reason of density dependent groundwater flow modeling for coastal aquifer is to determine the optimal freshwater pumping rate so that the saline water encroachment issue can be avoided. To understand the interaction between freshwater and brackish waterfront single phase contaminant transport codes are useful. These codes identify advective transport and dispersion as the main mechanism of saltwater transport to the well. These mechanisms are simulated by single phase transport and particle tracking codes. For mapping and quantifying saltwater intrusion to freshwater aquifer due to dispersion of salt, density dependent coupled flow and transport models are used. These models include fluid density gradients associated with solute concentration gradients in the simulation environment.

For groundwater modeling the most popular and widely accepted code is MODFLOW [48]. There are certain four MODFLOW based codes which are used for variable density groundwater flow modeling particularly for coastal aquifers. They are Sea Water Intrusion (SWI) package for MODFLOW-2000, SEAWAT, MODHMS, and MOCDENS3D. Among them SEAWAT, MOCDENS3D and MODHMS simulate advective and dispersive solute transport. Whereas Sea Water Intrusion Package for MODFLOW-2000 assumes non-dispersive, continuity of flow approach for simulating multiple density isosurfaces.

SEAWAT is an open-source code developed by United States Geological Survey (USGS) and it is readily available at USGS website. SEAWAT was developed coupling MODFLOW-88 and MT3DMS into a single program to solve variable-density groundwater flow and solute-transport equations [30]. MOCDENS3D code was developed based on the USGS finite difference code MOC3D [39]. Here density difference of fresh and saline water has been considered to simulate transient flow of fresh, brackish, and saline groundwater to coastal aquifers. MODHMS (©HGL) is kind of MODFLOW's capability enhancement code to simulate three dimensional saturated-unsaturated subsurface groundwater flow. This extension is also capable of simulating two-dimensional overland and one-dimensional channel flow. MODHMS has different simulation capabilities but the most important for coastal ecology modeling are flood prediction/mitigation and watershed/wetland management. Now coming to Sea Water Intrusion Package for MODFLOW-2000, it is capable of modeling density dependent three-dimensional salinity distribution for coastal aquifers [7]. The advantage of this model is it can simulate each aquifer with a single layer of cells. Therefore, it does not require vertical discretization of an aquifer. With SWI package saline water intrusion to coastal aquifer can be simulated through addition of only one input file of an existing MODFLOW model. SWI package can simulate continuously varying density flow, interface flow and stratified flow.

8.6 Discussion and Conclusions

Remote Sensing has become an inevitable tool for any reconnaissance study now a days. Application of it extends from earth observation, natural disaster prediction/monitoring (monitoring hurricane or any other coastal storm hazard), nation's security surveillance etc. ending up to planetary data collection and so on. GIS (commercial or open source) is the most widely used digital map making/holding platform. Remote Sensing coupled with GIS has very much utility in our daily life also. One of the most widely used service is Google Map (©Google) for motor vehicle navigation. Extending the service, we can even know our neighborhood from Google interactive map. Considering Google Earth (©Google) which is web-based GIS platform, one can upload shapefiles to real life coordinate system in this platform to make map of his own desire. Nowadays private satellite companies are there like SPACEX, Blue Origin, Virgin Galactic offering internet services through satellite, space tourism etc.

Now focusing down the discussion domain to coastal area surveillance and management in terms of its freshwater resource and ecological sustainability, a detailed discussion has already been made. Most of the dedicated air and space borne Remote Sensing platforms to monitor coastal zone of earth provide images which are helpful for addressing different environmental issues. The direct application of Remote Sensing in coastal ecology monitoring involves mapping of coastal marshes, mangroves, and sea grass systems and its spacio-temporal variation, thus having superficial knowledge on coastal ecology related issues. Remote Sensing also has direct application in capturing coastal geomorphology like sand dune migration, change of coastline due to relative sea level change, coastal land cover/land use monitoring & mapping etc. In the field of coastal hydrology/hydrogeology application of Remote Sensing extends up to delineation of river catchment area, flood prediction, sediment load discharge by river to sea etc. In all the above-mentioned context GIS can act as digital map making and map holding platform. GIS is also widely used as a special data repository. Measuring land subsidence and relative sea level change DEM data are very useful. Considering field based Remote Sensing survey, GPS has become an integral part of it to capture real life coordinates of the survey area for digital map making purpose. Remote Sensing and GIS have no direct application in groundwater modeling, management, and monitoring. However, there are some passive applications of Remote Sensing in groundwater exploration which has already been discussed. Satellite images can act as a base map for groundwater survey with GPS coordinate as location point of water level measurement. Digital map prepared in GIS platform can act as base map for groundwater numerical modeling using codes like MODFLOW. Not only addressing the environmental issues of coastal ecosystem a constant monitoring of it is required for economic purpose also. Nation can earn a significant amount of revenue from coastal resources like fishing and coastal tourism. With concluding remarks Remote Sensing and GIS have become an integral part of our life starting from scientific research to daily basis utility application like vehicle navigation, knowing the neighborhood, coastal storm hazard alarming & monitoring etc. In the upcoming days it will become more human interactive integrated with high-speed internet.

8.6.1 Case Study

With detailed discussion on application of Remote Sensing in coastal ecology monitoring and groundwater management here two case studies have been incorporated where Remote Sensing and GIS have been used as a tool to address the solution of the issue. **Case study 1** is from Nile Delta, Egypt where GIS has been adopted to manage/model groundwater resource of the study area. In this work GIS acted as base map for groundwater modeling and attribute tables acted as geodatabase to hold special data like hydraulic conductivity of the aquifer material, water level at well location etc. **Case study 2** is from Himmerfjärden, Sweden where ocean color Remote Sensing has been used for integrated coastal zone management. In this work Medium Resolution Imaging Spectrometer (MERIS) data has been selected/used to identify change in ocean color in and adjacent Baltic Sea of Sweden coast. Time series analysis of MERIS data were taken into consideration to identify different parameters which may be indicative to stress on coastal ecosystem and derivative maps were helpful for decision makers to identify the stress on the ecosystem and working towards its mitigation.

8.6.1.1 Case Study 1

"Application of GIS in groundwater management-Western Nile Delta"

The study area is located at Western Nile Delta region of Egypt as shown in Fig. 8.1. For the study area due to increasing water demand River Nile is no longer sufficient. Therefor it was required to exploit its groundwater resources. Most promising for that was the Quaternary aquifer, the Pliocene aquifer and the Moghra aquifer of Nile delta. To establish the conceptual model for the study area for groundwater modeling GIS was widely used to hold and compile special data. In this work GIS database was classified into three categories namely raster dataset, vector dataset and attribute table. Raster dataset included any preexisting georeferenced digital map of geology, soil, land use, topography, hydrogeology etc. of the study area. This also includes Remote Sensing data like satellite data of true color composite of the study area. Once all the available raster datasets were compiled then georeferenced base map for groundwater modeling was ready. Vector dataset included thematic information layer of geology, topography, land use and distribution of river & drains. Field data of observed groundwater level, hydraulic conductivity data, recharge areas were considered as part of attribute table. When all the available information was gathered then compiling all the data in GIS (ArcGIS) conceptual model were developed for numerical simulation. For numerical modeling a multi-layer aquifer system was built in a finite element grid. TRIWACO software was selected for numerical modeling purposes. The model was also capable of handling the variable density groundwater condition for coastal aquifers. Application of GIS in this modeling work acted as repository of time invariant special data mainly hydrogeological properties of aquifer (hydraulic conductivity, porosity etc.) and maps of land use, location of pumping wells, surface water bodies etc. GIS hydrogeological database in this modeling effort acted as database for holding field based hydrogeological parameters for special and time correlation of that with modeling result. It also proved helpful for hydrogeological data verification and validation for model calibration.

8.6.1.2 Case Study 2

"Remote Sensing based ocean color monitoring for coastal zone management-Himmerfjärden, Sweden"

Study area is located at just south of 60°N around Southern Stockholm Archipelago, Sweden at an opening into the Baltic Sea as shown in Fig. 8.2. Geomorphologically the area is low with a mean depth of 17 m, and it consists of sequence of basins divided by sills. The study area of Himmerfjärden is also characterized by very low freshwater input and sills restrict any additional tidal influence of Baltic Sea to the study area. Problem identified in the study area was algal filamentous cyanobacteria blooms like *Aphani zomenon sp. and Pseudanabaena limnetica* and occasional bloom of *Nodularia spumigena* at summer. However, the presence of *Nodularia spumigena* is more prominent in the Baltic Sea. These filamentous cyanobacteria are capable of

Fig. 8.1 Satellite image map of the study area demarcated by red border. ©Google Earth. Dawoud, M.A., Darwish, M.M. & El-Kady, M.M. "GIS-Based Groundwater Management Model for Western Nile Delta". Water Resour Manage 19, 585–604 (2005). https://doi.org/10.1007/s11269-005-5603-z

fixing nitrogen, which gives them a competitive advantage in comparison to other phytoplankton present in the ecosystem. Their growth is also related to availability of phosphorus. Optical property of coastal water body was taken under consideration to identify growth of algal bloom, presence of Colored Dissolved Organic Matter (CDOM) which is also termed as humic substances and Suspended Particulate Matter (SPM). Collectively it is named as Total Suspended Matter (TSM). Optical property of the coastal water body was evaluated by loss in light intensity, which is measured as diffuse attenuation coefficient (K_d, 490). This optical property evaluation was very helpful for identifying above mentioned substances since CDOM absorbs blue part of the visible part of the spectrum, inorganic suspended matter scatters light thus increasing the water-leaving radiance, phytoplankton absorbs light in the blue and in the red part of the visible part of the spectrum and scatter light. All these optical phenomena can be monitored by space born sea sensing satellite image. To sense the change of ecosystem response chlorophyll-α derived biomass was also taken under consideration. For Remote Sensing Data, Medium Resolution Imaging Spectrometer (MERIS) was selected. The spectral, spatial, and temporal resolution of MERIS is better than most other ocean color sensors in operation at present. Ocean color bands were carefully chosen for MERIS which were informative to measure light attenuation and quantify chlorophyll-α, SPM and CDOM concentration in the sea water. Secchi depth was used as the main indicator for eutrophication in the study area. Secchi depth for open Baltic Sea was measured by SeaWiFS & MODIS data and were compared with coastal water bodies of the study area.

In this work Integrated Project Science and Policy Integration for Coastal System Assessment (SPICOSA) framework was chosen for decision making on Integrated Coastal Zone Management (ICZM) for European countries. This method was

Fig. 8.2 Location of Himmerfjärden Bay, Sweden. © Google Earth. Kratzer, S. E., Harvey, E. T., Philipson, P. "The use of ocean color remote sensing in integrated coastal zone management—A case study from Himmerfjärden, Sweden" Marine Policy, Volume 43, 2014, Pages 29–39, ISSN 0308-597X, https://doi.org/10.1016/j.marpol.2013.03.023

executed at eighteen locations in Europe and this study area was one of them. The focus of the SPICOSA method was to develop a Systems Approach Framework (SAF) for making coastal zone assessment policies. SAF methodology considers coastal ecosystem dynamics and examines the potential consequences of different policy scenarios, at different spacio-temporal scales. The Coastal Zone System Information Feedback Loop (CZFBL) developed within SPICOSA considers different drivers, pressures, state, impacts and responses within an ecosystem with prognostic approach.

Considering above mentioned framework once all the required information was gathered from the satellite data it was identified that accumulation of nitrogen was the main reason for ecological stress of the study area as well as for entire Baltic Sea. Other interpretations were that Baltic Sea color was dominated by dissolved organic matter. Also, a clear optical signal was obtained due to growth of phytoplankton. Phytoplankton growth was correlated with nutrient input from land as well as nutrients coming from sea bottom. After analyzing all the Remote Sensing data, conclusions were made that inorganic suspended matter can be used as an indicator to determine the extent of coastal water which can extend up to 15–20 km from the coastline. Change in the ocean color of Baltic Sea was attributed not only due to phytoplankton growth but also related to changes in CDOM or SPM concentrations. Outcome of this research was deliverable maps of chlorophyll-α, TSM and CDOM absorption available to end-users within few days of image registration. Water quality maps were accessible to end-users by ArcGIS Server solution. These maps proved themselves the best possible method to capture special dynamics of above-mentioned parameters to understand Baltic Sea coastal ecology health and came up with best possible solution for its mitigation, protection, and sustainability.

References

1. Abdulla FA, Lettenmaier DP, Wood EF, Smith JA (1996) Application of a macroscale hydrological model to estimate the water balance of the Arkansas-Red River Basin. J Geophys Res
2. Al-Tahir A, Baban SMJ, Ramlal B (2006) Utilizing emerging geo-imaging technologies for the management of tropical coastal environments. West Indian J Eng 29:11–22
3. Allen RG, Jensen ME, Wright JL, Bunnan RD (1989) Operational estimates of reference evapotranspiration. Agron J 81:650–662
4. Archer L, Neal JC, Bates PD, House JI (2018) Comparing TanDEM-X data with frequently used DEMs for flood inundation modeling. Water Resour Res 54(12):10205–10222. https://doi.org/10.1029/2018WR023688
5. Arrow KJ, Cropper ML, Eads GC, Hahn RW, Lave LB, Noll RG, Portney PR, Russell M, Schmalensee R, Smith VK et al (1996) Is there a role for benefit-cost analysis in environmental, health, and safety regulation? Science 272:221–222
6. Aswani S, Mumby PJ, Baker AC, Christie P et al (2015) Scientific frontiers in the management of coral reefs. Front Marine Sci 2(50):1–13. https://doi.org/10.3389/fmars.2015.00050
7. Bakker M (2003) A dupuit formulation for modeling seawater intrusion in regional aquifer systems. Water Resour Res 39(5):1131. https://doi.org/10.1029/2002WR001710
8. Barbier EB, Hacker S, Kennedy C, Koch E, Stier A, Sillman B (2011) The value of estuarine and coastal ecosystem services. Ecol Monogr 81:169–193
9. Barbier EB, Lee KD (2014) Economics of the marine seascape. Int Rev Environ Resour Econ 7:35–65
10. Barras J, Beville S, Britsch D, Hartley S, Hawes S, Johnston J, Kemp P et al (2003) Historical and projected coastal Louisiana land changes: 1978–2050. USGS Open File Rep 03–334:39
11. Bobba AG, Bukata RP, Jerome JH (1992) Digitally processed satellite data as a tool in detecting potential groundwater flow systems. J Hydrol 131:25–62
12. Brock J, Sallenger A (2000) Airborne topographic LIDAR mapping for coastal science. U.S. Geo logical Survey, Open-File Report 01–46
13. Bruckler L, Witono H, Stengel P (1988) Near surface soil moisture estimation from microwave measurements. Remote Sens Environ 26:101–121
14. Burke LA, Kura Y, Kassem K, Revenga C, Spalding M, McAllister D (2001) Pilot analysis of global ecosystems: coastal ecosystems. World Resour Inst, Washington, DC
15. Campbell JB (2007) Introduction to remote sensing. The Guilford Press, New York
16. Carruthers RM, Greenbaum D, Jackson PD, Mtetwa S, Peart RJ, Shedlock SL (1993) Geological and geophysical characterization of lineaments in south-east zimbabwe and implications for groundwater exploration. Br Geol Survey, NERC, Techn. Rep. WC/9317
17. Chen PY, Chen CC, Chu L, McCarl B (2015) Evaluating the economic damage of climate change on global coral reefs. Glob Environ Change 30:12–20. https://doi.org/10.1016/j.gloenvcha.2014.10.011. ISSN 0959-3780
18. Chust G, Ibon Galparsoro I, Ángel Borja A, Javier Franco J, Adolfo Uriarte A (2008) Coastal and estuarine habitat mapping, using LIDAR height and intensity and multi-spectral imagery, Estuarine. Coast Shelf Sci 78(4):633–643. https://doi.org/10.1016/j.ecss.2008.02.003
19. Collier CG (1996) Applications of weather radar systems. A guide to uses of radar data in meteorology and hydrology, 2nd edn. Praxis Publishing Ltd., Chichester and Wiley, London, 390 pp
20. Devlin M, Bricker S, Painting S (2011) Comparison of five methods for assessing impacts of nutrient enrichment using Estuarine case studies. Biogeochemistry 106:177–205
21. Dey B, Moore H, Gregory AF (1977) The use of satellite imagery for monitoring ice break-up along the Mackenzie River, NWT. Arctic 30(4):234–242
22. Drabbe J, Ghyben BW (1889) Nota in verband met de voorgenomen putboring nabij Amsterdam. Tijdschrift van het Koninklijk Instituut van Ingenieurs, pp 8–22
23. Dregne HE (1986) Desertification of arid lands. In: El-Baz F, Hassan MHA (eds) Physics of desertification. Dordrecht: Martinus Nijhoff, p 473

24. Dubayah R (1992) Estimating net solar radiation using Landsat Thematic Mapper and digital elevation data. Water Resour Res 28(9):2469–2484
25. Eeman S, Leinse A, Raats PAC, Van der Zee SEATM (2011) Analysis of the thickness of a freshwater lens and of the transition zone between this lens and upwelling saline water. Adv Water Resour 34:291–302
26. Etikala B, Adimalla N, Madhav S, Somagouni SG, Keshava Kiran Kumar PL (2021) Salinity problems in groundwater and management strategies in arid and semi-arid regions. Groundw Geochem 42–56. https://doi.org/10.1002/9781119709732.ch3
27. Fassoni-Andrade AC, Durand F, Moreira D, Azevedo A, dos Santos VF, Funi C, Laraque A (2021) Comprehensive bathymetry and intertidal topography of the Amazon estuary. Earth Syst Sci Data 13(5):2275–2291. https://doi.org/10.5194/essd-13-2275-2021
28. Ganguly S, Tiwari S, Bhan U, Mittal S, Rai S et al (2015) Melting of sea ice inexplicable for recent global eustatic sea level rise. J Earth Sci Clim Change 6:245. https://doi.org/10.4172/2157-7617.1000245
29. De Groot R, Brander L, Van Der Ploeg S, Costanza R, Bernard F, Braat L, Christie M, Crossman N, Ghermandi A, Hein L et al (2012) Global estimates of the value of ecosystems and their services in monetary units. Ecosystem 1:50–61
30. Guo W, Langevin CD (2002) User's guide to SEAWAT: a computer program for simulation of three-dimensional variable-density ground-water flow. U.S. Geol Surv Open-File Rep 01–434, 79 p
31. Haralick RM, Shapiro LG (1985) Survey: image segmentation techniques. Comput Vis Graph Image Process 29:100–132
32. Herzberg A (1901) Die Wasserversorgung einger Nordsee bader. J Gasbeleucht Wasserversorg 44(815–819):842–844
33. Jarihani AA, Callow JN, McVicar TR, Van Niel TG, Larsen JR (2015) Satellite-derived digital elevation model (DEM) selection, preparation and correction for hydrodynamic modelling in large, low-gradient and data-sparse catchments. J Hydrol 524:489–506. https://doi.org/10.1016/j.jhydrol.2015.02.049
34. Jensen JR (2007) Remote sensing of the environment: an Earth resource perspective. Prentice Hall, New Jersey
35. Jia M, Zhang Y, Wang Z, Song K, Ren C (2014) Mapping the distribution of mangrove species in the core zone of Mai Po Marshes nature reserve Hong Kong using hyperspectral data and high-resolution data. Int J Appl Earth Obs 33:226–231
36. Joss J, Lee R (1995) The application of radar-gauge comparisons to operational precipitation profile corrections. J Appl Met 34:2612–2330
37. Klemas V, Dobson JE, Ferguson RL, Haddad KD (1993) A coastal land cover classification system for the NOAA coast watch change analysis project. J Coastal Res 9(3):862–872
38. Klemas V (2007) Remote sensing of coastal wetlands and estuaries. In: Proceedings of the coastal Zone 07. NOAA Coastal Services Center. Charleston, South Carolina
39. Konikow LF, Goode DJ, Hornberger GZ (1996) A three-dimensional method-of-characteristics solute-transport model (MOC3D). US Geol Surv Water-Resour Investig Rep 96–4267:87
40. Kostianoy AG, Lavrova OY, Strochkov AY (2022a) Satellite instrumentation and technique for monitoring of seawater quality. In: Di Mauro A, Scozzari A, Soldovieri F (eds) Instrumentation and measurement technologies for water cycle management. Springer Nature, Cham, Switzerland, pp 79–109. https://doi.org/10.1007/978-3-031-08262-7_5
41. Kostianoy AG, Lavrova OY (2022b) Satellite instrumentation and technique for oil pollution monitoring of the seas. In: Di Mauro A, Scozzari A, Soldovieri F (eds) Instrumentation and measurement technologies for water cycle management. Springer Nature, Cham, Switzerland, pp 53–77. https://doi.org/10.1007/978-3-031-08262-7_4
42. Liu L, Zhang T, Wahr J (2010) InSAR measurements of surface deformation over permafrost on the North Slope of Alaska. J Geophys Res 115:F03023. https://doi.org/10.1029/2009JF001547
43. Longhurst A, Sathyendranath S, Platt T, Caverhill C (1995) An estimate of global primary production in the ocean from satellite data. J Plank Res 17:1245–2127

44. Lu Z, Kwoun O, Rykhus R (2007) Interferometric synthetic aperture radar (InSAR): its past, present and future. Photogramm Eng Remote Sens 73:217
45. Lyon JG, McCarthy J (1995) Wetland and environmental applications of GIS. Lewis Publishers, New York
46. Martin S (2004) An introduction to remote sensing. Cambridge University Press, Cambridge
47. McCarthy MJ, Colna KE, El-Mezayen MM et al (20017) Satellite remote sensing for coastal management: a review of successful applications. Environ Manag 60:323–339. https://doi.org/10.1007/s00267-017-0880-x
48. McDonalds MG, Harbaugh AW (1988) A modular three-dimensional finite difference ground water flow model. USGS Open File Report, p 875
49. Michael HA, Mulligan AE, Harvey CF (2005) Seasonal oscillations in water exchange between aquifer and the coastal ocean. Nature 436:1145–1148
50. Milon WJ, Alvarez S (2019) The elusive quest for valuation of coastal and marine ecosystem services. Water 11:1518. https://doi.org/10.3390/w11071518
51. Moore GK (1979) What is a picture worth? A history of remote sensing. Hydrol Sci Bull 24(4):477–485. https://doi.org/10.1080/02626667909491887
52. Motz LH (1992) Salt water upconing in an aquifer overlain by a Leaky confining bed. Ground Water 30(2)
53. Muhadi NA, Abdullah AF, Bejo SK, Mahadi MR, Mijic A (2020) The use of LiDAR-derived DEM in flood applications: a review. Remote Sens 12:2308
54. Neumann B, Vafeidis A, Zimmermann J, Nicholls R (2015) Future coastal population growth and exposure to sea-level rise and coastal flooding-a global assessment. Plos One 10:e0118571
55. Oliveira AM, Fleischmann AS, Paiva RCD (2021) On the contribution of remote sensing-based calibration to model hydrological and hydraulic processes in tropical regions. J Hydrol 597. https://doi.org/10.1016/j.jhydrol.2021.126184
56. Purkis SJ (2005) A reef-up approach to classifying coral habitats from IKONOS imagery. IEEE T Geosci Remote 43:1375–1390
57. Schaeffer BA, Hagy JD, Conmy RN, Lehrter JC, Stumpf RP (2012) An approach to developing numeric water quality criteria for coastal waters using the SeaWiFS satellite data record. Environ Sci Technol 46:916–922
58. Selman M, Sugg Z, Greenhalgh S, Diaz R (2008) Eutrophication and hypoxia in coastal areas: a global assessment of the state of knowledge. World Resources Institute, Washington, DC
59. Stephens PR, Cihlar J (1982) Mapping soil erosion in New Zealand and Canada. In: Johannsen CJ, Sanders JL (eds) Remote sensing for resource management. Soil Conservation Society of America, Ankeny, Iowa, pp 232–242
60. Stewart RH (2008) Introduction to physical oceanography. Texas A&M University. https://sites.google.com/tamu.edu/ocean-world/home
61. Thompson A, Schroeder T, Brando VE, Schaffelke B (2014) Coral community responses to declining water quality: Whitsunday islands great barrier Reef Australia. Coral Reefs 33(4):923–938
62. Wang Y, Traber M, Milstead B, Stevens S (2006) Terrestrial and submerged aquatic vegetation mapping in Fire Island National Seashore using high spatial resolution remote sensing data. Mar Geodesy 30(1):77–95
63. Ward RC, Robinson M (1989) Principles of hydrology. McGraw-Hill (UK), 365 pp
64. Wehr A, Lohr U (1999) Airborne laser scanning—An introduction and overview. ISPRS J Photogramm 54:68–82
65. Xiaojun Y (ed) (2009) Remote sensing and geospatial technologies for coastal ecosystem assessment and management. Springer, Berlin, Heidelberg. https://doi.org/10.1007/978-3-540-88183-4
66. Yang X, Damen MCJ, Van Zuidam R (1999) Use of thematic mapper imagery and a geographic information system for geomorphologic mapping in a large deltaic lowland environment. Int J Remote Sens 20:568–591

Chapter 9
Hydrogeomorphological Mapping of Groundwater Potential Zones Using Multi-influence Factor (MIF) and GIS Techniques: A Case Study of Vishav Watershed, Western Himalayas

Suhail A. Lone and Ghulam Jeelani

Abstract In western Himalayas groundwater is one of the most significant source of freshwater playing an important role in sustaining the social and economic stability of the region. However, this resource is under serious threat of depletion and contamination as a result of global climatic variability and anthropogenic pressures. Thus, evaluating potential groundwater sites is critical for long-term water management and water quality standards. Multi-influence factor (MIF) and geospatial techniques were used in this study to assess potential groundwater sites in the Vishav watershed. Various thematic layers including geology, geomorphology, soil, landuse/landcover (lulc), lineaments, drainage network, slope, and rainfall were created in a GIS environment using different satellite data sets (e.g. Digital Elevation Model, Landsat ETM+, and Google Earth). According to our findings, more than 80% of the basin area is classified as having medium to good groundwater potential, while 20% is classified as having low potential. Because of the lower slope and unconsolidated sediments, lower parts of the basin have been identified as significantly potential areas. According to the findings, geology, geomorphology, and lineaments play important roles in groundwater occurrence and distribution. Our research will not only improve understanding and knowledge of groundwater occurrence and distribution, but it will also assist local and regional administrators in ensuring sustainable groundwater management throughout the western Himalayas.

Keywords Remote sensing · Geology · Soil · Groundwater · Satellite data

S. A. Lone (✉) · G. Jeelani
Department of Earth Sciences, University of Kashmir, Srinagar 190006, India
e-mail: geosuhail55@gmail.com

G. Jeelani
e-mail: geojeelani@gmail.com

E. Balaji et al. (eds.), *Emerging Technologies for Water Supply, Conservation and Management*, Springer Water, https://doi.org/10.1007/978-3-031-35279-9_9

9.1 Introduction

Water towers in Asia provide an important source of fresh water for millions of people who live in upstream and downstream areas of various watersheds [27, 32–34]. However, with recent climate change and a scarcity of potable surface water resources for domestic and agricultural use, groundwater has become increasingly important [13, 16, 23, 26, 28, 30, 31, 36, 48]. As a result, not only monitoring and conservation are critical, but the delineation of groundwater potential zones is highly critical under changing climatic scenarios [6, 31]. Groundwater exploration in hard rock areas includes the following factors: mass removal rate, fracture degree, porosity, slope, drainage, landscape, land use/land (LULC), and climatic conditions [13, 31, 37]. Groundwater is mostly confined to secondary fissures and weathered layers in these areas [4, 17, 19, 40, 52]. The systematic mixing of numerous datasets allows for the rapid and cost-effective allocation of groundwater potential zones in such regions. As a result, a combined approach for groundwater delineation is critical. Remote sensing and Geographic Information Systems (GIS), as well as several statistical techniques, such as the analytical hierarchical process (AHP), have widespread application in the management of various natural resources [1, 7–9, 11, 12, 39, 41–47, 49–51]. The use of geospatial tools to delineate potential groundwater zones has proven to be extremely effective [23, 48]. In recent years, the significant application of geospatial techniques in conjunction with secondary datasets has facilitated the collection of basic information about groundwater resources all over the world [2, 3, 10, 14, 21–23, 28, 31, 35, 53, 55].

In Vishav watershed groundwater is dominantly used for physical, chemical, and biological activities. Because of the basin's oval shape, precipitation from higher elevations flows directly and quickly into rivers and streams. Water resources and demand are unevenly distributed in UJB, both spatially and temporally. At present water shortage is a critical issue as a result of global climate change and anthropogenic problems [25, 28–30]. Significant mining of streambed sediments in different watersheds have caused depletion of groundwater, resulting in reduced spring runoff and/or depletion of springs and watertable depth [18]. As a result, groundwater exploration is increasing, resulting in over-exploitation and serious environmental issues such as declining groundwater levels, groundwater pollution, and deterioration of water quality. Groundwater is pumped from shallow springs, dug wells, and drilled wells in the Vishav watershed for drinking and irrigation purposes. As a result, a systematic understanding of groundwater resources in the watershed is required for better local and regional planning and management. Keeping in view its significance, the primary goal of this research is to use geospatial methods (remote sensing, GIS) and multi-influencer techniques to delineate and map potential groundwater zones in the Vishav watershed of western Himalayas.

9.2 Study Area

Vishav, a high altitude watershed in western Himalayas, lies in Pir-Panjal between the geographical coordinates of latitudes 33°33′ and 33°34′ N and longitudes 74°40′ and 75°00′ E, with an areal extent of of 1283 km^2 (Fig. 9.1).

The climate in the Vishav watershed is classified as mesotrophic/temperate by Koppens classification system. The Vishav watershed receives 1400 mm of precipitation $year^{-1}$ with highest in winter and early spring [15, 24]. The highest and lowest precipitation were measured in March (~180 mm) and November (~30 mm), with mean highest and lowest temperature acquired in July (~24 °C) and January (~1.2 °C, [15]). The vegetation of Vishav watershed varies from lush wetlands and terrestrial vegetation in the lowlands to lush evergreen conifers in highlands. Vishav watershed/sub-basin is endowed with many large and small sized moraine-dammed lakes ranging in size from 0.20 to 1.82 km^2, with Kousarnag being the basin's largest moraine dammed lake contributes maximum proportion of water to Vishav River. Brahmasar, Chirsar, and Indrasa are among the other lakes in the watershed. In Vishav watershed, the watertable depth ranges from 5 to 7 mbgl (below ground level) along river valleys to 20–30 mbgl at higher altitudes [17].

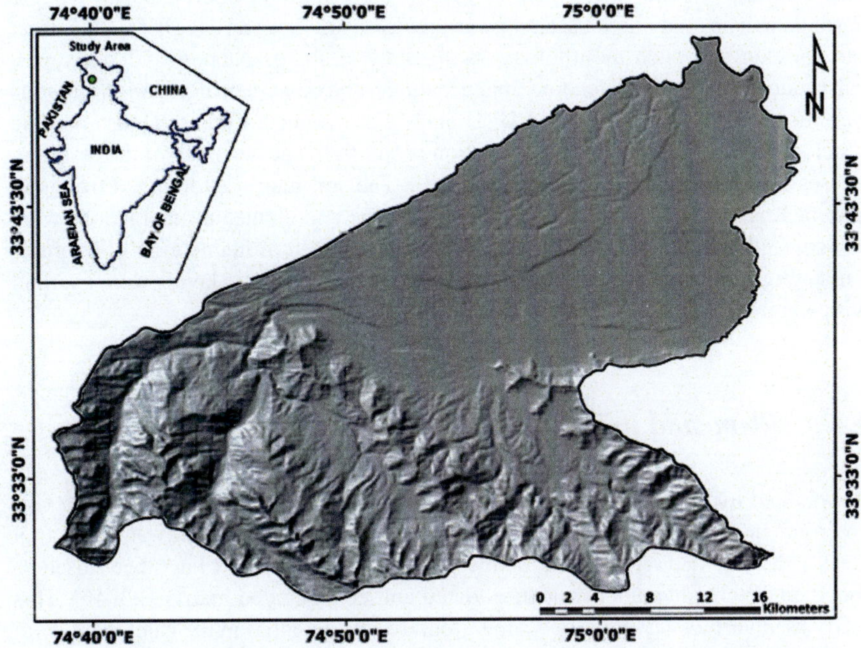

Fig. 9.1 Study area map (Vishav sub-basin). Hill shade used as base map was generated from Aster Digital Elevation Model of 30 m resolution in ArcGIS 10.7.1. The map towards the left corner depicts the position of Vishav sub-basin in India

9.3 Data Collection

The Landsat TM, ETM+ data and ASTER Digital Elevation Model was used in the current study for the identification and mapping of various attributes required for zonation. Least snow covered images were used for more accurate results. The satellite data was cross-checked with field data collected with a Global Positioning System and a Survey of India (SOI) Topo-sheet. The lithology of the sub-basin was digitized from geological map of western Himalayas and groundwater table data was obtained from a Central Ground Water Board report (CGWB 2018). Geological structures were delineated using Geometic_2015 from satellite data. Finally, all the thematic layers generated were merged in ArcGIS 10.7.1 Precipitation data was acquired from Weather Online Website.

9.4 Methodology Adopted

Remote sensing and GIS combined with field validation was used to delineate and map potential groundwater zones in Vishav watershed corroborated with Google Earth data. The study area basemap was created using an ASTER DEM. To extract various elements such as lithology, topography, drainage, slopes, and soil types, False color composition, standard image features, and on-screen digitization methods were used. Slope, drainage and LULC maps were created with arc Hydro and the raster surface tool and image classification properties. The accuracy of the resulting classes was evaluated using field based data. The soil map was digitized from soil map of Kashmir. The factors influencing recharge and elemental interactions were investigated. The importance of each item was assigned to item weights. To create a map of groundwater potential map of watershed, all thematic layers were overlaid with weighted overlay properties in ArcMap.

9.4.1 Elemental Relationship

A weighted overlay method based on the incorporation of influencer layers in GIS environment has been demonstrated to be useful in defining groundwater potential areas [31, 43]. In various parts of the globe, similar methods have been used to ascertain potential zones in unconsolidated and consolidated matrix [20, 46]. This study used various factors influencing groundwater potential in the catchment, with the hypothesis being that there is a relationship between several influencing factors that affect the sub-recharge basin's capacity. Major and minor relationships were revealed. The sum of all weights for each factor determines an element's characteristic weights. Higher-weighted elements have a greater impact on groundwater recharge and vice-versa. To validate potential groundwater zones, spatial assimilation and surveys were carried out in a GIS environment. We investigated the hydrological and geographic characteristics of the Vishav watershed and identified seven key factors influencing replenishment potential.

9.5 Results and Discussions

9.5.1 Geology

One of the most important factors influencing groundwater existence, movement, and storage within a watershed is geology. The specific rock type determines the type, depth, weathering rate, and pertrophysical properties of hydration discontinuities. Different geological conditions also influence the formation and properties of water-bearing zones with readily mobile water. The lithology of a specific area has a significant impact on the quantitative and qualitative occurrence of groundwater [23, 31, 38, 48]. In Vishav, Panjal Traps are the dominant litho-facies in the lower part of the sub-basin, followed by the Karewa deposit, recent alluvium, and Triassic-Jurassic limestone [54] (Fig. 9.2a). The pertrophysical properties of confined and unconfined aquifers is influenced by basin lithology. Due to secondary porosity and permeability, groundwater zoning maps indicated that optimal groundwater potential areas included limestone and alluvium. However, each rock unit in the basin behaves differently in terms of defining and governing groundwater recharge and runoff.

9.5.2 Geomorphology

The topography of a specific area greatly influences its groundwater potential and distribution. River sediments, Piedmont sediments, moderately rugged glacial landscapes/rugged hills, V-shaped valleys, lakes, and denuded hills are among the key physiographic units identified and mapped in the watershed (Fig. 9.2b). The major geomorphological features of the sub-basin are river sediments, moderately rugged hills, and glacial landscapes. Recent alluvial and Piedmont sediments in low-lying areas are significant sites because their non-consolidated nature allows surface water to infiltrate into groundwater, but large topography and high ridge gradients can lead to groundwater contamination. Surface runoff is more common in areas with low porosity than in areas with high porosity [23, 31]. However, when secondary pore systems are well developed and interconnected, these hard rocky landforms serve as excellent groundwater reservoirs.

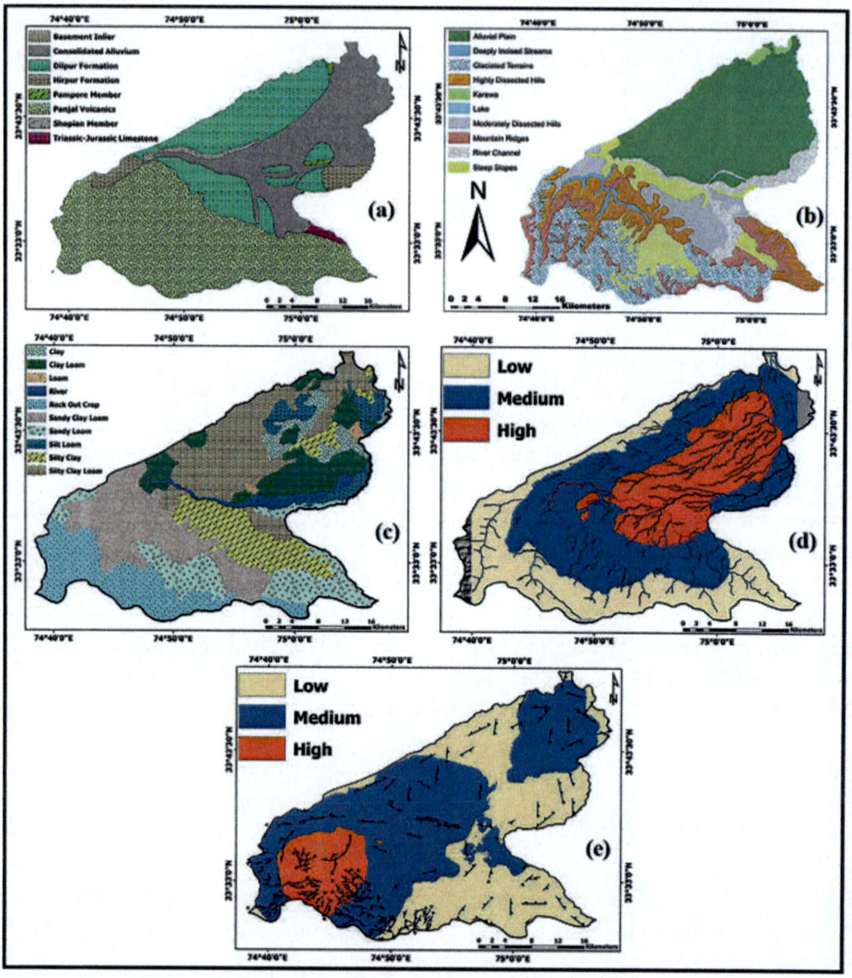

Fig. 9.2 Maps showing the **a** Geology. **b** Geomorphology. **c** Soil. **d** Drainage density and **e** Lineament density of Vishav sub-basin. All these maps were generated in ArcGIS 10.7.1

9.5.3 Soil

Because of its important pertrophysical properties, soil is a key factor influencing groundwater potential zones in a given area. The descending movement and subsequent penetration of water into different soil layers is highly dependent on porous nature of soil. Because of the predominance of unconsolidated material (sand, sandy loam, silty clay) carried down by various streams, soil maps of the study area clearly show higher infiltration in the lower areas. Water intrusion, on the other hand, is reduced at higher elevations due to steeper slopes and less land cover. Figure 9.2c depicts various soil types mapped in the Vishav sub-watershed.

9.5.4 Drainage Density

The drainage density is total drainage-length/unit area calculated as:

$$D_d = \frac{\sum_{i=0}^{i=n} Si}{A} \tag{9.1}$$

$\sum_{i=0}^{i=n} Si$, Drainage length (L) A, unit area (L^2).

Drainage density is proportional to intrinsic permeability. Permeability is primarily responsible for controlling water movement. Unconsolidated sediments dominate the Vishav sub-basin (Karewa and Alluvium). Karst landforms, such as numerous springs and Sack Valleys, predominate in the region. It resulted in well-developed secondary pores that were interconnected. Drainage densities exceeding 2.5 km/km^2 are typical of the best groundwater recharge zones. Upstream and midstream flow rates range from 0.50 to 1.5 km/km^2. This study identified two arrangements with different drainage densities (Fig. 9.2d). Fine drainage network was observed and delineated in the southern part of the Vishav sub-basin, but there were loosely structured zones towards eastern, western, and northern flanks. Lithology and structures primarily control sub-basin drainage textures and patterns in the north-western Vishav sub-basin (Fig. 9.2d). In the lower parts a dendritic drainage pattern was commonly observed.

9.5.5 Lineament Density

Lineaments are very important because they provide the subsurface flowpaths for groundwater movement therefore its existence defines a permeable zone. The lineament density is total lineament-length/unit area as:

$$L^d = \frac{\sum_{i=0}^{i=n} Li}{A} \tag{9.2}$$

$\sum_{i=0}^{i=n} Li$; Lineament-length (L) and A is the unit area (L^2).

Highly dense lineament illustrates high porosity, indicating a favorable groundwater area (Fig. 9.2e). Because of the interconnectivity of pores, runoff easily infiltrates in high density lineament areas, suggesting a considerable potential for groundwater development. These areas act as good reservoirs of groundwater, and exit for springs.

9.5.6 Slope

Another important element employed to identify groundwater potential zones is slope. Steep gradient slopes are more responsible for faster runoff, faster erosion, and less infiltration. Slope maps of the study area were created in an ArcGIS environment using ASTER DEM. The basin is divided into six classes, as shown on the map. Built-up and cultivated land areas with slope values between 0 and 5 cover the basin's lower reaches (Figs. 9.1 and 9.3a). Slope values of 5–13 to 13–20 are considered moderate and are covered by natural plantations such as forests. Arid and glacial environments dominate the values above 20 in the sub-northwest basin's and north. The area's slope is a key factor in determining groundwater availability and occurrence. Flatter, lower-sloping areas retain surface runoff (precipitation) for a longer period of time, allowing for a faster rate of recharge than steeper slopes with faster runoff [13, 31].

9.5.7 Land Use/Land Cover (LULC)

Climate, geomorphology, agro-ecology, and human development all have an impact on a region's LULC (Land Use/Land Cover). This is an another essential parameter affecting the potential and existence of groundwater in the watershed. This research was divided into seven categories: human development, forestry, agriculture, horticulture, wastelands, water, and snow.

These classes were delineated using a supervised classification in ArcMap 10.7.1 (Fig. 9.3b). Accuracy assessments of the resulting classes were performed against field data. The overall accuracy of maximum likelihood classification is >80% with a kappa coefficient of 0.80. The LULC classes dominating Vishav watershed includes: lush green forests, cultivated land, wastelands, and built-up. Due to unconsolidated nature and high infiltration rates, forests and agricultural/horticultural areas are the most suitable groundwater potential areas. Plant cultivation is primarily controlled by water availability. Therefore, the presence of a sufficient amount of groundwater in the area increases the degree of cultivation.

9.5.8 Rainfall Distribution

Precipitation is critical in the direct recharge of groundwater. It determines the amount of available water in the sub-soil (Fig. 9.3c). The annual precipitation in the Vishav watershed is approximately 1440 mm. 71% of total precipitation is accounted for by the mid-latitude circulation and rest is from Indian Summer Monsoon (ISM). The Vishav River's annual flow is primarily influenced by meltwater from winter snowfall and seasonal precipitation. The watershed relies on winter precipitation.

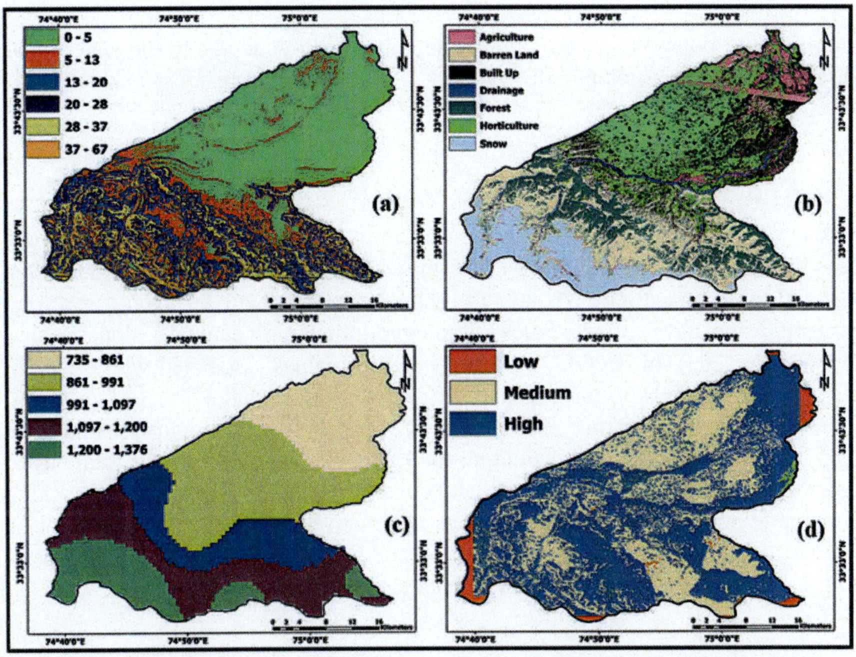

Fig. 9.3 Maps showing the **a** slop. **b** Landuse/landcover. **c** Rainfall and **d** Groundwater Potential of Vishav sub-basin. Using different modules all these maps were generated in ArcGIS 10.7.1

The distribution of precipitation along the basin's slope influences runoff infiltration and thus increases the likelihood for groundwater existence.

9.6 Groundwater Prospecting and Validation

In the current study, a groundwater map with potential zones was created in ArcGIS 10.7.1 using a combination of different thematic layers (Fig. 9.3d). Potential zones were evaluated and classified as low, medium, or high based on their spatial distribution (Fig. 9.3d). 0.80 km^2, 412 km^2, and 648 km^2 in the Vishav sub-basin are classified as low, medium, and high potential zones, accounting for 1.2%, 47%, and 52% of the total study area, respectively. The study also found promising potential groundwater zones in the basin's lower reaches due to the presence of large amounts of sand, gentle slopes, and agricultural land. Furthermore, regional drainage densities show that groundwater recharge is heavily dependent on river flow. The generated groundwater potential map was validated using data from the Central Groundwater Board [5] on groundwater table depth in different seasons, and areas with low to medium groundwater potential were found to have very deep underground water.

Areas of high potential, on the other hand, have shallow depths to the water table and little seasonal variation.

9.7 Conclusion

Geospatial and MIF techniques were used in the current study to create a groundwater potential zoning map for the Vishav sub-basin. We created thematic layers for various elements using physical properties (lithology, drainage, lineaments, slopes, soils, precipitation, etc.). In ArcGIS, properly weighted values associated with thematic layers were overlaid to create a map of groundwater potential. According to the findings, 0.80 km^2, 412 km^2, and 648 km^2 are classified as having low, medium, or high groundwater potential, accounting for 1.2%, 47%, and 52% of the total basin, respectively. Because of the reduced land cover and steep slopes, higher elevations were the least effective. The findings can help with the long-term management of Himalayan groundwater resources. This method has the potential to be widely used for groundwater exploration in mountain catchments.

Acknowledgements The authors gratefully acknowledge United State Geological Survey (USGS) (https://earthexplorer.usgs.gov/) for free access to satellite data.

References

1. Achu AL, Thomas J, Reghunath R (2020) Multi-criteria decision analysis for delineation of groundwater potential zones in a tropical river basin using remote sensing, GIS and analytical hierarchy process (AHP). Groundw Sustain Dev 10:100365
2. Bhuvaneswaran C, Ganesh A, Nevedita S (2015) Spatial analysis of groundwater potential zones using remote sensing, GIS and MIF techniques in uppar Odai sub-watershed, Nandiyar, Cauvery basin, Tamilnadu. Int J Curr Res 7(09):20765–20774
3. Chatterjee RS, Bhattacharya AK (1995) Delineation of the drainage pattern of a coal basin related inference using satellite remote sensing techniques. Asia Pac Remote Sens J 1:107–114
4. Chowdhury A, Jha MK, Chowdary VM (2010) Delineation of groundwater recharge zones and identification of artificial recharge sites in West Medinipur district, West Bengal, using RS, GIS and MCDM techniques. Environ Earth Sci 59(6):1209–1222
5. Central Ground Water Board (2018) Ground water yearbook 2016–17. http://cgwb.gov.in/Regions/GW-year-Books/GWYB-%202016-17/J&K.pdf
6. Dar IA, Sankar K, Dar MA (2011) Deciphering groundwater potential zones in hard rock terrain using geospatial technology. Environ Monit Assess 173(1):597–610
7. Das S, Behera SC, Kar A, Narendra P, Guha S (1997) Hydrogeomorphological mapping in ground water exploration using remotely sensed data—A case study in keonjhar district, orissa. J Indian Soc Remote Sens 25(4):247–259
8. Edet AE, Okereke CS, Teme SC, Esu EO (1998) Application of remote-sensing data to groundwater exploration: a case study of the Cross River State, southeastern Nigeria. Hydrogeol J 6(3):394–404

9. Etikala B, Golla V, Li P, Renati S (2019a) Deciphering groundwater potential zones using MIF technique and GIS: a study from Tirupati area, Chittoor District, Andhra Pradesh, India. HydroResearch 1:1–7
10. Etikala B, Veeraswamy G, Narsimha A, Rajasekhar M, Siddi Raju R (2019b) Assessing groundwater potential zones using remote sensing, GIS and MIF techniques: Renigunta area, Andhra Pradesh, India. Indian J Geophys Union 2019:469–477
11. Golla V, Badapalli PK, Etikal B, Sivakumar VL, Telkar SK (2021) Delineation of groundwater potential zones in semi-arid region (Ananatapuram) using geospatial techniques. Mater Today: Proc. https://doi.org/10.1016/j.matpr.2021.02.2393
12. Harinarayana P, Krishna G, Balasubramanian GS (2000) Remote sensing a data for groundwater development and management in Keralapura watersheds of Cauvery basin, Karnataka. In: Proceedings of the national seminar on earth resources. Mangalore University; (abs) Jones, HC The iron ore deposits of Bihar and Orissa. Mem. GSI, vol 63, no 22, pp 167–302
13. Hussein AA, Govindu V, Nigusse AGM (2017) Evaluation of groundwater potential using geospatial techniques. Appl Water Sci 7(5):2447–2461
14. Jaan O, Lone S, Malik R, Lone A, Wasim M (2016) Morphotectonic and morphometric analysis of Vishav basin left bank tributary of Jhelum River SW Kashmir Valley India. Int J Econ Environ Geol 17–26
15. Jeelani G, Deshpande RD, Shah RA, Hassan W (2017) Influence of southwest monsoons in Kashmir valley, western Himalayas. Isot Environ Health Stud 53:400–412
16. Jeelani G, Lone SA (2019) Establishing the effect of aridity on the stable isotopes (O and D) of precipitation in cold desert. Ladakh. Int J Water Res Arid Environ 8:161–168
17. Jeelani G, Lone SA, Nisa AU, Mukherjee A, Deshpande RD (2020) Sources and processes of groundwater arsenic mobilization in upper Jhelum basin, western Himalayas. J Hydrol 591:125292
18. Jeelani G, Lone SA, Lone A, Deshpande RD (2021b) Groundwater resource protection and spring restoration in Upper Jhelum Basin (UJB), western Himalayas. Groundw Sustain Dev 15:100685
19. Jeelani G, Lone SA, Nisa AU, Deshpande RD, Padhya V (2021a) Use of stable water isotopes to identify and estimate the sources of groundwater recharge in an alluvial aquifer of Upper Jhelum Basin (UJB), western Himalayas. Hydrol Sci J 66(16):2330–2339
20. Krishnamurthy J, Mani A, Jayaraman V, Manivel M (2000) Groundwater resources development in hard rock terrain-an approach using remote sensing and GIS techniques. Int J Appl Earth Obs Geoinf 2(3–4):204–215
21. Leblanc M, Leduc C, Razack M, Lemoalle J, Dagorne D, Mofor L (2003) Applications of remote sensing and GIS for groundwater modelling of large semiarid areas: example of the Lake Chad Basin, Africa. Int Assoc Hydrol Sci Publ 278:186–192
22. Lone S (2016) Mapping the glacial-geomorphological landforms in East Liddar Valley, NW Himalaya Kashmir India. J Geogr Nat Disast S 6:2167–587
23. Lone SA, Lone AA, Jeelani G (2016) Characterization of groundwater potential of Sindh watershed Western Himalayas. J Res Dev 16:29–40
24. Lone SA, Jeelani G, Deshpande RD, Shah RA (2017) Evaluating the sensitivity of glacier to climate by using stable water isotopes and remote sensing. Environ Earth Sci 76(17):1–9
25. Lone SA, Jeelani G, Deshpande RD, Mukherjee A (2019) Stable isotope ($\delta^{18}O$ and δD) dynamics of precipitation in a high altitude Himalayan cold desert and its surroundings in Indus river basin, Ladakh. Atmos Res 221:46–57
26. Lone SA, Jeelani G, Mukherjee A, Coomar P (2020) Geogenic groundwater arsenic in high altitude bedrock aquifers of upper Indus river basin (UIRB), Ladakh. Appl Geochem 113:104497
27. Lone SA, Jeelani G (2020) Stable water isotopic evidence for the moisture source and composition of surface runoff in Ladakh, upper Indus river basin (UIRB). In: Proceedings of the 5th international electronic conference on water sciences, MDPI: Basel, Switzerland. https://doi.org/10.3390/ECWS-5-07903

28. Lone SA (2020) Hydrogeochemical and isotopic evidence of groundwater evolution and recharge in Indus Basin Ladakh (J&K India). Doctoral dissertation. University of Kashmir. http://hdl.handle.net/10603/314053
29. Lone SA, Jeelani G, Deshpande RD, Mukherjee A, Jasechko S, Lone A (2021) Meltwaters dominate groundwater recharge in cold arid desert of upper Indus River Basin (UIRB), western Himalayas. Sci Total Environ 786:147514
30. Lone SA, Jeelani G, Mukherjee A, Coomar P (2021a) Arsenic fate in upper Indus river basin (UIRB) aquifers: controls of hydrochemical processes, provenances and water-aquifer matrix interaction. Sci Total Environ 795:148734
31. Lone SA, Jeelani G (2022) Appraising the groundwater potential of Liddar sub-Basin (Western Himalayas) using geospatial techniques. In: Balamurugan P, Baliram PC, Kirubakaran M, Anand B, Nagavinothini R (eds) Climate change impact on groundwater resources: human health risk assessment in arid and semi-arid regions. Springer, pp 139–154. https://doi.org/10.1007/978-3-031-04707-7
32. Lone SA, Jeelani G, Padhya V, Deshpande RD (2022) Identifying and estimating the sources of river flow in the cold arid desert environment of Upper Indus River Basin (UIRB), western Himalayas. Sci Total Environ 154964
33. Lone SA, Jeelani G, Alam A, Bhat MS, Farooq H (2022a) Effects of changing climate on the water resources of upper Jhelum basin (UJB), India. In: Mukherjee A (ed) Riverine systems: understanding the hydrological, hydrosocial and hydroheritage connections. Springer Capital Publishing, New Delhi, pp 139–154
34. Lone SA, Jeelani G (2022a) Assessing the hydrological processes controlling spatio-temporal patterns of snowmelt in high altitude arid-climatic environment of Upper Indus River Basin (UIRB), India 1032504. In: Frontiers in hydrology, future of water, San Juan, Puerto Rico
35. Lone SA, Jeelani G (2022b) Quantifying the dominance of source waters to river flow in high altitude arid climatic environment of Upper Indus River Basin (UIRB), western Himalayas 1032514. In: Frontiers in hydrology, future of water, San Juan, Puerto Rico
36. Lone SA, Jeelani G, Deshpande RD, Bhat MS, Padhya V (2023) Assessing the hydrological controls on spatio-temporal patterns of streamwater in glacierized mountainous Upper Indus River Basin (UIRB), western Himalayas. J Hydrol 129310
37. Lone SA, Jeelani G (2023) Spatio-temporal dynamics of Groundwater recharge in Dras sub-basin of upper Indus River Basin, Western Himalayas. Impacts of urbanization on hydrological systems in India. Springer International Publishing, Cham, pp 313–322
38. Middlemiss CS (1910) Revision of the Silurian-Trias sequence in Kashmir: records of the geological survey of India, vol 40
39. Muralidhar M, Raju KRK, Raju KSVP, Prasad JR (2000) Remote sensing applications for the evaluation of water resources in rainfed area, Warangal district, Andhra Pradesh. Indian Mineral 34(2):33–40
40. Murthy KSR (2000) Ground water potential in a semi-arid region of Andhra Pradesh-a geographical information system approach. Int J Remote Sens 21(9):1867–1884
41. Magesh NS, Chandrasekar N, Soundranayagam JP (2011) Morphometric evaluation of Papanasam and Manimuthar watersheds, parts of Western Ghats, Tirunelveli district, Tamil Nadu, India: a GIS approach. Environ Earth Sci 64(2):373–381
42. Salama RB, Tapley I, Ishii T, Hawkes G (1994) Identification of areas of recharge and discharge using Landsat-TM satellite imagery and aerial photography mapping techniques. J Hydrol 162(1–2):119–141
43. Saraf AK, Choudhury PR (1998) Integrated remote sensing and GIS for groundwater exploration and identification of artificial recharge sites. Int J Remote Sens 19(10):1825–1841
44. Sener E, Davraz A, Ozcelik M (2005) An integration of GIS and remote sensing in groundwater investigations: a case study in Burdur, Turkey. Hydrogeol J 13(5):826–834
45. Shaban A, Khawlie M, Abdallah C (2006) Use of remote sensing and GIS to determine recharge potential zones: the case of Occidental Lebanon. Hydrogeol J 14(4):433–443
46. Shashikkumar MC, Metilda LD (2012) GIS approach for the demarcation of groundwater recharge potential zones in hard rock terrain. Int J Earth Sci Eng 5(1):117–120

47. Subagunasekar M, Sashikkumar MC (2012) GIS for the assessment of the groundwater recharge potential zone in Karunkulam block, Thoothukudi district, Tamil Nadu, India. Int J Curr Sci 159–162
48. Shah RA, Lone SA (2019) Hydrogeomorphological mapping using geospatial techniques for assessing the groundwater potential of Rambiara river basin, western Himalayas. Appl Water Sci 9(3):1–11
49. Sikdar PK, Chakraborty S, Adhya E, Paul PK (2004) Land use/land cover changes and groundwater potential zoning in and around Raniganj coal mining area, Bardhaman District, West Bengal: a GIS and remote sensing approach. J Spat Hydrol 4(2):1–24
50. Thomas A, Sharma PK, Sharma MK, Sood A (1999) Hydrogeomorphological mapping in assessing ground water by using remote sensing data—A case study in lehra gaga block, Sangrur district, Punjab. J Indian Soc Remote Sens 27(1):31–42
51. Tiwari A, Rai B (1996) Hydromorphogeological mapping for groundwater prospecting using landsat-MSS images-a case study of part of Dhanbad District, Bihar. J Indian Soc Remote Sen 24(4):281–285
52. Tweed SO, Leblanc M, Webb JA, Lubczynski MW (2007) Remote sensing and GIS for mapping groundwater recharge and discharge areas in salinity prone catchments, southeastern Australia. Hydrogeol J 15(1):75–96
53. Vittala SS, Govindaiah S, Gowda HH (2005) Evaluation of groundwater potential zones in the sub-watersheds of North Pennar river basin around Pavagada, Karnataka, India using remote sensing and GIS techniques. J Indian Soc Remote Sens 33(4):483–493
54. Wadia DN (1975) Geology of India. Tata McGraw Hill Co, New Delhi, 344 pp.
55. Yeh HF, Lee CH, Hsu KC, Chang PH (2009) GIS for the assessment of the groundwater recharge potential zone. Environ Geol 58(1):185–195

Chapter 10
GIS-Based Disaster Risk Analysis of Floods Using Certainty Factor (CF) and Its Ensemble with Deep Learning Neural Network (DLNN): A Case Study of Dima Hasao District of Assam, India

Sk Ajim Ali, Farhana Parvin, and Rukhsar Anjum

Abstract Assam, a north-eastern state of India is vulnerable to floods and other natural calamities due to its wide-ranging network of river system that have a detrimental effect on the state's overall growth. Every year, during the monsoon season, the Brahmaputra and Barak Rivers, which receive water from more than 50 tributaries, produce severe floods. Dima Hasao was severely affected district in the first wave of flood in the year 2022 as communication issues starting in mid-May 2022. This research is aimed to assess disaster risk of floods in Dima Hasao district of Assam. So, this study considered various geo-physical factors and mapped for examining flood risk using certainty factor (CF) and its ensemble with deep learning neural networks (DLNN). Initially, sixteen flood conditioning factors and a numbers of historical flood locations were taken into consideration, and then the dataset were divided into training (70%) and testing datasets (30%). The findings revealed that the northern parts of the study area is where floods are most likely to occur, while the southern part is where they are least likely to occur. Model validation results using Receiver Operating Characteristic (ROC) curve further revealed that DLNN-CF is the most performant model with Area Under Curve (AUC) value of 0.974 in compare to stand-alone CF model with AUC = 0.963. For managing and mitigating future floods in the study region, this study could be useful to the decision makers and land use planners.

Keywords Flood risk assessment · GIS · Certainty factor · Ensemble learning · Deep learning neural network · Assam

---

S. A. Ali (✉) · F. Parvin (✉) · R. Anjum
Department of Geography, Faculty of Science, Aligarh Muslim University, Aligarh 202002, India
e-mail: skajimali.saa@gmail.com; skajimali@myamu.ac.in

F. Parvin
e-mail: farhanaparvin93@gmail.com; fparvin@myamu.ac.in

R. Anjum
e-mail: rukhsar.amu3@gmail.com

E. Balaji et al. (eds.), *Emerging Technologies for Water Supply, Conservation and Management*, Springer Water, https://doi.org/10.1007/978-3-031-35279-9_10

10.1 Introduction

In addition to being a deadly natural hazard, flooding can also result from rainfall and water overflow, which have much more dangerous effects [33, 53]. Therefore, the current study focused on this occurrence to analyse floods risk in southern district of Assam. Our daily lives and economy depend heavily on water. But water causes devastating impacts in form of floods, flash floods, landslides and mass movements [16, 56]. According to estimates, natural disasters in Asian countries account for more than 90% of all losses involving people and resources. Floods and landslides account for the majority of these losses [17]. The frequency of floods has grown during the past 20 years and approximately 110 million people have been impacted by it [29, 44]. In various areas of southern Assam (a state in northeast India), there have been landslides this year that have been triggered by floods, resulting in at least 30 fatalities and other losses of life [17, 43].

Flood is the most dangerous natural disasters that have caused property damage and fatalities in the mountainous valley regions. The combination of flood and other natural hazards makes it particularly perilous. Topographical, meteorological, geological, and other geo-environmental elements all interact to cause floods. Although it is hard to completely prevent flood due to its complexity, its damages can be minimised by identifying risky regions and doing a catastrophe risk analysis (DRA). The key measure in catastrophe risk mitigation is flood management [2, 6].

The early warning system (EWS), hazard mitigation, emergency response services, and flood planning for risk reduction all heavily rely on the flood susceptibility index and mapping. Prior research emphasized on assessment of flood vulnerable regions around the globe [3, 4, 20, 41, 46]. The space of old methods is constricted for flood risk assessment because of inadequate data [40, 62]. For example, the arithmetic simulation approach, run-off modeling, watermarks method, etc. are unsatisfactory for flood risk analysis [1, 10, 25, 65]. Applying these methods and models is tiring, time taking, and costly, and the mandatory data for defined regions, but data scarcity is a major problem for many developing countries [17]. Therefore, remote sensing images and GIS-assisted hybrid models are currently preferred for disaster risk analysis [12, 32, 59, 67].

Different methods for analysing flood risk have been presented in recent literatures; they can be divided into knowledge-driven and data-driven methods [17, 28]. Data-driven approaches prefer statistical methods, while knowledge-driven methods are subjective [68]. Soft computing has advanced statistical models, which have higher accuracy than traditional techniques. Examples of these models include ANNs, LR, SVM, RF, NB, ANFIS, and other tree-based classification models [8, 34, 36, 38, 63]. However, rather of analysing just one model, the recent advances is to ensemble multi-models in order to propose a more complex model and gain better accuracy. Numerous investigations have shown that distinct models have been hybridised [9, 24, 47].

The Dima Hasao district of Assam in India is vulnerable to devastating floods brought on by ongoing rains and stream overflow. The Brahmaputra valley drains the

Indian state of Assam and is extremely susceptible to calamities like landslides and floods [21, 55]. However, no research on floods and slope instability has been done in this region. As a result, in this study, the locations most likely to flood were identified using the certainty factor (CF) in combination with a DLNN. Every model has inherent drawbacks, and the accuracy of the predictions depends on the availability of input data and model construction. Although there is not universal agreement over which method is optimal for each location, combining many models could result in improved accuracy and robustness [17]. Literatures show that combining several models can lead to the best model structure for hazard susceptibility [14, 27].

In order to stress the aforementioned point, the current work sought to evaluate flood modelling utilising a combination of deep machine learning algorithms and bivariate statistics. The findings of this study will serve as a significant body of literature for the assessment of flood risk, notably for the Dima Hasao district in southern Assam. Additionally, the current analysis has identified flood-prone areas, which will aid planners in determining the best areal extension for flood risk mitigation and sustainable management plans in the study area.

10.2 Objectives

Following are the key objectives of the current study:

- To evaluate major factors responsible for disaster risk of floods in Dima Hasao district of Assam
- To collect information on past flood events and locations in in Dima Hasao district and develop models for future flood risk areas using machine learning algorithms
- To identify vulnerable areas prone to future disaster risk of floods in Dima Hasao district
- To assess and identify best models for flood risk using a model comparative analysis.

10.3 Geographical Profile of the Study Area

Assam is well-known for its frequent multi-hazard occurrences, including earthquakes, landslides, floods, and flash floods. Geographically, this state of India is separated the Barak Valley region, the Himalayan foothills, and the Brahmaputra Valley. The study region, Dima Hasao district, once was called as North Cachar Hills (Fig. 10.1). Assam is primarily a flat region; however the Dima Hasao and Cachar districts in the south contain some low and raised hills. While there are significant quantities of bituminous coal in the northern half, the Cachar area is known for its limestone reserves, which are predominantly sedimentary rocks [17].

Approximately, 75% geographical area of Dima Hasao district is covered by various kinds of forest. Different types of forest make this area one of north-east

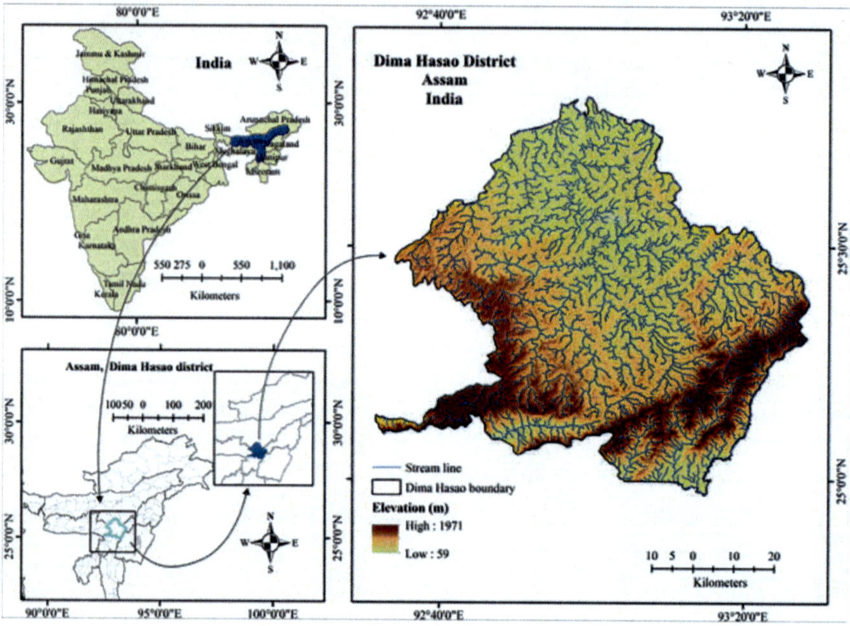

Fig. 10.1 Location of the study area

India's biodiversity zones. The tropical monsoon climate predominates in the study area, which experiences cloudy skies and heavy rainfall between the months of April and October. The highest temperature (25–38 °C) recorded throughout the summer, and annual average temperature range between 4 and 35 °C. In the Dima Hasao district, yearly rainfall ranges from 1200 to 2200 mm. Heavy rains are typically seen in June and July, which can result in flooding, erosion, and other natural disasters. Floods are brought into this area by the torrential rain and discharge water. Therefore, this present study focused on a crucial area of research for lowering the district of Dima Hasao's disaster risk due to flooding.

10.4 Materials and Methods

10.4.1 Materials

10.4.1.1 Data Source

Understanding the cause of floods in the studied region is essential for creating engineering measurements. It was discovered after a thorough investigation of the data on floods in the study area that significant rainfall triggers a downslope flow

Table 10.1 Data extraction sources and their spatial resolution

Data	Data source	Data extracted	Spatial resolution	Resolution of processed data
Topographic data (DEM)	Earth data—NASA (https://www.earthdata.nasa.gov/)	*As shown in* Fig. 10.2 (A. Land elevation; B. Slope; C. Aspect; I. General curvature; J. Plan curvature; K. Profile curvature; M. Flow accumulation)	30 m	20 m
Elevation and slope	Calculated from DEM	*As shown in* Fig. 10.2 (D. CI; E. TRI; F. SPI; L. LS factor)	30 m	20 m
Drainage network and flow accumulation	Calculated from DEM	*As shown in* Fig. 10.2 (F. SPI; H. TWI)	30 m	20 m
Climatic data	India Meteorological Department (https://mausam.imd.gov.in/)	*As shown in* Fig. 10.2 (P. Rainfall)	NA	20 m
MODIS land cover data	MODIS Land Cover (https://modis.gsfc.nasa.gov/)	*As shown in* Fig. 10.2 (O. LULC)	250 m	20 m
Flood-induced landslide locations	Google Earth pro and field report	*As shown in* Fig. 10.3	NA	NA

Source Data extracted and prepared by authors

of rainwater that leads to large-scale flooding. To analyse the catastrophe event and create the flood susceptible index (FSI), the key conditioning elements listed in the following table were applied (Table 10.1).

10.4.1.2 Flood Conditioning Factors Selection

In general, risk mapping call for assessments of the interactions among certain causative factors which have a direct or indirect impact on vulnerability and spatial occurrence of hazard incidences [42, 50]. The creation of the database using recognised conditioning factors and the integration of spatial data are essential steps in risk mapping [49, 58]. Therefore, a number of important factors that are required for the detection of flood risk areas were selected. The selected factors in the present analysis are land elevation, slope, aspect, CI, TRI, SPI, TPI, TWI, general curvature, plan curvature, profile curvature, LS factor, flow accumulation, Downslope Distance Gradient (DDG), LULC, and rainfall (Fig. 10.2a–p).

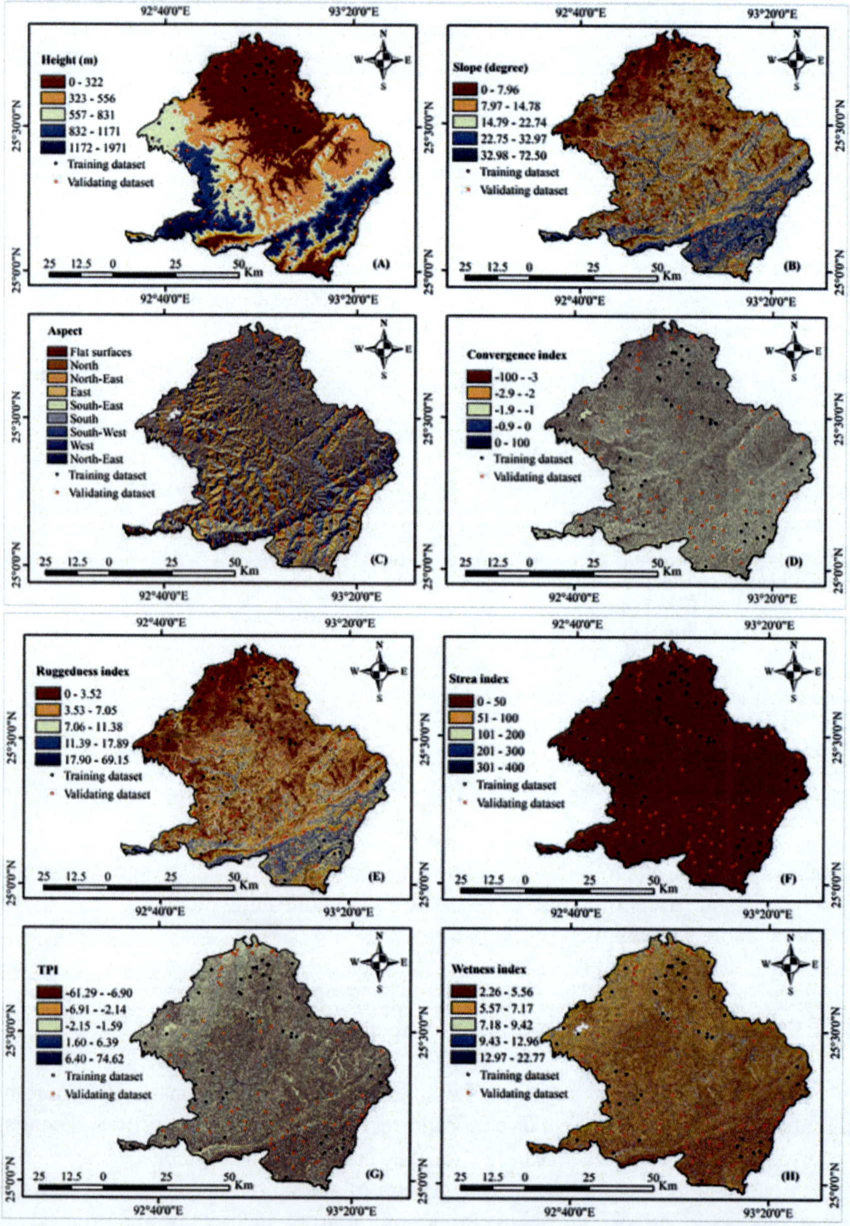

Fig. 10.2 Selected factors. **a** Land elevation; **b** Slope; **c** Aspect; **d** CI; **e** TRI; **f** SPI; **g** TPI; **h** TWI) selected factors (**i** General curvature; **j** Plan curvature; **k** Profile curvature; **l** LS factor) selected factors (**m** Flow accumulation; **n** DDG; **o** LULC; **p** Rainfall)

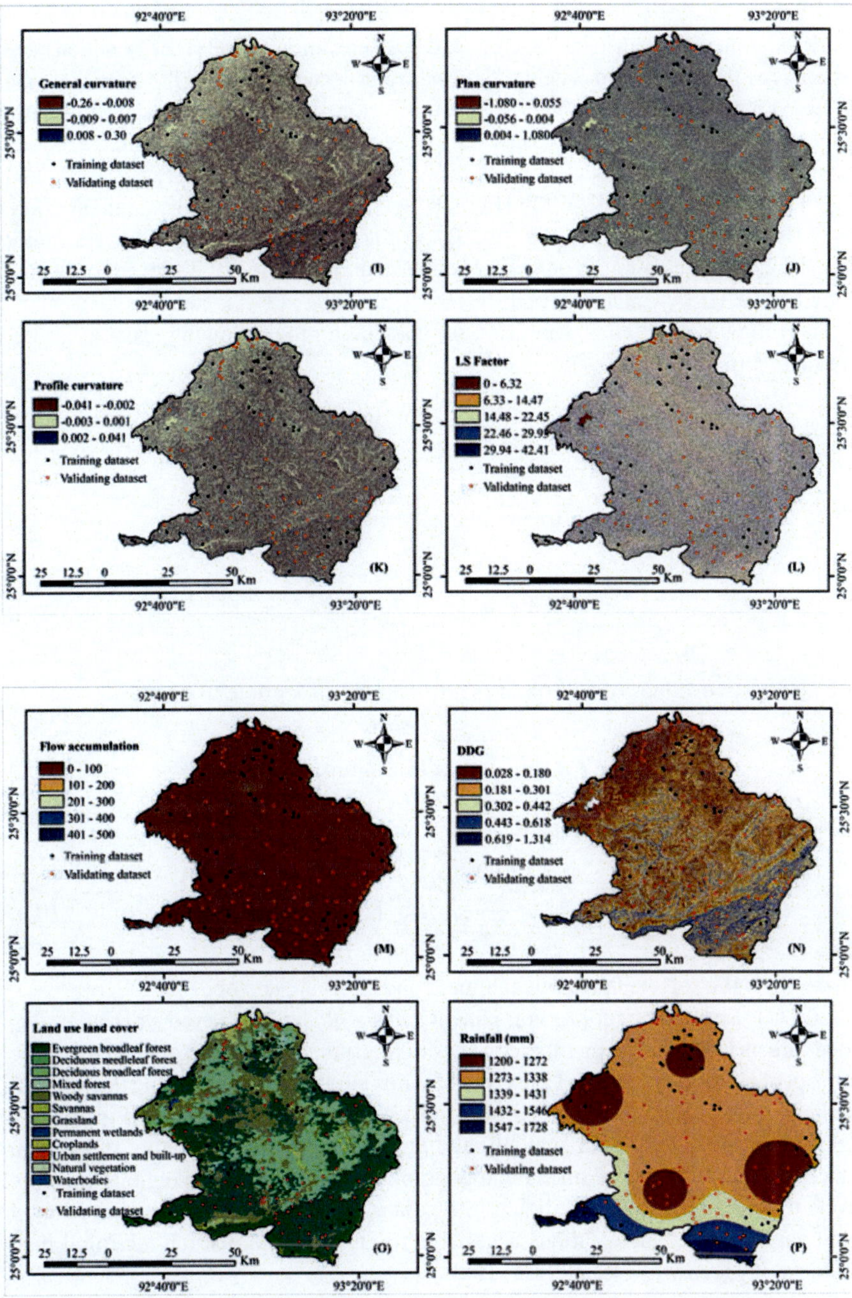

Fig. 10.2 (continued)

SRTM data were considered to create a digital elevation model (DEM), which have a spatial resolution of 30 m. The raw data was reprojected into the UTM zone from the WGS 1984 projection. For greater spatial resolution, the 30 m DEM was additionally reduced into 20 m, and the necessary data were produced. the predictor maps of land elevation (Fig. 10.2a), slope (Fig. 10.2b), aspect (Fig. 10.2c), CI (Fig. 10.2d), TRI (Fig. 10.2e), SPI (Fig. 10.2f), TPI (Fig. 10.2g), TWI (Fig. 10.2h), general curvature (Fig. 10.2i), plan curvature (Fig. 10.2j), profile curvature (Fig. 10.2k), LS factor (Fig. 10.2l), flow accumulation (Fig. 10.2m), and DDG (Fig. 10.2n), were produced using SAGA GIS v-2.2 and ArcGIS v-10.5.

The following formulas were used for calculating the following specific factors from DEM (Eqs. 10.1–10.4):

$$TWI = In\left(\frac{A}{\tan\beta}\right) \tag{10.1}$$

where A is the part of the upstream area and $tan\beta$ is the slope

$$TPI = E_0 - \sum_{i=1}^{n} E_0/n \tag{10.2}$$

The negative values of TPI point to areas much susceptible to flood.

$$TRI = \sqrt{(\max H^2 - \min H^2)} \tag{10.3}$$

where $\max H$ and $min H$ are the maximum and minimum values.

$$SPI = \left(\frac{C_a}{22.13}\right)^{0.6} \times \left(\frac{\tan\beta}{0.0896}\right)^{1.3} \tag{10.4}$$

where C_a is the selected area of catchment and $tan\beta$ is the slope.

LULC is a crucial triggering component that should be considered when assessing flood threats [39]. The dismembered zones of a certain area were explained using the concept of land cover [5]. The LULC map for this study was created using the MODIS land cover data from 2018, which was divided into twelve categories (Fig. 10.2o). Rainfall episodes vary both spatially and temporally due to the nature of rainfall, which changes with the weather and topography [52]. High rainfall regions are more prone to flooding [57, 64]. Utilizing interpolation method of rainfall data collected from various rain gauge stations accessible in Dima Hasao district, a rainfall map for the study area was created (Fig. 10.2p).

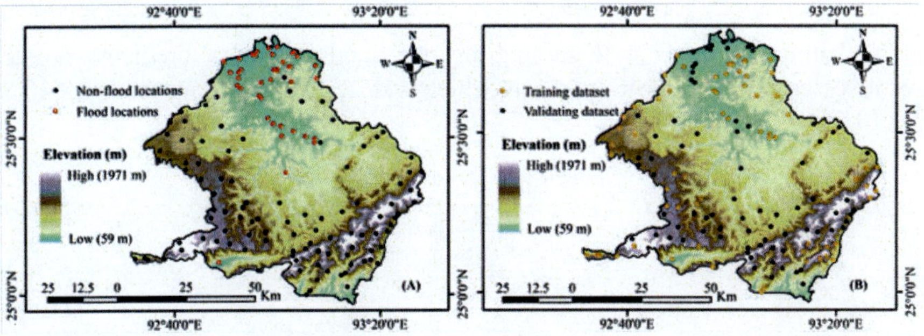

Fig. 10.3 Inventory map of flood. **a** The flood and non-flood sites. **b** Dataset used for model training and validating

10.4.1.3 Preparation of Flood Inventory

Using several methods, such as fieldwork, previous records, and the interpretation of satellite images, an inventory map was created to depict the locations of floods that have occurred in the past or are now occurring. It is used to display the geographic distribution of both flood-prone and flood-free areas in the study area. Additionally, it shows potential floodplains. As far as previous literatures are concerned, it is crucial to prepare an inventory first before moving further analysis [51]. In order to create an inventory map, different scholars favoured different approaches like field surveys, published reports, previous images, and real-time images [8, 60, 61].

Because other studies also favoured the use of earlier imagery when creating an inventory map, the present study used filed-based report and real-time Google earth engines to prepare flood inventory of the study area [45, 48]. Therefore, ten years of historical imagery were considered to identify 100 flood areas. It is also necessary to have some sites where no floods have ever happened for training and validation. To prevent personal bias, an equal number of non-flood areas were picked at random. The whole dataset were divided into 70:30 ratios for DLNN training and testing (Fig. 10.3).

10.4.2 Methods

10.4.2.1 Feature Selection Using Information Gain Ratio (IGR)

The IGR approach was used to assess the applicability of the input parameters. In the instance of predicting flood susceptibility, the IGR approach is a feature analysis technique that can identify significant input variables [22]. If IGR value is higher, it is considered as greater impact of the flood. To enhance the precision of the output, the factors which have no bearing on the incidence of floods should be eliminated from the modelling process [17]. Regarding this, the flood predictors' predictive power was assessed using the IGR approach, which was determined using the Eqs. (10.5 and 10.6).

$$IGR\,(T, E) = \frac{\text{Entropy}\,(T) - \text{Entropy}\,(T, E)}{\text{SplitEntropy}\,(T, E)} \tag{10.5}$$

$$\text{Entropy}\,(T) = -\sum_{i=1}^{2} \frac{(X_i, E)}{|T|} \log_2 \frac{n(X_i, E)}{|T|} \tag{10.6}$$

where T is training data collected from n input samples, n $n(X_i, E)$ is the No. of samples T have its place to the class label X_i (floods, non-floods)

10.4.2.2 Certainty Factor (CF)

The CF is often rule-based systems that are used to give specific elements in decision analysis weight [30]. CF studies the interaction between a variety of variables that can affect an occurrence and the uncertainty of the input data [23]. Two functions—confidence true (ct) and a measure of certainty false (cf) form the foundation of the notion of the certainty factor. When assessing flood risk, these can be described by the Eqs. (10.7 and 10.8):

$$ct = \begin{cases} \text{if } r(F) = +1 \\ \frac{\max[r(F/e), r(F)] - r(F)}{\max[+1, -1] - r(F)} \end{cases} \tag{10.7}$$

$$\text{Otherwise, } cf = \begin{cases} \text{if } r(F) = -1 \\ \frac{\min[r(F/e), r(F)] - r(F)}{\min[+1, -1] - r(F)} \end{cases} \tag{10.8}$$

where $r(F)$ is the risk of floods 'F' being true, and $r(F/e)$ is the risk of floods 'F' is true based on validation 'e'.

The ct and cf value range from -1 to $+1$, indicating categorically false and true, while the middle values, as shown in the table below, signify varied degrees of certainty (Table 10.2).

$+1$ and -1 value of CF indicates floods in that area is organized and unorganized by this class interval. These are CF's extremely high and low values. The likelihood of a flood is indicated by an increasing plus and minus value of CF, respectively. Each factor class's weight is represented by its CF. A factor class's observed likelihood of being absolutely false or true for the occurrence of flooding is expressed by the weight. Thus, the CF was normalized to decide the weight of factor's class (F_{wg}) using Eq. (10.9).

Table 10.2 Degrees of certainty used in CF

Degrees of certainty	Value of CF
Absolutely/surely not	−1
Nearly not	−0.80
Undoubtedly not	−0.60
Could be not	−0.40
Unidentified	−0.20 to +0.20
Could be	+0.40
Possibly	+0.60
Nearly	+0.80
Absolutely/surely	+1

Source Adopted from Costache et al. [17]

$$F_{wg} = \frac{\frac{f_i+1}{1-(-1)}}{\sum_{p=1}^{n} \frac{f_i+1}{1-(-1)}} \quad (10.9)$$

where f_i is the CF value of a factor class 'p'. It is similar to probabilistic reasoning.

10.4.2.3 Deep Learning Neural Network (DLNN)

DLNN is typically applied to unstructured data to improve results. In recent years, numerous studies have used deep learning approaches [7, 26]. For calculating the model, training data structures at increasingly higher levels is beneficial. A more advanced type of artificial neural network (ANN) that makes use of several deep networks is known as DLNN. The primary distinction between DLNN and ANN is in terms of network complexity.

Identifying flood risk areas in the present study, a DLNN model was applied because flood risk would be calculated at certain areas. The chosen causative factors formed input datasets in the first layer of the model, travelled over numerous hidden layers, and risk category was determined in the output layer. The sigmoid function was used to estimate B(Q = 1/p), which has values of 0 and 1, from the binary classification for the output layer. One neuron from each class I makes up the output layer, which calculates an estimate of B(Q = 1/p). When all values are added together, they equal to 1. The common function softmax was considered (Eq. 10.10).

$$\mathrm{softmax}(b_i) = \frac{\exp(b)}{\sum_k \exp(b_i)} \quad (10.10)$$

Now, the multi-layer deep neural network can be presented with the help of mathematical formulation as shows in Eqs. (10.11–10.14):

For $f = 1, \ldots, h$ (hidden layers),

$$x^{(f)}(a) = y^{(f)} + V^{(f)} p^{(f-1)}(a) \quad (10.11)$$

$$p^{(f)}(a) = \varnothing(x^{(f)}(a)) \quad (10.12)$$

For $f = O + 1$ (output layer),

$$x^{(O+1)}(a) = y^{(O+1)} V^{(O+1)} p^{(O)}(a) \quad (10.13)$$

$$p^{(O+1)}(a) = \Psi\left(x^{(O+1)}(a)\right) = I(a, \theta) \quad (10.14)$$

where, activation function is indicated by $\varnothing$ and Ψ. In each step, $V^{(f)}$ is the structure connected with many neurons in row layers f and column layers $f - 1$.

Setting all functions are crucial throughout the development of the DLNN model. The type and quantity of training and validating datasets were taken into consideration when setting the hidden layers and processing elements in this case.

10.4.2.4 Accuracy Assessment and Model Validation

Statistical Measures

The sensitivity (S_t), specificity (S_f), overall accuracy (O_v), F-score (F_s), and Kappa (Kp) statistics were considered in order to evaluate the performance of the applied models. The mathematical expressions of such measures are given (Eqs. 10.15–10.19):

$$S_t = \frac{TP_F}{TP_F + FN_F} \quad (10.15)$$

$$S_f = \frac{TN_F}{FP_F + TN_F} \quad (10.16)$$

$$O_v = \frac{TP_F + TN_F}{TP_F + FP_F} + TN_F + FN_F \quad (10.17)$$

$$F_s = 2 * \frac{p * r}{p + r} \therefore \begin{bmatrix} p = \frac{TP_F}{TP_F + FP_F} \\ r = \frac{TP_F}{TP_F + FN_F} \end{bmatrix} \quad (10.18)$$

$$Kp = \frac{F_a - F_e}{1 - F_e} \quad (10.19)$$

where TP_F and TN_F are the No. of flood pixels exactly classified; FP_F and FN_F are the No. of flood pixels inaccurately classified; F_a and F_e are the observed and predicted flood pixels, correspondingly, p and r are precision and recall, respectively.

ROC Curve

A receiver operating characteristic (ROC) curve has emerged as one of the most often used tools for comparing and evaluating model performance for determining the accuracy of flood vulnerability assessments [47]. AUC can be taken as a measurement to assess the overall effectiveness of a model, whereas ROC is an effective way to outline the value of prediction systems. The AUC's value ranges from 0.5 to 1. Better prediction ability is shown by a higher AUC value. AUC values above 0.7 indicate great predictive capability, while values between 0.7 and 0.8, 0.9, and 1.0 indicate poor predictive capability [13].

10.5 Results

10.5.1 Result of Feature Selection

Calculating IGR values was the first stage in the workflow designed to determine whether a flood causative factor has an impact in flood occurring or not. The resulting IGR values are displayed in Fig. 10.4 as a result of using Weka 3.8. Since all IGR values were greater than 0, the ensuing study took into account all flood predictors.

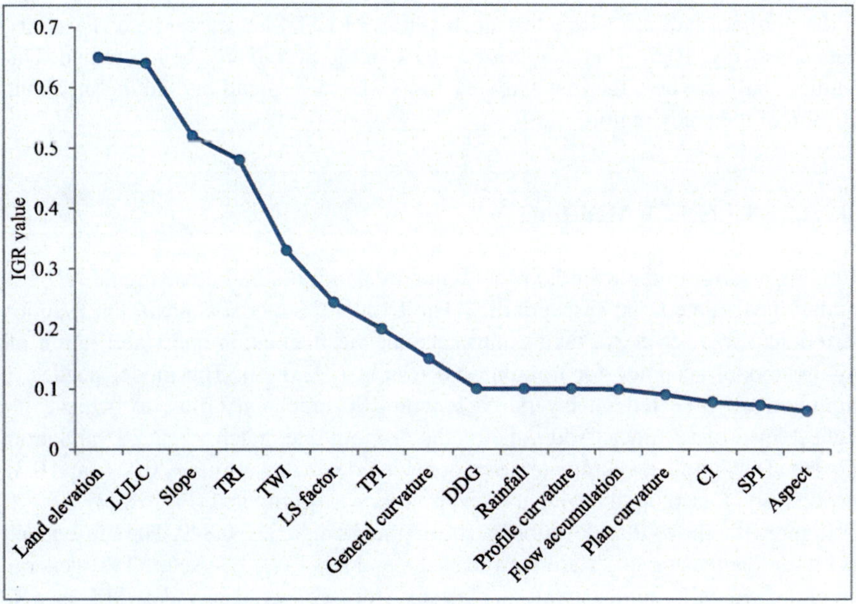

Fig. 10.4 IGR value of all flood conditioning predictors

10.5.2 *Flood Risk Assessment*

10.5.2.1 CF Modelling

For the chosen flood causative factors, the values of each class under CF were determined using the expression described in the methodology section of CF. In order to carry out the computations for the 91 classes GIS environment was utilized. In land use category, the urban settlement and built-up class was given the highest rating of CF, which is equal to 0.721. For the Downslope Degree class, the same result was achieved between 0.619 and 1.314. The category of TWI between 9.43 and 12.96 scored a CF of 0.660, whereas the grassland in land use classification obtained a CF value of 0.689 (Table 10.3). It should be noted that a lowest CF value of −1 characterises 13 different types of flood predictors.

The Flood Susceptibility Index (FSI) was calculated using CF values once they had been derived, using both the stand-alone CF approach and a hybrid method that combined CF with DLNN.

The stand-alone CF approach was applied by entering the CF values into ArcGIS v-10.5's Raster Calculator and adding them up. The class of extremely high FSI_{CF} values occurs in the northern portion of the study region and makes up roughly 6.64% of the entire area under investigation (Fig. 10.5). High FSI_{CF} values are also found in the Nordic portion of the study region, where they are found on roughly 15.13% of the entire area under study. The research area's southernmost extreme as well as the northern half are where the mean values of LSI_{CF} appear to be consistently distributed (Fig. 10.5). They comprise 27.01% of the overall area being studied. The southern half is where the low and very low FSI_{CF} are spread out, taking up about 51.19% of the study region.

10.5.2.2 DLNN-CF Modelling

Running a script in the RStudio programme made it possible to use the DLNN-CF model to calculate flood susceptibility. The DLNN-CF model's maximum training period size was set at 30, taking into account the fluctuation and stabilisation of highest model accuracy for training and testing (i.e. 0.99). The model needed a minimum of three hidden layers, each with 100 hidden neurons, to achieve its highest level of accuracy. Additionally, the dropout rate, batch size, and validation rate are also significant parameter to be considered which were set as 0.1, 4, and 0.3, respectively. Based on that weight of each factor was computed (Fig. 10.6).

Figure 10.6 shows that elevation had the highest weight i.e., 0.849. The subsequent factors in decreasing order are land use, slope, TRI, TWI, LS factor, TPI, general curvature, rainfall, profile curvature, plan curvature, convergence index, SPI, aspect, flow accumulation, and downslope with weight of 0.819, 0.790, 0.740, 0.642, 0.520, 0.471, 0.448, 0.388, 0.380, 0.07, 0.19, 0.11, 0.09, 0.08, and 0.01, respectively. The flood susceptibility index ($FSI_{DLNN\text{-}CF}$) was generated using the raster calculator in

Table 10.3 Flood triggering factors and calculated value of CF

Flood predictors	Category	CF value
Elevation	0–322	0.419
	323–556	−0.511
	557–831	−1.00
	832–1171	−1.00
	1172–1971	−1.00
Slope	0–7.96	0.629
	7.97–14.78	−0.141
	14.79–22.74	−0.721
	22.75–32.97	−0.93
	32.98–72.50	−1.00
Aspect	Flat surfaces	−1.00
	North	−0.023
	North-East	0.004
	East	0.189
	South-East	−0.159
	South	0.051
	South-West	−0.206
	West	−0.182
	North-East	0.205
Convergence index	(−100) – (−3)	0.039
	(−2.9) – (−2)	−0.422
	(−1.9) – (−1)	−0.441
	(−0.9) – 0	−0.201
	0–100	−0.006
TRI	0–3.52	0.491
	3.53–7.05	−0.711
	7.06–11.38	−0.909
	11.39–17.89	−1.00
	17.90–69.15	−1.00
SPI	0–50	−0.059
	51–100	0.605
	101–200	0.067
	201–300	0.203
	301–400	−1.00
TPI	(−61.29) – (−6.90)	−0.860
	(−6.91) – (−2.14)	−0.629
	(−2.15) – 1.59	0.40

(continued)

Table 10.3 (continued)

Flood predictors	Category	CF value
	1.60–6.39	−0.466
	6.40–74.62	−1.00
TWI	2.26–5.56	−0.889
	5.67–7.17	0.009
	7.18–9.42	0.440
	9.43–12.96	0.660
	12.97–22.77	0.455
General curvature	(−0.26) – (−0.008)	−0.65
	(−0.009) – 0.007	0.301
	0.008 – 0.300	−0.736
Plan curvature	(−1.080) – (−0.055)	0.450
	−0.056 – 0.004	−0.169
	0.004 – 1.080	0.490
Profile curvature	(−0.041) – (−0.002)	−0.609
	(−0.001) – 0.001	0.250
	0.002–0.041	−0.439
LS factor	0–6.32	0.559
	6.33–14.47	0.540
	14.48–22.45	0.461
	22.46–29.93	−0.620
	29.94–42.41	−1.00
Flow accumulation	0–100	−0.056
	101–200	0.452
	201–300	0.140
	301–400	−0.008
	401–500	−1.00
DDG	0.028–0.180	−0.509
	0.181–0.301	−0.110
	0.302–0.442	0.319
	0.443–0.618	−0.337
	0.619–1.314	0.721
Land use	Evergreen broadleaf forest	−0.903
	Deciduous needleleaf forest	−1.00
	Deciduous broadleaf forest	−0.681
	Mixed forest	−1.00
	Woody savannas	−0.269
	Savannas	0.451

(continued)

Table 10.3 (continued)

Flood predictors	Category	CF value
	Grassland	0.689
	Permanent wetlands	0.624
	Croplands	0.590
	Urban settlement and built-up	0.721
	Natural vegetation	−0.509
	Waterbodies	−1.00
Rainfall	1200–1272	−0.461
	1273–1338	−0.879
	1339–1431	−0.074
	1432–1546	0.248
	1547–1728	0.410

Source Calculated by authors

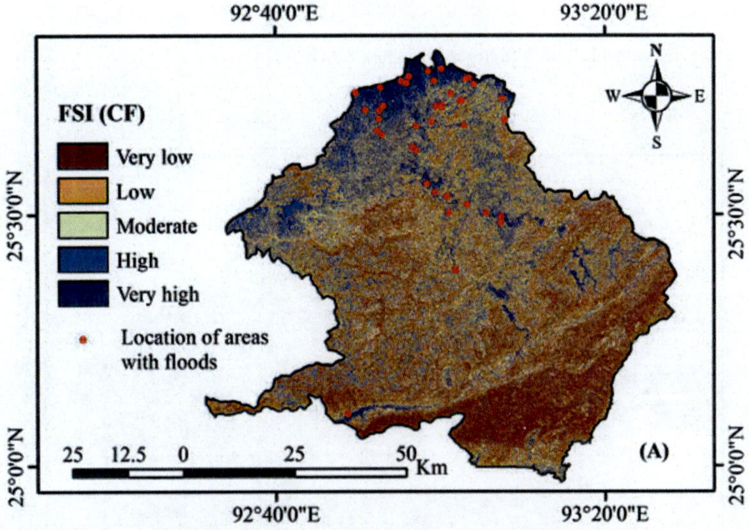

Fig. 10.5 Flood susceptibility index using FSI_{CF} model

ArcGIS v-10.5 taking the weights of the components into account. The northern part of the research region has the very high values of $FSI_{DLNN\text{-}CF}$, which make up roughly 6.45% of the total area. 14.74% of the research region has high levels of $FSI_{DLNN\text{-}CF}$, whereas 21.43% of it has intermediate susceptibility. The low and very low values of $FSI_{DLNN\text{-}CF}$ cover approximately 57.36% of the entire study region which are situated mainly in the southern parts (Fig. 10.7).

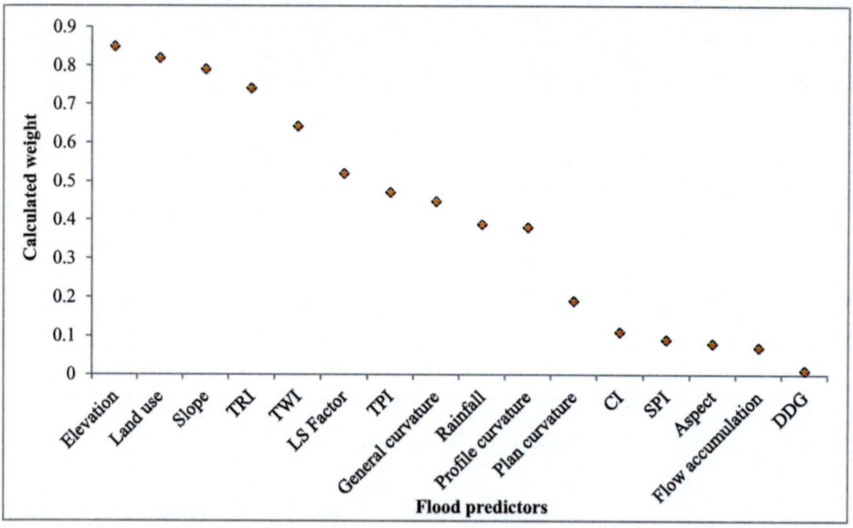

Fig. 10.6 Weight of each flood predictor estimated by DLNN-CF

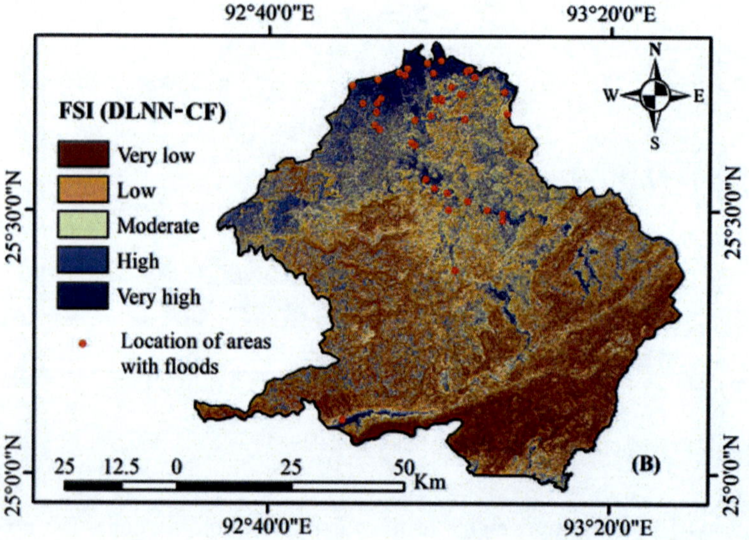

Fig. 10.7 Flood susceptibility index using $FSI_{DLNN\text{-}CF}$ model

10.5.3 Results of Model Validation

The results were validated using ROC Curve and other statistical methods. Both the success rate (SR) and prediction rate (PR) using the training and validating dataset were constructed from the perspective of ROC Curves (Fig. 10.8). According to the SR, the $FSI_{DLNN\text{-}CF}$ recorded the highest AUC-ROC values (0.982), while the FSI_{CF} recorded the lowest values (0.961). In terms of PR, the $FSI_{DLNN\text{-}CF}$ had the highest AUC-ROC value (0.969), and the FSI_{CF} had the lowest value (0.938). Therefore, it can be concluded that the ROC curve illustrates a strong performances of the both used models.

In addition to using ROC Curve, other statistical methods were also used to assess the results' validity. In compared to FSI_{CF} (0.901), the $FSI_{DLNN\text{-}CF}$ model thus achieved the greatest accuracy score of 0.923 for the training dataset. Using a validating dataset, the outcome showed that $FSI_{DLNN\text{-}CF}$ and FSI_{CF} both had accuracy of 0.889 and 0.878, respectively (Table 10.4).

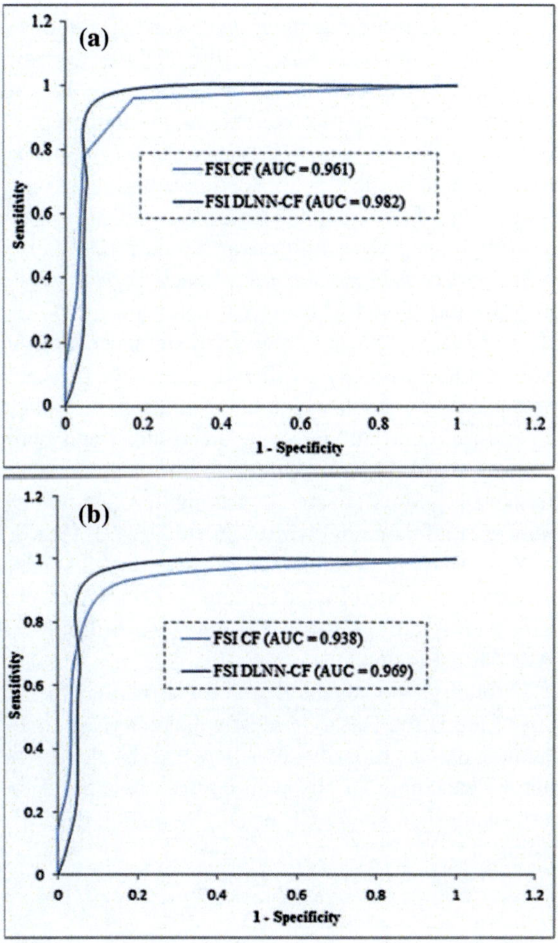

Fig. 10.8 Showing ROC curve of the applied models. **a** SP. **b** PR

Table 10.4 Statistical measures tested for evaluating performance of the applied models

	Models	S_t	S_f	O_v	k-index
Training	CF	0.901	0.901	0.901	0.80
	DLNN-CF	0.923	0.923	0.923	0.83
Validating	CF	0.878	0.878	0.878	0.77
	DLNN-CF	0.889	0.889	0.889	0.80

Source Calculated by authors

10.6 Discussion

Floods are currently among the record hazardous occurrences, and when they combine through other hazards like landslides, mass migration, and erosion, these risks become disastrous for the regions where they occur [11, 31, 35, 66]. Because of enormous amount of precipitation, the study region is one of vulnerable to water related disaster in Asia [37]. Therefore, for the purpose of protecting property and human life, identification of flood potential areas are very crucial [17, 54]. In order to determine FSI in a southern district of Assam state, two models namely, FSI_{CF} and $FSI_{DLNN\text{-}CF}$ were utilised in the current study (Dima Hasao). Therefore, the sophisticated machine learning model, DLNN, was integrated with the statistical bivariate model (CF). The great precision of the used models, as underlined by SR and PR, served as a sign of their dependability in the current study. In the case of PR, $FSI_{DLNN\text{-}CF}$ achieved the highest AUC of 0.969. It is amazing that the alternative model, which obtained an AUC value of 0.938, likewise came quite close. In the current study, the stand-alone and optimised DLNN model used by Costache et al. [17] for susceptibility analysis of landslides caused by floods in Barak valley, Southern Assam was applied, and found that the DLNN-CF model was as most efficient model with AUC of 0.974. Hence, it can be said that adopting the DLNN ensemble structure can produce findings with more accuracy. This claim is further corroborated by the finding by [19], who were able to get an AUC of 0.959 using their Iterative Classifier Optimizer—Multilayer Perceptron—Frequency Ratio (ICO-MLP-FR) model.

Through this analysis, it was found that all the models agree that the northern part of the Dima Hasao is mostly characterised by high risk of flood, whereas the southern part is primarily defined by less risk flood. The current study also concluded that the ensemble DLNN with CF to assess flood vulnerability is proved as more performant than single CF or DLNN model, earlier studies also found the similar results [15, 17–19].

10.7 Conclusions

The great exposure of such natural calamity highlights the need for in-depth studies on flood susceptibility modelling. In this study, an ensemble models (DLNN-CF) and a stand-alone bivariate model (CF) were proposed and tested to detect the flood vulnerable areas in Dima Hasao, a southern district of Assam, India. For model training and testing, total 200 flood and non-flood locations were considered as input data, along with 16 flood causative factors. Elevation was the most significant flood conditioning element, per the machine learning modelling. All of the models showed that more than 21% of the study area is located in areas with high and very high flood susceptibility, with these areas being particularly prevalent in the region's north. Additionally, the DLNN-CF model was shown to be the most effective one, with an AUC-ROC Curve above 0.966.

The use of the ensemble model to predict the flood vulnerability zones for Dima Hasao district is the key basis of novelty that distinguishes this study from previous ones in the literature. The authorities in charge of disaster risk reduction may also find a lot of information to use from this study.

10.8 Recommendations

Some flood monitoring approaches date back to ancient times. These approaches include creating flood-ways, terracing hillsides to limit flow downhill, and plantation to retain extra water. Among other options include constructing levees, dikes, dams, reservoirs, or retention ponds to hold excess water during floods.

Recently with advancement of satellite technology, GIS tools, and artificial intelligence, the flooding events are predicted to calculate the probability that occur and to estimate environmental, social, and economic implications under existing conditions. The present study also highlighted how models based GIS tools and machine learning algorithms can enable to detect flood-prone areas.

However, still now flood events devastatingly affect many terrains throughout the world, which indicates that more researches and practical approaches are required to overcome such issue and fulfil the existing research gap. In this regard, the following recommendations are be utilized:

(i) A nature based flood solution is encouraged. Land conservation, permeable pavements, coastal dunes, greenways, restoring and protecting wetlands, and storm water parks are some recommendation for nature based flood management and mitigation. In case of urban flood, Nature based solutions (NBS) favour storm water preservation and percolation, help in flood mitigation and enhancement of water quality. Figure 10.9 shows some examples of NBS for floods.

(ii) It is important to use a participative method to establish an acceptable level of risk.

Fig. 10.9 Examples of Nature based solutions for flood management and risk mitigation. *Source* The World Bank https://www.worldbank.org

(iii) Impact assessment should be prepared and detailed impacts outlined.
(iv) To comply with the degree of risk assessment, appropriate measures should be evaluated and implemented. This procedure helps the public accept different measures that may increase or decrease risk exposure and can improve community involvement in the developed flood problem solutions.
(v) A grid-based rainfall-runoff model is also suggested in this regard. In this model, rainfall-runoff relationships are described mathematically of a catchment area, drainage basin or watershed. Using this model, storm water runoff can be managed to protect water quality and promote sustainable development.

Acknowledgements We thankfully acknowledge the reviewers and editorial team for their valuable time, productive comments and suggestions for improving the overall quality of the manuscript.

Ethical Approval NA.

Consent to Participate NA.

Consent to Publish NA.

Authors Contributions **Sk Ajim Ali:** Conceptualization, model preparation, methodology development, Writing-original draft, Software, Formal analysis, Visualization. **Farhana Parvin** and **Rukhsar Anjum**: Formal analysis; Data curation, Writing, Review and editing.

Competing Interests There are no conflicts of interest to declare.

Availability of Data and Materials The data that support the findings of this study are available from the corresponding authors, [Sk Ajim Ali, skajimali.saa@gmail.com and Farhana Parvin, farhanaparvin93@gmail.com], upon reasonable request.

References

1. Abdessamed D, Abderrazak B (2019) Coupling HEC-RAS and HEC-HMS in rainfall–runoff modeling and evaluating floodplain inundation maps in arid environments: case study of Ain Sefra city, Ksour Mountain. SW of Algeria. Environ Earth Sci 78(19):1–17
2. Abu El-Magd SA, Orabi HO, Ali SA, Parvin F, Pham QB (2021) An integrated approach for evaluating the flash flood risk and potential erosion using the hydrologic indices and morpho-tectonic parameters. Environ Earth Sci 80(20):1–17
3. Ajibade FO, Ajibade TF, Idowu TE, Nwogwu NA, Adelodun B, Lasisi KH et al (2021) Flood-prone area mapping using GIS-based analytical hierarchy frameworks for Ibadan city, Nigeria. J Multi-Criteria Decis Anal 28(5–6):283–295
4. Ali SA, Khatun R, Ahmad A, Ahmad SN (2019) Application of GIS-based analytic hierarchy process and frequency ratio model to flood vulnerable mapping and risk area estimation at Sundarban region, India. Model Earth Syst Environ 5(3):1083–1102
5. Ali SA, Parvin F, Pham QB, Khedher KM, Dehbozorgi M, Rabby YW et al (2022) An ensemble random forest tree with SVM, ANN, NBT, and LMT for landslide susceptibility mapping in the Rangit River watershed, India. Nat Hazards 1–33
6. Ali SA, Parvin F, Pham QB, Vojtek M, Vojteková J, Costache R et al (2020) GIS-based comparative assessment of flood susceptibility mapping using hybrid multi-criteria decision-making approach, naïve Bayes tree, bivariate statistics and logistic regression: a case of Topľa basin, Slovakia. Ecol Indic 117:106620
7. Ali SA, Parvin F, Vojteková J, Costache R, Linh NTT, Pham QB et al (2021) GIS-based landslide susceptibility modeling: a comparison between fuzzy multi-criteria and machine learning algorithms. Geosci Front 12(2):857–876
8. Al-Juaidi AE, Nassar AM, Al-Juaidi OE (2018) Evaluation of flood susceptibility mapping using logistic regression and GIS conditioning factors. Arab J Geosci 11(24):1–10
9. Band SS, Janizadeh S, Chandra Pal S, Saha A, Chakrabortty R, Melesse AM, Mosavi A (2020) Flash flood susceptibility modeling using new approaches of hybrid and ensemble tree-based machine learning algorithms. Remote Sens 12(21):3568
10. Ben Khalfallah C, Trudel M, Leblanc SG (2020) Calibration of a hydrodynamic model (HEC-RAS) using UAV imagery to delineate high-water mark of a flood event. In: AGU fall meeting abstracts, vol 2020, pp H011–0004
11. Bordbar M, Aghamohammadi H, Pourghasemi HR, Azizi Z (2022) Multi-hazard spatial modeling via ensembles of machine learning and meta-heuristic techniques. Sci Rep 12(1):1–17
12. Bui DT, Tsangaratos P, Ngo PTT, Pham TD, Pham BT (2019) Flash flood susceptibility modeling using an optimized fuzzy rule based feature selection technique and tree based ensemble methods. Sci Total Environ 668:1038–1054
13. Cao J, Zhang Z, Du J, Zhang L, Song Y, Sun G (2020) Multi-geohazards susceptibility mapping based on machine learning—A case study in Jiuzhaigou, China. Nat Hazards 102(3):851–871
14. Chen W, Chen X, Peng J, Panahi M, Lee S (2021) Landslide susceptibility modeling based on ANFIS with teaching-learning-based optimization and Satin bowerbird optimizer. Geosci Front 12(1):93–107
15. Chowdhuri I, Pal SC, Arabameri A, Saha A, Chakrabortty R, Blaschke T et al (2020) Implementation of artificial intelligence based ensemble models for gully erosion susceptibility assessment. Remote Sens 12(21):3620
16. Coscarelli R, Aguilar E, Petrucci O, Vicente-Serrano SM, Zimbo F (2021) The potential role of climate indices to explain floods, mass-movement events and wildfires in Southern Italy. Climate 9(11):156
17. Costache R, Ali SA, Parvin F, Pham QB, Arabameri A, Nguyen H et al (2021a) Detection of areas prone to flood-induced landslides risk using certainty factor and its hybridization with FAHP, XGBoost and deep learning neural network. Geocarto Int 1–36
18. Costache R, Arabameri A, Blaschke T, Pham QB, Pham BT, Pandey M et al (2021) Flash-flood potential mapping using deep learning, alternating decision trees and data provided by remote sensing sensors. Sensors 21(1):280

19. Costache R, Arabameri A, Costache I, Crăciun A, Islam ARMT, Abba SI et al (2022) Flood susceptibility evaluation through deep learning optimizer ensembles and GIS techniques. J Environ Manag 316:115316
20. Costache R, Pham QB, Arabameri A, Diaconu DC, Costache I, Crăciun A et al (2021b) Flash-flood propagation susceptibility estimation using weights of evidence and their novel ensembles with multicriteria decision making and machine learning. Geocarto Int 1–33
21. Devrani R, Srivastava P, Kumar R, Kasana P (2022) Characterization and assessment of flood inundated areas of lower Brahmaputra River Basin using multitemporal synthetic Aperture Radar data: a case study from NE India. Geol J 57(2):622–646
22. Dodangeh E, Panahi M, Rezaie F, Lee S, Bui DT, Lee CW, Pradhan B (2020) Novel hybrid intelligence models for flood-susceptibility prediction: Meta optimization of the GMDH and SVR models with the genetic algorithm and harmony search. J Hydrol 590:125423
23. Elkano M, Galar M, Sanz JA, Schiavo PF, Pereira S Jr, Dimuro GP et al (2018) Consensus via penalty functions for decision making in ensembles in fuzzy rule-based classification systems. Appl Soft Comput 67:728–740
24. Fang L, Huang J, Cai J, Nitivattananon V (2022) Hybrid approach for flood susceptibility assessment in a flood-prone mountainous catchment in China. J Hydrol 612:128091
25. Firozjaii AM, Saghafi HR (2020) Review on chemical enhanced oil recovery using polymer flooding: fundamentals, experimental and numerical simulation. Petroleum 6(2):115–122
26. Ghorbanzadeh O, Blaschke T, Gholamnia K, Meena SR, Tiede D, Aryal J (2019) Evaluation of different machine learning methods and deep-learning convolutional neural networks for landslide detection. Remote Sens 11(2):196
27. Ghorbanzadeh O, Rostamzadeh H, Blaschke T, Gholaminia K, Aryal J (2018) A new GIS-based data mining technique using an adaptive neuro-fuzzy inference system (ANFIS) and k-fold cross-validation approach for land subsidence susceptibility mapping. Nat Hazards 94(2):497–517
28. Gideon D, Mustafa FB (2022) The application of a data-driven method for spatial analysis and prediction of gully erosion susceptibility. In: Methodological approaches in physical geography. Springer, Cham, pp 27–51
29. Guo Y, Wu Y, Wen B, Huang W, Ju K, Gao Y, Li S (2020) Floods in China, COVID-19, and climate change. Lancet Planet Health 4(10):e443–e444
30. Hassanzad M, Orooji A, Valinejadi A, Velayati A (2017) A fuzzy rule-based expert system for diagnosing cystic fibrosis. Electron Phys 9(12):5974
31. Hitouri S, Varasano A, Mohajane M, Ijlil S, Essahlaoui N, Ali SA et al (2022) Hybrid machine learning approach for gully erosion mapping susceptibility at a watershed scale. ISPRS Int J Geo Inf 11(7):401
32. Kalantar B, Ueda N, Saeidi V, Janizadeh S, Shabani F, Ahmadi K, Shabani F (2021) Deep neural network utilizing remote sensing datasets for flood hazard susceptibility mapping in Brisbane, Australia. Remote Sens 13(13):2638
33. Khan AN (2011) Analysis of flood causes and associated socio-economic damages in the Hindukush region. Nat Hazards 59(3):1239–1260
34. Khoirunisa N, Ku CY, Liu CY (2021) A GIS-based artificial neural network model for flood susceptibility assessment. Int J Environ Res Public Health 18(3):1072
35. Leroy SA, Gracheva R, Medvedev A (2022) Natural hazards and disasters around the Caspian Sea. Nat Hazards 1–44
36. Liu J, Wang J, Xiong J, Cheng W, Li Y, Cao Y et al (2022) Assessment of flood susceptibility mapping using support vector machine, logistic regression and their ensemble techniques in the Belt and Road region. Geocarto Int 1–30
37. Ludescher J, Martin M, Boers N, Bunde A, Ciemer C, Fan J et al (2021) Network-based forecasting of climate phenomena. Proc Natl Acad Sci 118(47):e1922872118
38. Mohammadi A, Kamran KV, Karimzadeh S, Shahabi H, Al-Ansari N (2020) Flood detection and susceptibility mapping using sentinel-1 time series, alternating decision trees, and bag-adtree models. Complexity

39. Mojaddadi H, Pradhan B, Nampak H, Ahmad N, Ghazali AHB (2017) Ensemble machine-learning-based geospatial approach for flood risk assessment using multi-sensor remote-sensing data and GIS. Geomat Nat Haz Risk 8(2):1080–1102
40. Muhadi NA, Abdullah AF, Bejo SK, Mahadi MR, Mijic A (2020) The use of LiDAR-derived DEM in flood applications: a review. Remote Sens 12(14):2308
41. Mukherjee F, Singh D (2020) Detecting flood prone areas in Harris County: a GIS based analysis. GeoJournal 85(3):647–663
42. Nguyen TT, Bonetti J, Rogers K, Woodroffe CD (2016) Indicator-based assessment of climate-change impacts on coasts: a review of concepts, methodological approaches and vulnerability indices. Ocean Coast Manag 123:18–43
43. Pandey K, Vishwakarma DK (2019) 7 Flash floods cause and remedial measures for their control in hilly regions. Appl Agric Pract Mitigating Clim Change 2:77
44. Park E, Ho HL, Tran DD, Yang X, Alcantara E, Merino E, Son VH (2020) Dramatic decrease of flood frequency in the Mekong Delta due to river-bed mining and dyke construction. Sci Total Environ 723:138066
45. Parvin F, Ali SA, Calka B, Bielecka E, Linh NTT, Pham QB (2022) Urban flood vulnerability assessment in a densely urbanized city using multi-factor analysis and machine learning algorithms. Theor Appl Climatol 1–21
46. Pathan AI, Girish Agnihotri P, Said S, Patel D (2022) AHP and TOPSIS based flood risk assessment-a case study of the Navsari City, Gujarat, India. Environ Monitor Assess 194(7):1–37
47. Pham BT, Luu C, Van Phong T, Nguyen HD, Van Le H, Tran TQ et al (2021) Flood risk assessment using hybrid artificial intelligence models integrated with multi-criteria decision analysis in Quang Nam Province, Vietnam. J Hydrol 592:125815
48. Pham QB, Ali SA, Bielecka E, Calka B, Orych A, Parvin F, Łupikasza E (2022) Flood vulnerability and buildings' flood exposure assessment in a densely urbanised city: comparative analysis of three scenarios using a neural network approach. Nat Hazards 1–39
49. Pourghasemi HR, Beheshtirad M, Pradhan B (2016) A comparative assessment of prediction capabilities of modified analytical hierarchy process (M-AHP) and Mamdani fuzzy logic models using Netcad-GIS for forest fire susceptibility mapping. Geomat Nat Haz Risk 7(2):861–885
50. Pradhan B (2011) Use of GIS-based fuzzy logic relations and its cross application to produce landslide susceptibility maps in three test areas in Malaysia. Environ Earth Sci 63(2):329–349
51. Rahmati O, Pourghasemi HR, Zeinivand H (2016) Flood susceptibility mapping using frequency ratio and weights-of-evidence models in the Golastan Province, Iran. Geocarto Int 31(1):42–70
52. Ran Q, Fu W, Liu Y, Li T, Shi K, Sivakumar B (2018) Evaluation of quantitative precipitation predictions by ECMWF, CMA, and UKMO for flood forecasting: application to two basins in China. Nat Hazard Rev 19(2):05018003
53. Safiah Yusmah MY, Bracken LJ, Sahdan Z, Norhaslina H, Melasutra MD, Ghaffarianhoseini A et al (2020) Understanding urban flood vulnerability and resilience: a case study of Kuantan, Pahang, Malaysia. Nat Hazards 101(2):551–571
54. Saha AK, Agrawal S (2020) Mapping and assessment of flood risk in Prayagraj district, India: a GIS and remote sensing study. Nanotechnol Environ Eng 5(2):1–18
55. Saikia L (2022) Flood Hazard of the Brahmaputra River in Assam: current mitigation approaches, challenges and sustainable solution options. In: Riverine systems. Springer, Cham, pp 99–111
56. Sati VP (2013) Extreme weather related disasters: a case study of two flashfloods hit areas of Badrinath and Kedarnath Valleys, Uttarakhand Himalaya, India. J Earth Sci Eng 3(8):562
57. Shen Y, Morsy MM, Huxley C, Tahvildari N, Goodall JL (2019) Flood risk assessment and increased resilience for coastal urban watersheds under the combined impact of storm tide and heavy rainfall. J Hydrol 579:124159
58. Tehrany MS, Pradhan B, Mansor S, Ahmad N (2015) Flood susceptibility assessment using GIS-based support vector machine model with different kernel types. Catena 125:91–101

59. Tien Bui D, Khosravi K, Shahabi H, Daggupati P, Adamowski JF, Melesse AM et al (2019) Flood spatial modeling in northern Iran using remote sensing and GIS: a comparison between evidential belief functions and its ensemble with a multivariate logistic regression model. Remote Sens 11(13):1589
60. Tripathy P, Malladi T (2022) Global Flood Mapper: a novel Google Earth Engine application for rapid flood mapping using Sentinel-1 SAR. Nat Hazards 1–23
61. Veh G, Korup O, von Specht S, Roessner S, Walz A (2019) Unchanged frequency of moraine-dammed glacial lake outburst floods in the Himalaya. Nat Clim Chang 9(5):379–383
62. Wagenaar D, Curran A, Balbi M, Bhardwaj A, Soden R, Hartato E et al (2020) Invited perspectives: how machine learning will change flood risk and impact assessment. Nat Hazard 20(4):1149–1161
63. Wang Y, Hong H, Chen W, Li S, Panahi M, Khosravi K et al (2019) Flood susceptibility mapping in Dingnan County (China) using adaptive neuro-fuzzy inference system with biogeography based optimization and imperialistic competitive algorithm. J Environ Manag 247:712–729
64. Wei L, Hu KH, Hu XD (2018) Rainfall occurrence and its relation to flood damage in China from 2000 to 2015. J Mt Sci 15(11):2492–2504
65. Wright DB, Yu G, England JF (2020) Six decades of rainfall and flood frequency analysis using stochastic storm transposition: review, progress, and prospects. J Hydrol 585:124816
66. Yanilmaz S, Baskak D, Yucesan M, Gul M (2021) Extension of FEMA and SMUG models with Bayesian best-worst method for disaster risk reduction. Int J Disaster Risk Reduct 66:102631
67. Yousefi S, Avand M, Yariyan P, Pourghasemi HR, Keesstra S, Tavangar S, Tabibian S (2020) A novel GIS-based ensemble technique for rangeland downward trend mapping as an ecological indicator change. Ecol Ind 117:106591
68. Yu L, Zhou C, Wang Y, Cao Y, Peres DJ (2022) Coupling data-and knowledge-driven methods for landslide susceptibility mapping in human-modified environments: a case study from Wanzhou County, three Gorges Reservoir Area, China. Remote Sens 14(3):774

Chapter 11
Geospatial and Analytical Hierarchical Techniques to Assess the Groundwater Potential Areas in Kanyakumari District, Tamil Nadu, India

Annmaria K. George, M. Suresh Gandhi, P. Muthukumar, and S. Selvam

Abstract The utilization of surface and groundwater resources are increasing every year. Intensive groundwater extraction, growing urban complexes and industries are causing enormous pressure on the existing groundwater resources in the Kanyakumari district. This study define the Groundwater potential zones (GWPZ) of Kanyakumari district, Tamil Nadu, India by the application of Geographic Information System (GIS) and Analytical Hierarchy Process (AHP). Slope, geology, soil texture, lineament, drainage, rainfall and land use maps of the study area were collected from different data sources. These parameters were then converted to raster datasets using Arc GIS software's raster conversion function. On the basis of contribution from each parameter towards the groundwater potential, the weights were assigned to each map and then combined to produce the final map. The resulting GWPZ map was divided into different classes that range from very high to very low. Areas of high and very-high aquifer recharge potentials were localized in the southern parts, spanning 19.02% (321 km^2) and 18.7% (317 km^2) respectively and areas of low and very low aquifer recharge potentials were distributed in the northern part about 27.91% (471 km^2) and 9.06% (153 km^2) respectively. Moderate potential zones were found to be 25.19% (425 km^2) and is spread unevenly over the study area. The obtained results were compared with well data collected from various wells in the district and similarities were found. Applying these decision-making models

A. K. George (✉) · M. S. Gandhi
Department of Geology, University of Madras, Guindy Campus, Chennai 600025, India
e-mail: annmariakgeorge@gmail.com

M. S. Gandhi
e-mail: msureshgandhi@gmail.com

P. Muthukumar · S. Selvam
Department of Geology, V.O. Chidambaram College, Thoothukudi, Tamil Nadu 628008, India
e-mail: muthukumargeo24@gmail.com

S. Selvam
e-mail: geoselvam10@gmail.com

E. Balaji et al. (eds.), *Emerging Technologies for Water Supply, Conservation and Management*, Springer Water, https://doi.org/10.1007/978-3-031-35279-9_11

not only helps to identify and predict vulnerabilities, but also helps policy makers to mitigate risk and plan solutions.

Keywords Groundwater · GIS · AHP · Kanyakumari

11.1 Introduction

Groundwater is an abundant source of freshwater found beneath the ground in the tiny gaps between the rocks, soils and sediments. The area under the ground where the groundwater is stored is called an aquifer. The origin and movement of groundwater is mainly controlled by several properties of aquifers such as thickness, rock formation, structure and permeability. Rocks with high porosity and permeability in an area with lot of rain can be a good aquifer. The other factors that influence groundwater flow is differences in pressure or overlying weight of the rock. Groundwater will flow away from the area with greatest pressure above it (the greatest weight of the overlying rock) and towards the area with the least pressure (lowest weight of the overlying rock). Depending on the type of material through which the water flows, the flow can be much slower or faster. Groundwater is variable with seasons. When the recharge is less, due to dry season, the water table decline and conversely, when the recharge is high by excess rainfall, the water table rises. Approximately 85% of domestic water needs in rural areas, 50% of urban water needs, and more than 50% of irrigation water needs are met from groundwater sources. Evaluating groundwater potential zones enable us to reduce water scarcity and improve the irrigation practices. With proper groundwater management we can make sure that our groundwater benefits many generations to come. The areas surrounding substantial and economically usable quantity of groundwater resources are termed as groundwater potential zones.

To learn more about the availability and control of groundwater prospecting, it is necessary to find areas with very high groundwater potential and areas with very low groundwater potential. Several methods have been identified for detecting potential groundwater areas. Geological, geophysical, and hydrogeological approaches were considered as the early conventional techniques for groundwater prospect. Now a days, GIS and Remote sensing is treated as a well-known method for groundwater studies across the world. The purpose of this research is to assess the groundwater likely zones in the Kanyakumari district of Tamil Nadu. A total of seven influencing factors were used for the study. The final output map depicts the groundwater potential zones, which are classified as poor, good, high, and very high classes. Therefore, through this method, vulnerable areas of the groundwater can be identified and appropriate remedial actions can be taken.

11.1.1 Review on Groundwater Potential Studies Using GIS and RS

Population growth and climate change can increase the demand on groundwater resources and hence there is chances of drought occurrence [1]. The distribution, quantity and chemical quality of water are enormously affected by the human activities [2]. Groundwater can be replenished annually, but its availability is not constant in time and space [3]. Recognition of GWPZs is necessary for water resource estimation, water quality protection, delineation of vulnerable zones and management of environment [4, 5]. Elements such as slope, lithology, geomorphology, drainage, soil type, land use/land cover (LULC), lineament, rainfall and distance to river have a substantial impact on groundwater determination [6]. It is not an easy task to locate the water beneath the earth surface as there is no direct method for it. Since aquifers containing groundwater is restricted to certain regions and is geographically variable, many researchers are interested in mapping the different areas where groundwater may be present.

There are various methods and tools available for determining groundwater probable zones in a specific area. Among them, GIS and RS are the most practical and affordable tools that require little manual labour. Rajesh et al. [7] considered RS and GIS as a valuable tool for observing and monitoring groundwater resources within a watershed area in short stretch of time. Several researchers have applied methods such as frequency ratio [8, 9, 10], multi-criteria decision analysis [11, 12, 13], weights—of—evidence [9, 14], (Pourghasemi et al. 2015), multi-influence factor [15], logistic regression [9], certainty factor [10], decision tree [16], artificial neural network model [17, 18], Shannon's entropy [19], machine learning techniques which include random forest [20] and so on for the identification of potential groundwater zones. Remote sensing and GIS are proved to be a powerful tool that can be used for fast estimation of natural resources [21]. MCDA technique is a source of bias due to expert opinion. The most popular and widely applied method for defining groundwater prospecting zones is AHP. AHP method was proposed by Saaty in 1980 and it is useful in solving socio-economic decision-making problems. AHP demarcate groundwater potential zones because it is an easy, quick, dependable, and efficient method [22]. It states that whether one criterion is more significant than the other or less important than the other and at given level of priority. Data can be transformed into useful information by the integration of GIS and AHP and it helps the managers and policymakers [8, 22]. For many years, researchers have used GIS and AHP in groundwater studies (e.g., [16, 23–30]). Even though remote sensing provides a wide range of applications in groundwater studies, it cannot provide the direct observation of the underground, hence the accuracy of study should be improved. The future research should focus on improving data accuracy, developing more reasonable evaluation models and multi method fusion.

11.1.2 *Groundwater Potential Studies in Kanyakumari District*

The following studies on groundwater potential zones have been reported from Kanyakumari district, Tamil Nadu. Geoelectric investigations was carried out by Sankar et al. [31] in order to access the groundwater potentially of Kanyakumari district. In their study, potential zones were classified based on the resistivity range and thickness of the subsurface layers. Application of RS and GIS was used by Suja Rose and Krishnan [32] to delineate the groundwater potential zones in Kanyakumari and Nambiyar basin, India. They concluded that Kanyakumari basin has more groundwater potential compared to Nambiyar basin. Perumal et al. [33] studied the hydrogeology along the coastal regions from Kanyakumari to Colachel. The study provides a detailed description about the landforms and groundwater occurrences of the area. Srinivas et al. [34] used Vertical Electrical Sounding (VES) to designate groundwater probable zones throughout the Kanyakumari district's coastal stretch. The areas affected by over exploitation and saline water intrusion were also demarcated by analysing the resistivity values. Satellite imagery with GIS applications were used for mapping the groundwater potential zones along the Kodayar basin, Tamil Nadu by Danee Joyce and Santhi [35]. In their study various thematic layers were prepared from satellite imageries, topographic maps and conventional datasets and these layers were assigned with proper weights and integrated in GIS platform to obtain the final groundwater potential map of the basin. The present study identifies the groundwater potential zones in the Kanyakumari district using GIS based AHP Technique and also recommend suitable methods for the wise use of the groundwater in the district.

11.2 Materials and Methods

11.2.1 *Study Area*

Kanyakumari is the southernmost district of Indian peninsula with an area of 1682 km^2. The district is bounded on the west by the Arabian Sea, and on the south and south-east by the Indian Ocean and the Gulf of Mannar, respectively. The north-east side is defined by the Tirunelveli district, while the north-west side is bounded by Kerala state. It stretches from 77°06′38″ to 77°31′58″ East and 8°04′26″ to 8°34′40″ North. The coastline is 62 km on the west (Arabian coast) and 6 km on the east (Bay of Mannar/Bay of Bengal) [36]. Thamirabarani, Karaiyar and Paralayar are the major rivers of the district [37]. Manakudy estuary is located in the southwest coast of the district and is enriched with different types of flora and fauna. Several land forms like marine uplands, shales, saltpans, water bodies like rivers and estuaries are found in the study area [38]. The area is having an undulating topography and elevation ranging from 0 to 1798 m. The region has a tropical climate in all years.

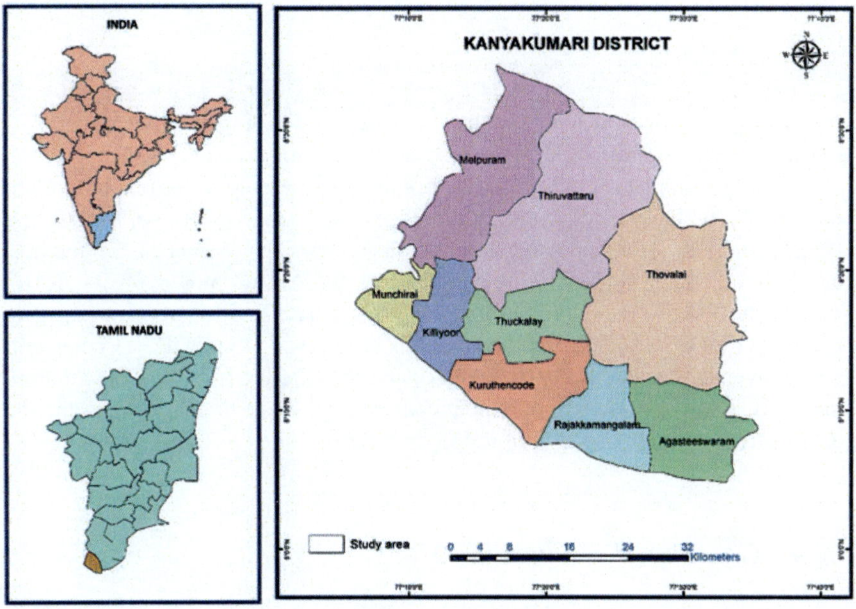

Fig. 11.1 Study area

Temperature over the region range from 22 to 35 °C in summer and 21–34 °C in winter. The average yearly rainfall ranges from about 826 to 1456 mm [39]. The soils of the district consist mainly of red loam varieties. The location of study area is shown in Fig. 11.1.

11.2.2 Geology and Hydrogeology of the Study Area

The coastline of Kanyakumari is characterised by laterite-capped sand dunes and short stretches of beach. The Western Ghats Mountain range occupies the north western parts of the district with a maximum elevation of 1658 m. The district constitutes about 80% hard rock and 20% sedimentary formations. Charnockites, khondalite, migmatite, laterites, sandstones, variegated clay, and river alluvium constitute the major lithology of the area. Unconsolidated to semi-consolidated formations and fractured crystalline rocks make up the region's aquifer system. In hard-rock regions, the nature and thickness of weathered zone vary both laterally and with depth influencing the groundwater recharge while groundwater is found under water table conditions in alluvial formation.

11.2.3 Data Collection

Different steps are involved in this study which include deciding groundwater factors, calculating weights through AHP, creating a spatial database, and validation of the results. There are diverse controlling factors on which groundwater potential of an area depends, which include: rainfall, drainage, lineament, soil, slope, LULC, and geology [40]. All the thematic layers were created by using toposheet as a reference. Study area falls under Survey of India toposheet nos. 58H/12, 58H/8, 58H/4, 58H/3 of scale 1:50,000. All the contributing parameters were compiled from various data sources, such as USGS Earth Explorer, Geological Survey of India, PWD, converted to raster layers with same cell size (10 * 10) and projected to UTM WGS 84 datum. Arc GIS 10.3 software was used in the preparation of all the thematic maps and the groundwater potential zone map of the study area.

11.2.4 Generation of Thematic Maps

The different thematic maps used in the study was collected from various sources. Digital Elevation Model (DEM) of the study area was downloaded from USGS Earth Explorer. From DEM, slope and drainage map can be derived. The rainfall and well yield data of the study area were gathered from the Public Work Department (PWD) Chennai. The obtained data was subjected to the interpolation technique in the Arc GIS software to obtain the rainfall map of the study area. The Geological Survey of India provided the lineament and geology map of the study area. Landsat 8 satellite image was used in the preparation of the LULC map. National Bureau of Soil Survey and Land Use Planning (NBSSLUP) is an organisation that provide the soil map of the study area. Figure 11.2. shows the methodology adopted for the study. Thematic layers were then categorised and rated based on their impact on groundwater (Table 11.1).

11.2.5 Weight Calculation for Groundwater Prospecting Parameters

11.2.5.1 Analytical Hierarchical Process

The AHP model is a frequently used approach for assigning a normalised weight to each thematic layer of groundwater prospecting factor [41]. The weightages were determined using expert opinion surveys. The weights are allotted for each parameter depending upon Saaty scale (1–9) value of relative importance. While assigning weightage to different themes, previous studies and field experiences are taken into consideration. According to the Saaty, "values of 9 specifies extreme importance, 8

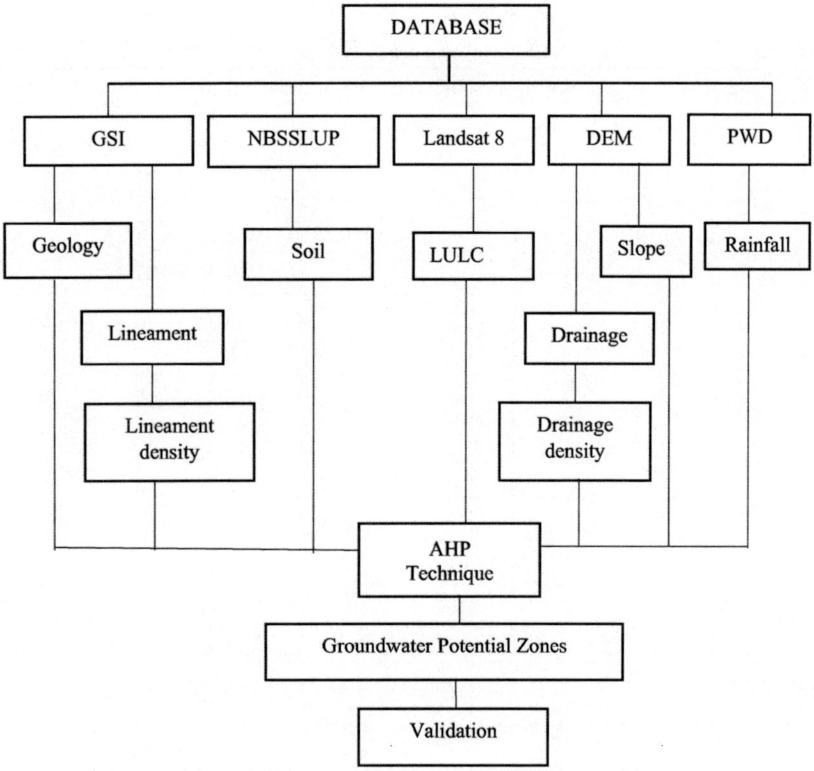

Fig. 11.2 Methodology adopted for the study is shown in the above flow chart

very, very strong, 7 very strong, 6 strong plus, 5 strong importance, 4 moderate plus, 3 moderate importance, 2 weak and 1 equal importance" [42]. Weights are assigned to the thematic layers in accordance with the classification based on their significance and water holding capacity. It is necessary to check the logical consistency of pairwise judgment. The consistency ratio is described as the proportion of the consistency index (CI) to the random consistency index (RI) (Eqs. 11.1 and 11.2). In order to obtain the normalised weights for the layers, the consistency index and consistency ratio values of the built pairwise matrix had to be satisfied. The CR value less than 0.1, indicates matrix is consistent.

$$\textbf{Consistency ratio (CR)} = \textbf{CI/RI} \tag{11.1}$$

$$\mathbf{CI} = \frac{\boldsymbol{\lambda}\mathbf{max} - \boldsymbol{n}}{\boldsymbol{n} - 1} \tag{11.2}$$

where λ max is the principal eigenvalue of the matrix and n is the number of factors considered for the groundwater potential [43]. In this study the consistency ratio (CR)

Table 11.1 Weights assigned to the parameters using AHP method

Sl. no	Theme	Weights	Influence (%)	Class	Rank
1	Rainfall	38	0.38	<50	1
				50–75	2
				75–108	3
				108–190	4
2	Geology	25	0.25	Laterite	5
				Active marine deposits	3
				Charnockite	3
				Khondalite	2
				Migmatite	1
				Paleo fluvial deposits	4
				Paleo marine deposirs	4
				Sandstone	5
3	Slope	13	0.13	0–6.47	5
				6.47–16.84	4
				16.84–29.15	3
				29.15–42.11	2
				42.11–82.61	1
4	Drainage density	9	0.09	0–46.95	5
				46.95–93.90	4
				93.90–140.86	3
				140.86–187.81	2
				187.81–234.76	1
5	Land use/land cover	6	0.06	Evergreen broad leaf forest	2
				Mixed forest	3
				Plantations	3
				Deciduous Broad leaf forest	2
				Water bodies	5
				Grassland	4
				Barren land	4
				Shrubland	1
				Cropland	5
				Built up	1
				Waste land	2
				Fallow land	2
				Agriculture land	5
6	Lineament density	5	0.05	0–18.21	1
				18.21–36.43	2
				36.43–54.65	3
				54.65–72.87	4
				72.87–91.09	5
7	Soil texture	4	0.04	Clayey soil	1
				Sandy soil	5
				Sandy loam soil	4
				Sandy clay loam	3
				Rock exposure	2
				Sandy clay	3
				Loamy sand	4

was found to be 0.05, and it indicates the weights used for the groundwater potential model were adequate and the comparisons of features were quite consistent.

11.2.5.2 Groundwater Potential Index

Thematic maps were selected and merged into a GIS framework to develop a groundwater potential index (GWPI). The following equation, which multiplies the rank and weight of each thematic layer, was used to calculate the GWPI [44] (Eq. 11.3):

$$\mathbf{GWPI} = \sum_{i=1}^{n}\sum_{j=1}^{m} Wj x Xi \tag{11.3}$$

where, Wj indicate the *j*th factor's normalised weight; Xi is the rank of the *i*th class of a thematic layer; There are m total theme layers, and there are n total classes in a thematic layer. The rank and weightage of each parameter were multiplied and added together in GIS software to create the groundwater prospect map of the district (Eq. 11.4).

$$\mathbf{GWPI} = \sum \begin{pmatrix} \textbf{Rainfall rank x weight} + \textbf{Geology rank} \times \textbf{weight} \\ + \textbf{Slope rank} \times \textbf{weight} + \textbf{Drainage density rank} \times \textbf{weight} \\ + \textbf{Land use rank} \times \textbf{weight} + \textbf{Lineament density rank} \times \textbf{weight} \\ + \textbf{Soil rank} \times \textbf{weight} \end{pmatrix} \tag{11.4}$$

The final output map was classed into several categories ranging from very high to very low potential zones.

11.3 Result and Discussions

11.3.1 Rainfall

Rainfall controls the groundwater table. Seasonal rainfall resultants in fluctuations of groundwater table. In order to construct the yearly rainfall map of the research region, data from Public Work Department was collected and processed using the IDW Interpolation technique (Fig. 11.3a). Regions with abundant rainfall are ranked highly, whereas those with less rainfall are ranked poorly. Potential of groundwater is maximum in areas with more rainfall and vice versa.

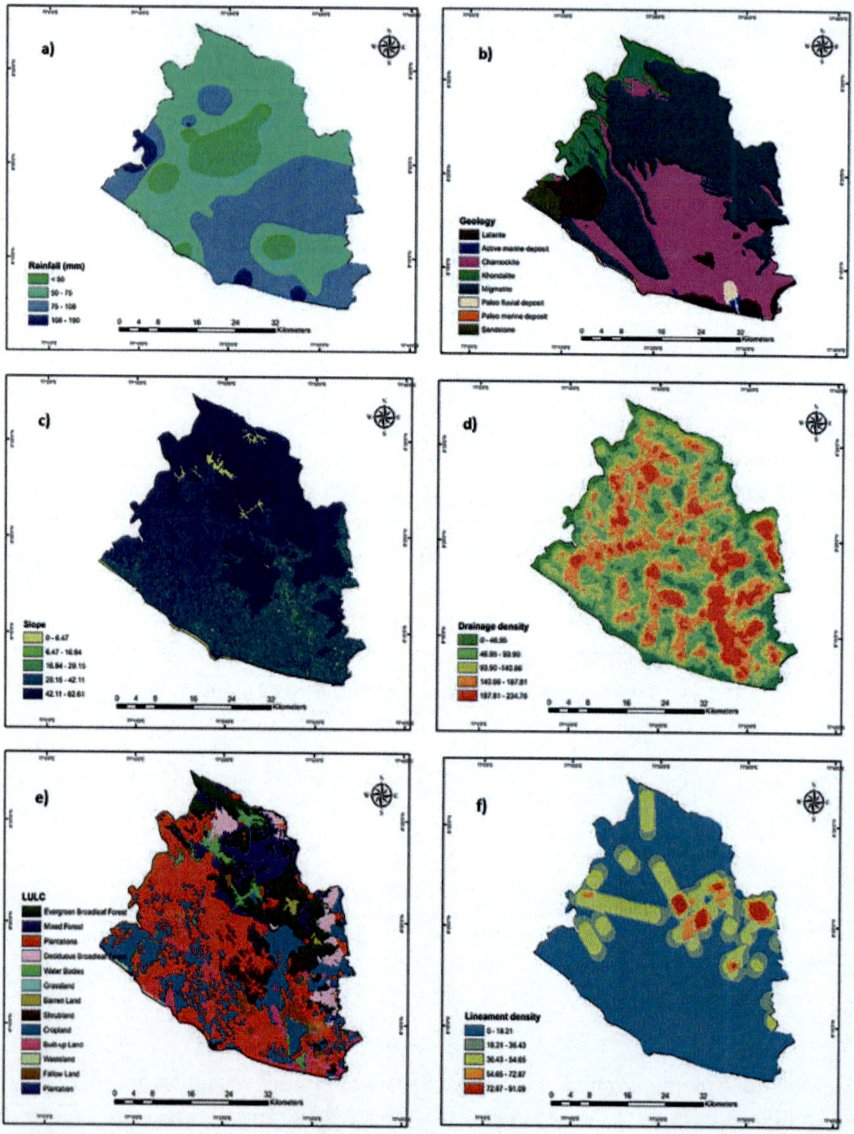

Fig. 11.3 **a–g** Represents the rainfall, geology, slope, drainage density, LULC, lineament density and soil map respectively of the study area

11.3.2 Geology

The entry of surface water into aquifers and its percolation are governed by the geological characteristics of rocks [45]. The geological map of the region was digitised from Geological Survey of India (Fig. 11.3b). Charnockite, Khondalite,

Fig. 11.3 (continued)

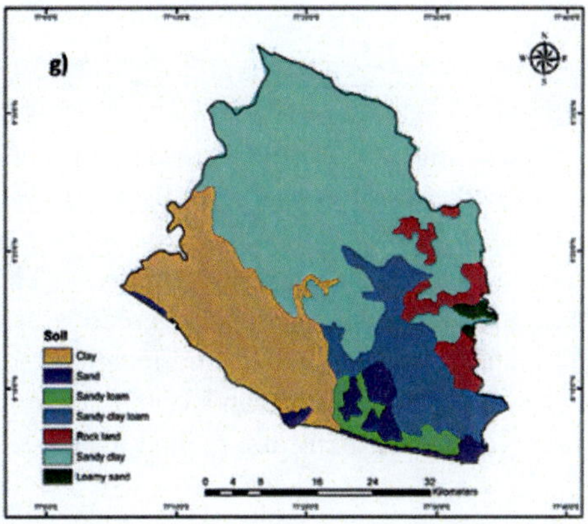

migmatite, laterite, marine sediments and sandstones constitute the major lithology of the study area. The formations with high porosity and less permeability are indications of high groundwater potentials. High rank is given to rock units having high water holding capacity like laterite and sandstone whereas low rank is assigned to hard rock such as migmatite and Khondalite.

11.3.3 Slope

Slope implies the terrain steepness. Groundwater recharge is greatest on flat or near-horizontal slopes, while runoff dominates in steeper areas. The slope map was created using the SRTM (Shuttle Radar Topography Mission) elevation data (Fig. 11.3c). The slope map of the region ranges between 0 and 83°. Gentle slope areas have high infiltration and low runoff, therefore assigned high rank. Slope with high values have high runoff and low infiltration property hence, it is assigned low rank.

11.3.4 Drainage Density

The distance of the stream to a unit area is defined as drainage density [46]. It is an acceptable tool for the investigation of landforms in terms of groundwater potential. DEM is used for preparing the drainage map of the study area (Fig. 11.3d). The drainage density is inversely proportional to the groundwater prospect [41]. Hence, the low drainage density areas are assigned high rank and low rank is given for high drainage density areas.

11.3.5 LULC

Land use/land cover influence various processes like runoff, infiltration, and groundwater recharge of an area [8]. LULC maps were generated from Landsat 8 (OLI) satellite imagery for the year 2021 (Fig. 11.3e). Supervised classification was performed to classify the different LULC patterns. The map describes that presence of settlement and wide leaf forest causes reduced water infiltration, but open area, irrigated land, and rainfed land are good locations for water percolation to recharge groundwater. Agricultural land and vegetation cover are ranked higher, while deciduous forests and settlements are ranked lower.

11.3.6 Lineament Density

Lineament represents a linear feature on the Earth's surface that indicates a fault. Lineament map of study area is gathered from the Geological survey of India. The density tool in the Arc GIS software is used to calculate lineament density. There is a significant correlation between a region's groundwater potential and lineament density. Areas with high lineament density have high groundwater potential and high porosity and permeability, so they are ranked high, while areas with low lineament density have low groundwater potential and rank low (Fig. 11.3f).

11.3.7 Soil

Water infiltration to the ground surface is mostly influenced by the soil [47]. The principal soil types in the research region are clay, sand, sandy loam, loamy sand, and rock land (Fig. 11.3g). Surface water can easily percolate and infiltrate into the subsurface due to very porous soils [23]. Considering the porosity and permeability factors, high rank is assigned to sandy type of soil which has high groundwater potential compared to clayey soils.

11.3.8 Groundwater Potential Zones

The GWPZ map of the study area is created and it shows distinct colours based on the groundwater availability (Fig. 11.4). The southern part of the district is more prone to groundwater occurrence than the northern part due to high rainfall, low drainage density, high lineament density, sandstone and laterite capping. Areas of high and very-high aquifer recharge potentials were localized in the southern parts, spanning 19.02% (321 km^2) and 18.7% (317 km^2) respectively. Areas of low and very low aquifer recharge potentials were distributed in the northern part about 27.91% (471 km^2) and 9.06% (153 km^2) respectively. Moderate potential zones were found to be 25.19% (425 km^2) and is randomly distributed over the study area (Table 11.2). Munchirai, Kuruthankode, Rajakkamangalam, Agatheeswaram, Thuckalay and Thovalai blocks falls in the moderate to very high potential zones whereas Melpuram, Thiruvattaru and Killiyoor blocks falls under very low to moderate zones. The resultant map is validated using the well data of the study area.

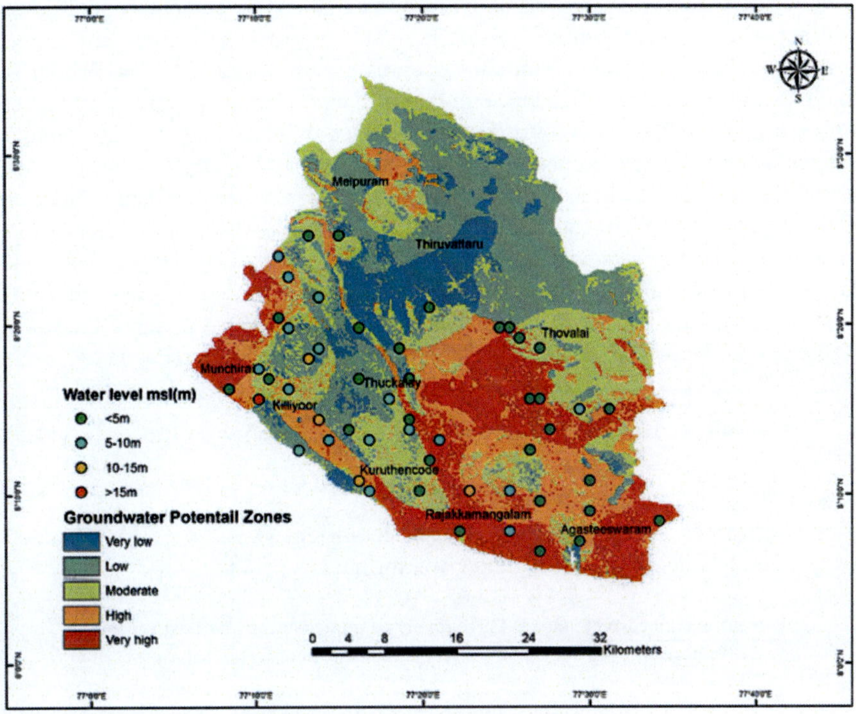

Fig. 11.4 Groundwater potential zones of the study area

Table 11.2 Groundwater potential zones of the study area

Sl. no	Groundwater potential zones	Area in km^2	Area in %
1	Very low	153	9.06
2	Low	471	27.91
3	Moderate	425	25.19
4	High	321	19.02
5	Very high	317	18.7

11.4 Validation of the Groundwater Potential Zone Map

Validation step enable us to evaluate the model. In order to verify groundwater potential zones, many researchers have used information from existing wells of their study area [22]. The topographic features, lithology and degree of weathering all have an impact on the yield characteristics of wells. For the purpose of finding the accuracy of the test, the obtained data was compared to actual groundwater yield of the district. Well yield is the amount of water that can be pumped from a given well per unit of time. Open well yields in the study region vary from 150 to 200 m^3/day at a drawdown of 1–3 m and a pumping period of 2–4 h. The district is chiefly composed of unconfined to semi-confined aquifers. The depth to water level ranges from 6 to10 m bgl and the discharge ranges from 10 to 20 m^3/day. Ground water exists in a phreatic condition in the sand dunes. The depth to water level is 4–8 m bgl, and the discharge is 30 to 80 m^3/day. The yield of large diameter wells ranges between 150 and 200 m^3/day. An accuracy assessment is performed to correlate the derived groundwater potential zone map with the observed field well data. For accuracy calculations, observation well data were used as a reference point. A confusion matrix or an error matrix was created to assess accuracy of the study (Table 11.3). The overall accuracy is calculated using the formula (Eq. 11.5).

$$\textbf{Overall accuracy} = \frac{\textbf{No.of Correct OWL}}{\textbf{Total No.of OWL}} = 38/53 = 71.6\% \tag{11.5}$$

where OWL is the Observation well locations.

A multivariate approach namely Kappa coefficient (K) is used for accuracy assessment. It is determined using the given formula (Eq. 11.6).

$$\mathbf{K} = \frac{\textbf{Percent overall correct value} - \textbf{Percent correct agreement to observed values}}{\textbf{Total no.of class} - \textbf{Percent correct agreement to observed values}} \tag{11.6}$$

$$\begin{aligned} \text{Over all accuracy} &= 38/53 = 71.6\% \\ \text{Kappa coefficient} &= 0.6\% \end{aligned}$$

Table 11.3 Error matrix of groundwater potential zones

Sl. no	GWPZ	Very high	High	Moderate	Low	Very low	Total	Correct samples
1	Very high	6	4	0	0	0	10	6
2	High	0	12	4	0	0	16	12
3	Moderate	0	2	13	2	0	17	13
4	Low	0	0	0	5	3	8	5
5	Very low	0	0	0	0	2	2	2
	Total	6	18	17	7	5	53	38

By substituting values in the above equations, it was found that, the overall accuracy is 71.6% with a Kappa coefficient of 0.6%. This finding suggests that a good correlation exist among the potential groundwater zones and the observed well data in the research area.

11.5 Conclusion

Assessing groundwater potential is an important step in effectively and efficiently using and managing resources. AHP approach is used in conjunction with GIS to estimate groundwater likely regions in Kanyakumari district, Tamil Nadu. Several parameters like, steepness of the terrain, rainfall, lithology, soil, land use, drainage and lineament are considered for the study. The AHP technique assigns weight to each influencing factor, which is then superimposed in Arc GIS software to make the final map. The final map depicts the district's very high, high, moderate, low, and very low groundwater zones. Regions with high and very-high aquifer recharge potentials were localized in the southern parts, spanning 19.02% (321 km^2) and 18.7% (317 km^2) respectively whereas, regions of low and very low aquifer recharge potentials were distributed in the northern part about 27.91% (471 km^2) and 9.06% (153 km^2) respectively. Moderate potential zones were found to be 25.19% (425 km^2) and is randomly distributed over the study area. Comparing to other blocks of the district, coastal blocks like Agatheeswaram, Rajakkamangalam and Munchirai have very high groundwater potential zones. The results obtained were then cross-checked using well yield data from the research region, and they showed a strong connection. The weighed overlay method was found to be very effective in mapping groundwater potential areas. The outcome of the study indicates the potential groundwater areas in the district and hence it will be useful for the future groundwater studies of the district.

11.6 Recommendations

The study of GWPZs is helpful in devising new production wells to meet increasing water demand. It is recommended that the potential area should be surveyed and verify the extractable amounts of water quantity. This study shows that zones with very high groundwater potential are mainly located near the coast. Coastal areas are always prone to various phenomenon like saline water intrusion, submarine groundwater discharge and coastal erosion. Hence a thorough study should be carried out for the wise use of water from the coastal region of the district. The study also helps in recommending sites for artificial recharges where groundwater is deficient.

Acknowledgements The authors are indebted to the financial assistance provided by DST-INSPIRE programme, Government of India. The first author is grateful to the faculties and Research scholars in V.O. Chidambaram College, Thoothukudi for their sincere support and also thankful to the Head of the department for his guidance and encouragement.

References

1. Dai A (2013) Increasing drought under global warming in observations and models. Nat Clim Chang 3(1):52–58
2. Thomas CW, Harvey JW, Franke OL, Alley WA (1998) Ground water and surface water a single resource. United States Geol Surv Circul 1139
3. Navane VS, Sahoo SN (2017) A review of integrated RS and GIS technique in groundwater potential zone mapping
4. Owolabi ST, Madi K, Kalumba AM, Orimoloye IR (2020) A groundwater potential zone mapping approach for semi-arid environments using remote sensing (RS), geographic information system (GIS), and analytical hierarchical process (AHP) techniques: a case study of Buffalo catchment, Eastern Cape, South Africa. Arab J Geosci 13(22):1–17
5. Waikar ML, Nilawar AP (2014) Identification of groundwater potential zone using remote sensing and GIS technique. Int J Innov Res Sci Eng Technol 3(5):12163–12174
6. Parameswari K, Padmini TK (2018) Assessment of groundwater potential in Tirukalukundram block of southern Chennai Metropolitan Area. Environ Dev Sustain 20(4):1535–1552
7. Rajesh J, Pande CB, Kadam SA, Gorantiwar SD, Shinde MG (2021) Exploration of groundwater potential zones using analytical hierarchical process (AHP) approach in the Godavari River basin of Maharashtra in India. Appl Water Sci 11(12):1–11
8. Guru B, Seshan K, Bera S (2017) Frequency ratio model for groundwater potential mapping and its sustainable management in cold desert, India. J King Saud Univ-Sci 29(3):333–347
9. Ozdemir A (2011) GIS-based groundwater spring potential mapping in the Sultan Mountains (Konya, Turkey) using frequency ratio, weights of evidence and logistic regression methods and their comparison. J Hydrol 411(3–4):290–308
10. Razandi Y, Pourghasemi HR, Neisani NS, Rahmati O (2015) Application of analytical hierarchy process, frequency ratio, and certainty factor models for groundwater potential mapping using GIS. Earth Sci Inf 8(4):867–883
11. Chowdhury A, Jha MK, Chowdary VM, Mal BC (2009) Integrated remote sensing and GIS-based approach for assessing groundwater potential in West Medinipur district, West Bengal, India. Int J Remote Sens 30(1):231–250
12. Mukherjee S, Mukherjee S, Garg RD, Bhardwaj A, Raju PLN (2013) Evaluation of topographic index in relation to terrain roughness and DEM grid spacing. J Earth Syst Sci 122(3):869–886
13. Rahmati O, Nazari Samani A, Mahdavi M, Pourghasemi HR, Zeinivand H (2015) Groundwater potential mapping at Kurdistan region of Iran using analytic hierarchy process and GIS. Arab J Geosci 8(9):7059–7071
14. Corsini A, Cervi F, Ronchetti F (2009) Weight of evidence and artificial neural networks for potential groundwater spring mapping: an application to the Mt. Modino area (Northern Apennines, Italy). Geomorphology 111(1–2):79–87
15. Etikala B, Golla V, Li P, Renati S (2019) Deciphering groundwater potential zones using MIF technique and GIS: a study from Tirupati area, Chittoor District, Andhra Pradesh, India. HydroResearch 1:1–7
16. Lee S, Lee CW (2015) Application of decision-tree model to groundwater productivity-potential mapping. Sustainability 7(10):13416–13432
17. Samantaray S, Sahoo A, Ghose DK (2019) Assessment of groundwater potential using neural network: a case study. In: International conference on intelligent computing and communication. Springer, Singapore, pp 655–664

18. Yadav S, Salui CL (2021) Artificial Neural Network for Identification of Groundwater Potential Zones in Part of Hugli District, West Bengal, India. In: Groundwater and society. Springer, Cham, pp 247–257
19. Naghibi SA, Pourghasemi HR, Pourtaghi ZS, Rezaei A (2015) Groundwater qanat potential mapping using frequency ratio and Shannon's entropy models in the Moghan watershed, Iran. Earth Sci Inform 8(1):171–186
20. Golkarian A, Naghibi SA, Kalantar B, Pradhan B (2018) Groundwater potential mapping using C5.0, random forest, and multivariate adaptive regression spline models in GIS. Environ Monitor Assess 190(3):1–16
21. Arulbalaji P, Padmalal D, Sreelash K (2019) GIS and AHP techniques-based delineation of groundwater potential zones: a case study from southern Western Ghats, India. Sci Rep 9(1):1–17
22. Ifediegwu SI (2022) Assessment of groundwater potential zones using GIS and AHP techniques: a case study of the Lafia district, Nasarawa State, Nigeria. Appl Water Sci 12(1):1–17
23. Abrar H, Legesse Kura A, Esayas Dube E, Likisa Beyene D (2021) AHP based analysis of groundwater potential in the western escarpment of the Ethiopian rift valley. Geol Ecol Landsc 1–14
24. Agarwal R, Garg PK (2016) Remote sensing and GIS based groundwater potential & recharge zones mapping using multi-criteria decision-making technique. Water Resour Manag 30(1):243–260
25. Kom KP, Gurugnanam B, Sunitha V (2022) Delineation of groundwater potential zones using GIS and AHP techniques in Coimbatore district, South India. Int J Energy Water Resour 1–25
26. Kumar A, Krishna AP (2018) Assessment of groundwater potential zones in coal mining impacted hard-rock terrain of India by integrating geospatial and analytic hierarchy process (AHP) approach. Geocarto Int 33(2):105–129
27. Mohammadi-Behzad HR, Charchi A, Kalantari N, Nejad AM, Vardanjani HK (2019) Delineation of groundwater potential zones using remote sensing (RS), geographical information system (GIS) and analytic hierarchy process (AHP) techniques: a case study in the Leylia-Keynow watershed, southwest of Iran. Carbonates Evaporites 34(4):1307–1319
28. Panahi MR, Mousavi SM, Rahimzadegan M (2017) Delineation of groundwater potential zones using remote sensing, GIS, and AHP technique in Tehran-Karaj plain, Iran. Environ Earth Sci 76(23):1–15
29. Saranya T, Saravanan S (2020) Groundwater potential zone mapping using analytical hierarchy process (AHP) and GIS for Kancheepuram District, Tamil Nadu, India. Model Earth Syst Environ 6(2):1105–1122
30. Shekhar S, Pandey AC (2015) Delineation of groundwater potential zone in hard rock terrain of India using remote sensing, geographical information system (GIS) and analytic hierarchy process (AHP) techniques. Geocarto Int 30(4):402–421
31. Sankar K, Jegatheesan MS, Balasubramanian A (1996) Geoelectrical resistivity studies in the Kanyakumari District, Tamil Nadu. J Appli Hydrol 9(1&2):83–90
32. Suja Rose RS, Krishnan N (2009) Spatial analysis of groundwater potential using remote sensing and GIS in the Kanyakumari and Nambiyar basins, India. J Indian Soc Remote Sens 37(4):681–692
33. Perumal SB, Thamarai P, Elango L (2010) Hydrogeological studies along the coastal area of Kanyakumari to colachel after Tsunami, South Tamil Nadu. IJEP 30:793–813
34. Srinivas Y, Oliver DH, Raj S, Chandrasekher N (2014) Delineation of groundwater potential zones along the coastal parts of Kanyakumari district, Tamilnadu. Indian J Geophys Union 18(3):356–362
35. CS DJ, Santhi H (2014) Mapping the groundwater potential zones in sub-basin of kodayar using satellite imagery with GIS application. IJESE 7(5):1973–1980
36. Ajithra AK, Venkatraman B, Jose MT, Chandrasekar S, Shanthi G (2017) Assessment of natural radioactivity and associated radiation indices in soil samples from the high background radiation area, Kanyakumari district, Tamil Nadu, India. Radiat Protect Environ 40(1):27

37. Vignesh KS, Anandakumar I, Ranjan R, Borah D (2021) Flood vulnerability assessment using an integrated approach of multi-criteria decision-making model and geospatial techniques. Model Earth Syst Environ 7(2):767–781
38. Chandrasekar N, Immanuel JL, Sahayam JD, Rajamanickam M, Saravanan S (2007) Appraisal of tsunami inundation and run-up along the coast of Kanyakumari District, India-GIS analysis. Oceanologia 49(3)
39. Balachandran A (2008) District groundwater brochure Kanyakumari district. Tamil Nadu
40. Das S, Pardeshi SD (2018) Integration of different influencing factors in GIS to delineate groundwater potential areas using IF and FR techniques: a study of Pravara basin, Maharashtra, India. Appl Water Sci 8:1–16
41. Melese T, Belay T (2022) Groundwater potential zone mapping using analytical hierarchy process and GIS in Muga Watershed, Abay Basin, Ethiopia. Glob Challenge 6(1):2100068
42. Meesariganda BR, Ishizaka A (2017) Mapping verbal AHP scale to numerical scale for cloud computing strategy selection. Appl Soft Comput 53:111–118
43. Ahmadi H, Kaya OA, Babadagi E, Savas T, Pekkan E (2020) GIS-based groundwater potentiality mapping using AHP and FR models in central Antalya, Turkey. Environ Sci Proc 5(1):11
44. Malczewski J (1999) GIS and multicriteria decision analysis. Wiley
45. Derdour A, Bouanani A, Kaid N, Mukdasai K, Algelany AM, Ahmad H, Menni Y, Ameur H (2022) Groundwater potentiality assessment of Ain Sefra region in upper Wadi namous basin, Algeria using integrated geospatial approaches. Sustainability 14(8):4450
46. Jhariya DC, Khan R, Mondal KC, Kumar T, Singh VK (2021) Assessment of groundwater potential zone using GIS-based multi-influencing factor (MIF), multi-criteria decision analysis (MCDA) and electrical resistivity survey techniques in Raipur city, Chhattisgarh, India. J Water Supply Res Technol AQUA 70(3):375–400
47. Ahmed R, Sajjad H (2018) Analyzing factors of groundwater potential and its relation with population in the lower barpani watershed, Assam, India. Nat Resour Res 27(4):503–515

Chapter 12
Application of Remote Sensing and GIS in Mapping Groundwater Potential Zones Through Fuzzy Integration in Kodavanar Watershed, A Part of Amaravathi Basin, Tamil Nadu, India

Thilagaraj Periasamy and Masilamani Palanisamy

Abstract Groundwater is an indispensable natural resource where its availability is shrinking in recent decades due to over exploitation. Accordingly, the rate of consumption and natural accumulation is inequitable in several parts of the country as the current rainfall has no immediate effect in recharging groundwater. So, it is necessary to demarcate the groundwater potential zones (GPZ) for recharging aquifers and conserving the low potential zones. This study aims to map groundwater potential zones in the semi-arid region of Kodavanar watershed, which is facing severe groundwater crisis and needs recharge structures in appropriate zones for conserving and managing groundwater. The morphometric parameters of various aspects are responsible for understanding the nature of the topographical, hydrological, and geomorphic structures along with lithology. Eventually, the advanced techniques of Geographic Information System (GIS) and remote sensing datasets have rendered the application of visualizing the GPZ. Thus, in the present study parameters such as geomorphology, infiltration ratio, aspect, geology, rainfall, slope, drainage density, land use/land cover, lineament density, relative relief, and Topographic Wetness Index (TWI) has been utilised. All the influencing factors were reclassified and integrated with the global weightage for mapping groundwater potential zones through fuzzy triangular membership function. The results infer that almost 16 and 32% of the Kodavanar watershed comes under very low and low potential class. Thus, the zones as identified from the study are suggested to the policy makers for immediate implementation of recharge structures for management and conservation of groundwater.

T. Periasamy · M. Palanisamy (✉)
Department of Geography, School of Earth Science, Bharathidasan University, Tiruchirappalli, Tamil Nadu 620 024, India
e-mail: masilamani@bdu.ac.in

T. Periasamy
e-mail: pthilagaraj23@gmail.com

E. Balaji et al. (eds.), *Emerging Technologies for Water Supply, Conservation and Management*, Springer Water, https://doi.org/10.1007/978-3-031-35279-9_12

Keywords Groundwater · Morphometric variables · Kodavanar watershed · Artificial recharge zones · Fuzzy triangular membership

12.1 Introduction

Groundwater resource is considered as a most significant natural resource which is stored in the critical zone (pore spaces and fractures) of geological formation beneath the earth surface [21]. However, it is dynamic in nature and significant contribution for drinking, irrigation and industrial requirements in comparison with the surface water [27]. In addition, the prime quality and quantity of resource in proper time and space is also significant [6, 26]. Globally, the demand for freshwater resource is increasing at an alarming rate and it is noted to be the severe problem in the developing countries [12, 22]. Eventually, in India, the dependency on groundwater resource is around 85% for domestic purpose in rural areas, 50% demand on urban and industrial purpose and 50% supply for irrigated area is totally depend on the groundwater [5, 15]. Thus the increasing demand of groundwater resource on domestic, industrial and irrigation purpose increases the thirst for groundwater potential zones identification [3, 24]. This will help in optimum utilization and conservation of the significant resource for the future [14, 17]. Accordingly, the study region of Kodavanar watershed is carefully chosen as a case study for assessing the groundwater potential zones as this region is affected with increased water scarcity due to exponential rise in population and socio-economic development. The Kodavanar watershed is a hard-rock terrain and is mainly depend on the seasonal rainfall throughout the year especially on the northeast monsoon season. The aquifers are recharged with the surface water accumulation during the pre-monsoon and are the major source all consumable use [2, 11]. Thus the study, attempted to identify the spatial variability of GPZ and artificial recharge zone for the conservation of the groundwater to cope up with the future needs.

In this technological era the modern techniques such as geographic information system (GIS) and remote sensing (RS) have been emerged as a significant tool in identifying the potential zones of groundwater [16]. Accordingly, many research articles have utilised various parameters like drainage density, lineament density, geomorphology, geology, lithology and soil for groundwater potential zone mapping through remote sensing and GIS techniques [13, 20, 23]. Further, most of the research work attempted in the hard rock terrain evident that the RS-GIS techniques is significant in integrating all the factors for mapping the potential zones [7, 10]. The results of individual factor involved in the assessment of groundwater potential zones are weighted and ranked using the Multi-criteria decision making (MCDM) method which is significant in qualitative assessment of the groundwater resource [8, 29, 30]. There have been many studies focused on various techniques in GIS which include multi-influencing factor (MIF), Analytical hierarchical process (AHP), Technique for order performance by similarity to ideal solution (TOPSIS) and frequency ratio (FR) technique for weighting the influencing factors [1, 9, 31], but only few studies have focused on the MCDM technique of Fuzzy Analytical hierarchical process (FAHP) which helps to tackle the problems of unequal distributions of weighting and ranking the influencing factors for mapping groundwater potential zones [18, 19, 25]. Thus in the present study, remote sensing data sets such as Advanced Land Observation Satellite-Phased Array type L-band Synthetic Aperture Radar (ALOS-PALSAR) DEM, Sentinel 2B and Persian-CCS-CDR along with the ancillary datasets from Survey of India Toposheet (SOI), Geological survey of India map (GSI) and Bhuvan web portal have been utilised to analyse and map the parameters like slope, geology, geomorphology, drainage density, rainfall, landuse/landcover, relative relief, aspect, lineament density, infiltration ratio and topographic wetness index. Then, all the parameters were reclassified and integrated with the global weightage for mapping groundwater potential zones through fuzzy triangular membership function as it is essential techniques in multi criteria decision and tackle the problems of unequal contributions of each sub-factor. Hence the vulnerable zones identified from the study are consistently monitored and conserved with the proper management plans at appropriate region for the implementation of recharge structure which helps in development of the groundwater resource in Kodavanar watershed.

12.2 Study Area

The case study of Kodavanar watershed, is selected as the study region which is facing severe groundwater issues and needs recharge structures in appropriate zones for conserving and managing groundwater. It is spatially positioned between 10°11′37″ and 10° 27′30″ north latitudes and 77° 37′46–78° 01′10″ east longitude as shown in the Fig. 12.1. The total geographical area of Kodavanar watershed is about 2254 km^2, in which its major portion is distributed over the Dindigul district, south India and the northern portion in the lower part of the watershed is distributed in Karur district, south India. The main stream of Kodavanar river originate in Pandrimalai and Sirumalai Hills lies in southern part of watershed and it flows from south to north in order to merge with the main stream of Amaravathi river in Karur district, south India. The highest elevation of hilly terrain of the watershed is marked at 1350 m and the lowest elevation of plain region is ranging around 300 to 100 m above sea level. Then the rainfall distribution ranges between 480 to 1880 mm and the black and red soil is the major soil type whereas loamy sand is the dominant soil type of the watershed. The dominant crops of the Kodavanar watershed include sugarcane, cotton, sorghum, rice and pulses.

12.3 Methodology

The present study attempted to map the groundwater potential zones through fuzzy integration using GIS and remote sensing. The groundwater potential zones of the Kodavanar watershed was mapped by utilizing the influencing parameters, which includes geology, geomorphology, rainfall, landuse/landcover, slope, infiltration

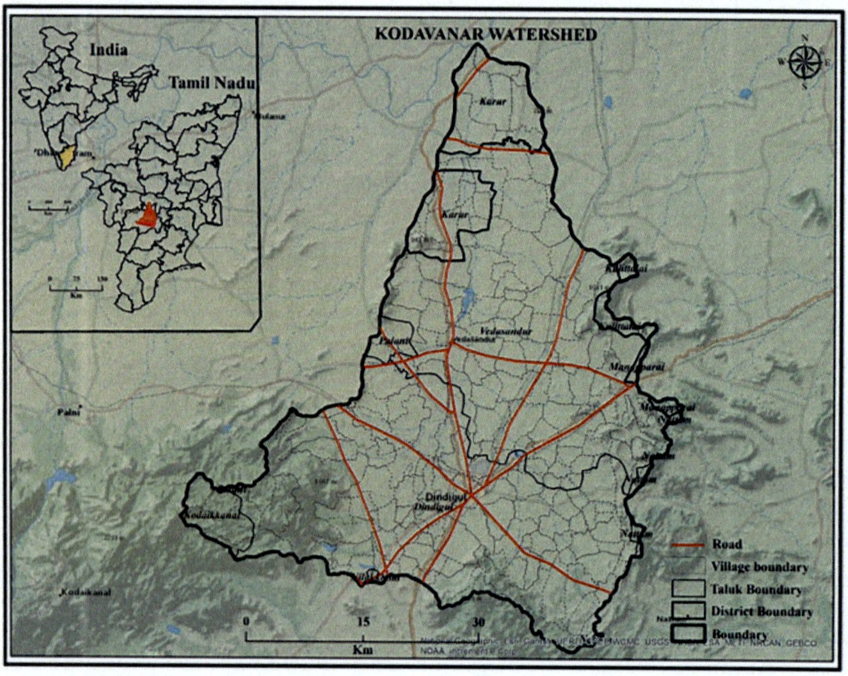

Fig. 12.1 Topographical visualization of case study region (Kodavanar watershed) with administrative boundaries

ratio, aspect, lineament density, relative relief, drainage density, and topographic wetness index shown in Figs. 12.2a–e and 12.3a–e. The analysis of the above-mentioned factors includes both the remote sensing and ancillary datasets. In which the remote sensing data sets such as Advanced Land Observation Satellite-Phased Array type L-band Synthetic Aperture Radar (ALOS-PALSAR) DEM, Sentinel 2B and Persian-CCS-CDR. The high resolution (12.5 m) terrain corrected ALOS PALSAR DEM was utilised to map the parameters like slope, aspect and topographic wetness index (TPI). Eventually, the stream network and watershed boundary was extracted to perform the morphometric analysis and the parameters like drainage density, relative relief and infiltration ratio are calculated for the study. Then, the Sentinel 2B of 10 m resolution is utilised for obtaining the landuse/landcover of six classes which includes built-up, crop land, vegetation, water body, barren rocky and barren land. The Persian-CCS-CDR data for the preparation of mean annual rainfall of the Kodavanar watershed. On the other hand the ancillary datasets such as Survey of India Toposheet (SOI), Geological survey of India (GSI) map and Bhuvan web portal. The SOI toposheet is utilised for the delineation of Kodavanar watershed boundary which act as study region for the present study. Then, the geology map of the study area is captured and classified by utilizing geological survey of India map. The Bhuvan web portal has been utilised for the computing the geomorphology and lineament density.

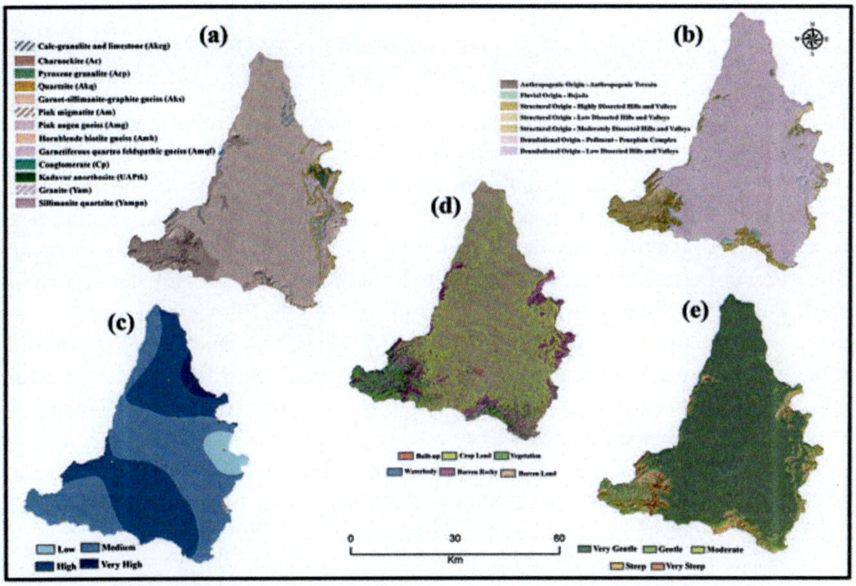

Fig. 12.2 **a** Geomorphology, **b** Geology, **c** rainfall, **d** LULC and **e** Slope

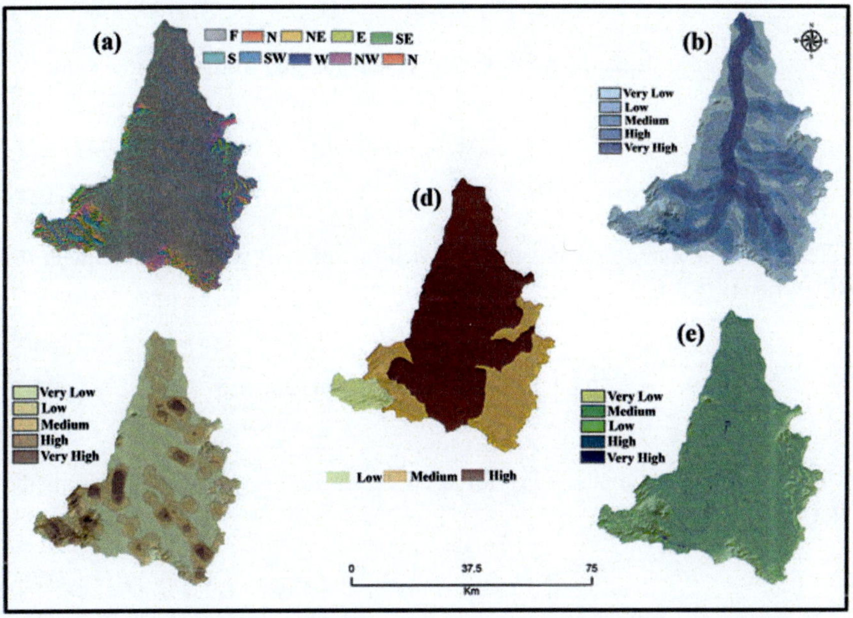

Fig. 12.3 **a** Aspect, **b** drainage density, **c** lineament density, **d** infiltration ratio and **e** TWI

12.3.1 Fuzzy Analytical Hierarchical Process (FAHP)

The factors involved in the computation of groundwater potential zones and their required datasets have been discussed above. Here, the method which is utilised to prepare the final result of potential zones of Kodavanar watershed will be discussed. The multi criteria decision making technique of Fuzzy Analytical Hierarchical Process (FAHP) is utilised in the present study for obtaining the ranks and weights (Table 12.1) of each factor responsible for the delineation of groundwater potential zones. This MCDM technique of FAHP is introduced by Buckley [4] which is the fusion set values of both the fuzzy and AHP [28]. In this study, for tackling the problems of unequal contributions of the each individual factor, the geometric mean approach has been employed. The steps and procedure involved in the assigning of fuzzy weightage is given below.

Initially, the parameters values are converted into triangular fuzzy elements for each factor and the pairwise comparison matrix will be constructed through (Eq. 12.1).

$$\tilde{A}_{ij} = (\tilde{a}_{ij})n \times n = \begin{pmatrix} (1,1,1) & (l_{12}, m_{12}, n_{12}) & (l_{1n}, m_{1n}, n_{1n}) \\ (l_{12}, m_{12}, n_{12})^{-1} & (1,1,1) & (l_{in}, m_{in}, n_{in}) \\ (l_{1n}, m_{1n}, n_{1n})^{-1} & (l_{in}, m_{in}, n_{in})^{-1} & (1,1,1) \end{pmatrix} \tag{12.1}$$

Through the Eq. 12.2, the fuzzy values were computed from the initial value.

$$\tilde{a}_{ij} = (l_{ij}, m_{ij}, n_{ij}) = \tilde{a}_{ij}^{-1} = \left(\frac{1}{n_{ij}}, \frac{1}{m_{ij}}, \frac{1}{l_{ij}}\right) \text{ where } i,\ j = 1, 2, \ldots n \text{ and } i \neq j. \tag{12.2}$$

The fuzzy geometric mean value for each factors (a_{ij}) was calculated with the help of Eq. 12.3 [4].

$$\tilde{r}_i = \left(\prod_{j=1}^{n} a_{ij}\right)^{1/n} \quad \text{Where, } i = 1, 2, \ldots..n \tag{12.3}$$

The fuzzified weightage ($\tilde{w}_i$) for each factor (a_{ij}) was calculated from the Eq. (12.4).

$$\tilde{w}_i = \tilde{r}_i \times (\tilde{r}_1 + \tilde{r}_2 + \ldots\ldots\tilde{r}_n)^{-1} \tag{12.4}$$

The Eq. 12.5 responsible for the De-fuzzification process and the numerical value (w_i) was calculated.

$$w_i = \left(\frac{l\tilde{w}_i + m\tilde{w}_i + n\tilde{w}_i}{3}\right) \tag{12.5}$$

Table 12.1 FAHP local & global weightage

Parameters	Class/ranges	Local/class weightage	Global/ parameters weightage
Slope	0–10	0.426	0.1
	(10–15)	0.31	
	15–25	0.114	
	25–35	0.09	
	>35	0.06	
Aspect	Flat (−1)	0.012	0.07
	North (0°–22.5°)	0.036	
	Northeast (22.5°–67.5°)	0.052	
	East (67.5°–112.5°)	0.095	
	Southeast (112.5°–157.5°)	0.087	
	South (157.5°–202.5°)	0.119	
	Southwest (202.5°–247.5°)	0.300	
	West (247.5°–292.5°)	0.205	
	Northwest (292.5°–337.5°)	0.059	
	North (337.5°–360°)	0.036	
TWI	>22	0.06	0.13
	15–22	0.09	
	11–15	0.114	
	8–11	0.31	
	<8	0.426	
Geomorphology	Anthropogenic Origin—Anthropogenic Terrain	0.04	0.08
	Fluvial Origin—Bajada	0.05	
	Structural Origin—Highly Dissected Hills and Valleys	0.21	
	Structural Origin—Low Dissected Hills and Valleys	0.19	
	Structural Origin—Moderately Dissected Hills and Valleys	0.28	
	Denudational Origin—Pediment—Peneplain Complex	0.12	
	Denudational Origin—Low Dissected Hills and Valleys	0.11	
Geology	Charnockite (Ac)	0.096	0.11
	Calc-granulite and limestone (Akcg)	0.087	

(continued)

Table 12.1 (continued)

Parameters	Class/ranges	Local/class weightage	Global/ parameters weightage
	Quartzite (Akq)	0.065	
	Garnet-sillimanite-graphite gneiss (Aks)	0.068	
	Pyroxene granulite (Acp)	0.047	
	Pink migmatite (Am)	0.064	
	Pink augen gneiss (Amg)	0.052	
	Hornblende biotite gneiss (Amh)	0.097	
	Garnetiferous quartzo feldspathic gneiss (Amqf)	0.099	
	Conglomerate (Cp)	0.079	
	Kadavur anorthosite (UAPtk)	0.068	
	Granite (Yam)	0.085	
	Sillimanite quartzite (Yampu)	0.093	
LULC	Vegetation	0.181	0.11
	Water	0.367	
	Build-up	0.12	
	Rocky land	0.02	
	Barren land	0.112	
	Crop land	0.2	
Rainfall	<500	0.15	0.12
	500–750	0.114	
	750–1000	0.31	
	>1000	0.426	
Drainage density	0–1.2	0.06	0.14
	1.2–2.4	0.09	
	2.4–3.6	0.114	
	3.6–4.8	0.31	
	4.8–6	0.426	
Lineament density	0–0.34	0.06	0.07
	0.34–0.99	0.09	
	0.99–1.57	0.114	
	1.57–2.11	0.31	
	2.11–2.69	0.426	
Infiltration ratio	7.82–8.55	0.114	0.07
	8.55–12.30	0.31	
	12.30–14.08	0.426	

Lastly, the values will be normalised through Eq. (12.6) for the final parameter weights.

$$N_i = \frac{w_i}{\sum_{i=1}^{n} w_i} \tag{12.6}$$

12.4 Results and Discussion

The groundwater potential zone result of the Kodavanar watershed is computed by utilizing factors such as geology, geomorphology, rainfall, landuse/landcover, slope, relative relief, aspect, lineament density, infiltration ratio drainage density, and topographic wetness index. The results of all the factor mentioned above is interpreted and discussed below.

12.4.1 Geology

The prevailing geology and groundwater resource of a region has a strong relationship as the infiltration and porosity are purely depending on the underlying geology which supports the groundwater potential in a particular zone. The geology of the Kodavanar watershed is distributed with Sillimanite quartzite (Yampu), Granite (Yam), Kadavur anorthosite (UAPtk), Conglomerate (Cp), Garnetiferous quartzo feldspathic gneiss (Amqf), Hornblende biotite gneiss (Amh), Pink augen gneiss (Amg), Pink migmatite (Am), Garnet-sillimanite-graphite gneiss (Aks), Quartzite (Akq), Pyroxene granulite (Acp), Charnockite (Ac) and Calc-granulite and limestone (Akcg) as shown in the Fig. 12.2a. In which Hornblende biotite gneiss (Amh) and Charnockite (Ac) is the predominant class covering almost 80% of the study region and is marks as the hard rock region.

12.4.2 Geomorphology

The geomorphology of the Kodavanar watershed is categorized based on the origin into four classes which includes denudational origin, structural origin fluvial origin and anthropogenic origin. The denudational origin is classified as low dissected hills and valleys and pediment—peneplain complex (Fig. 12.2b). Then structural origin is classified as highly, moderately and low dissected hills and valleys. Further, fluvial origin is classified as bajada and anthropogenic origin as anthropogenic terrain. In which denudational origin of pediment—peneplain complex and structural origin of highly dissected hills and valleys is the dominant class distributed in the Kodavanar watershed.

12.4.3 Rainfall

The rainfall of a region will have a direct impact on the availability of the groundwater resource as it is the significant sources which affect the quality and quantity of the ground water resource of a study region. Here, it has been classified as very high, high, medium and low. In which very high class is mostly distributed in the north eastern portion of the watershed as portrayed in Fig. 12.2c. The high rainfall class is distributed over the north eastern, south eastern and south western part of the study region. The medium and low rainfall class is marked over the middle portion and lowest south western portion of the study region.

12.4.4 Landuse/Landcover

The potential zones of groundwater are highly influenced by the LULC classes as some of the class supports infiltration and contribute to high potential zone and on the other hand some of the classes restrict the infiltration and reliable to the low potential zones. The LULC of the Kodavanar watershed has been categorised into six classes as built-up, crop land, vegetation, water body, barren rocky and barren land. It is noted from the result that the barren land and crop land are the predominant class present over the study region (Fig. 12.2d). The barren rocky class is distributed in the elevated portion of the study area bordering the watershed in the central and southern portion. The vegetation class is majorly distributed over the sirumalai and pandrimalai hills in the southern portion. The water body and the built-up are the minor class in which high concentration of the built-up class is accumulated more in the middle of southern portion which is the core Dindigul city region.

12.4.5 Slope

The degree of slope variation is a significant factor influencing the groundwater potential as the steep slope region are limited with holding of water for infiltration to take place, whereas the gentle slope supports accumulation of water and supports the infiltration process. The slope variation is categorized into five classes namely, very gentle, gentle, moderate, steep and very steep as in the Fig. 12.2e. The very gentle and gentle class of slope is the dominant class present in the study area. The moderate to steep and very steep class are mostly distributed in the elevated region of the watershed that borders the Kodavanar region.

12.4.6 *Aspect*

The aspect which constitute to the direction of slope has been classified as south west, north, east, north east, west, south east, south, north west and north (Fig. 12.3a). The north and north east class is the dominant class present over the study region.

12.4.7 *Drainage Density*

The value of drainage density is categorized as very high, high, medium, low and very low (Fig. 12.3b). The very high and high drainage density class is highly concentrated across the main steam order of the Kodavanar River that flows from south to north direction. The medium class is distributed over the adjacent portions of the high stream order of the watershed. Then the low and very low drainage class is mostly found in the foot hills region where there is presence of first and second order streams resulted in low drainage density.

12.4.8 *Lineament Density*

The lineament density is the most significant factor involved in the identification of groundwater potential zones as it constitutes the fracture, joints and underlying lithology that have direct relation with the percolation of groundwater resource. It is classified into five categories as very high, high, medium, low and very low as shown in the Fig. 12.3c. The very high and high lineament density class is mostly distributed over the hilly terrain situated in the southern and south western portion of the watershed. The medium and low classes are bordering the watershed in the northern and north eastern portion of Kodavanar watershed with minor hill portions. Then the very low class is dominant over the study region as most of the portions of the study area falls under the plain region.

12.4.9 *Infiltration Ratio*

The morphometric factors are helpful in understanding the hydrological behavior of the watershed. Whereas, the infiltration ratio is considered in reflecting the ranges of infiltration surface present over the study region. The infiltration ratio is classified into three classes as low, medium and high (Fig. 12.3d). In which the high infiltration class is dominant as most of the region of Kodavanar watershed falls under the plain region. Then the medium class is distributed in the south eastern and south western part which is the plateau and foothill region of the study region. The low infiltration class is marked over the border of south western part which is the high elevated region of the watershed.

12.4.10 Topographic Wetness Index

This index helps in identifying the degree of wetness associated with the watershed and has the indirect relationship with the groundwater potential areas as the wetness is high there will be least chance of infiltration in that particular zone. It is classified into five classes as very low, low, medium, high and very high (Fig. 12.3e). In which, the very low wetness index class are marked over the elevated region of sirumalai and pandrimalai hills. Then the low topographic wetness is the major class present over the watershed covering most of the plain region. Further, the medium and high class is minor distribution partially distributed over the southern regions which are nearer to the water bodies. The very high class is sparsely distributed over the watershed in exact match with the distribution of water bodies.

12.4.11 Groundwater Potential Zones

The individual factors responsible for the GPZ identification and their results have been interpreted and discussed above. Accordingly, the final result of groundwater potential zone of Kodavanar watershed will be discussed here. The results of every parameter involved in the demarcation of GPZ has been weighted and ranked (Table 12.1) using FAHP and is shown in Fig. 12.4. The GPZ of the Kodavanar watershed is categorized as very high, high, medium, low and very low.

The very low potential zone covers almost 16% of the study area distributed over the hilly terrain of the study region where the infiltration and percolation is highly limited. The low potential zone class covers almost 32% of the Kodavanar watershed distributed mostly over the northern, north eastern and north western portion where the lineament and drainage density is also low. Then the medium and high class covers about 36% of the study area distributed in the mid portion between low foot hill and low potential zones class. The very high class covers almost 13% of the study area (Table 12.2) distributed over the portion where the drainage and lineament density is high. Thus, it is interpreted from the study that very low and low potential zones cover nearly half (48%) of the study area which has to be monitored and appropriate management plans has to be implemented. Then, the medium and high class covers 36% of the Kodavanar watershed, which has to be maintained, and conserved using proper planning and mitigation plans for the development of the Kodavanar watershed. Further, in the present study some of the artificial zones has been identified which falls under medium and low GPZ. The suggested recharge zones are the place where the recharge structures can be planned to improve the quantity of groundwater resource over the region. However, the major objective of

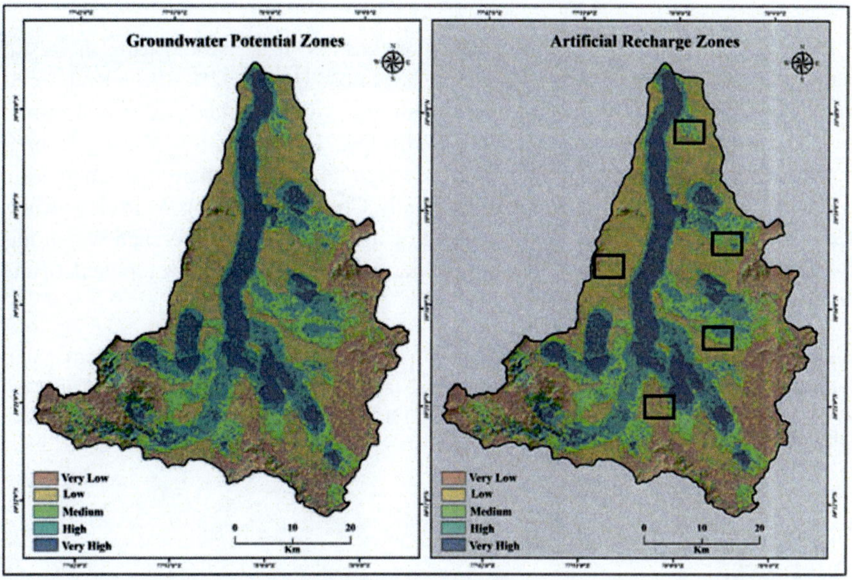

Fig. 12.4 FAHP based groundwater potential zone map

Table 12.2 Geographical area of each class in groundwater potential zones

Classes	Count	Area	Area (%)
VL	5171	364.316	16.1948
L	10,379	731.2387	32.50548
M	5924	417.3676	18.55308
H	6025	424.4834	18.8694
VH	4431	312.1803	13.87723

this study focuses on mapping groundwater potential zones; the future study on this Kodavanar has planned to identify the proper sites for implementing appropriate recharge structures.

12.5 Conclusion

The framework of the current study focuses on the mapping of the GPZ of the Kodavanar watershed through fuzzy integration. The factors involved in the identification of GPZ are analysed and each individual parameters are weighted and ranked with their specific significance through the fuzzy triangular membership function and the final result has been computes. Thus the application of GIS and RS is proved effective in delineating the potential zones of Kodavanar watershed through the multi criteria decision making method of fuzzy integration. It is marked from the result that the low &very low potential class cover almost 16 and 32% of the Kodavanar watershed where the infiltration and percolation is highly limited and in addition to this the lineament and drainage density is also low. These very low & low potential classes covers nearly 48% of study region which needs immediate implementation of Recharge/Injection Wells, Percolation Ponds, Check Dams, Dug Wells, Rainwater Harvesting. However the medium and high class 36% of the watershed which has to be monitored and conserved using proper planning for the development of the Kodavanar watershed.

Acknowledgements I would like to thank the Department of Geography and SAP DRS-II, Bharathidasan University for letting to utilise the lab facilities and their support. I would extend my gratitude to thank the University for providing the financial support through University Research Fellowship.

References

1. Abdul Rahaman S, Venkatesh R (2020) Application of remote sensing and google earth engine for monitoring environmental degradation in the nilgiri biosphere reserve and its ecosystem of western ghats, india. In: The international archives of the photogrammetry, remote sensing and spatial information sciences, XLIII-B3-2020, pp 933–940. https://doi.org/10.5194/isprs-archives-XLIII-B3-2020-933-2020
2. Balasubramani K et al (2021) Revealing the socio-economic vulnerability and multi-hazard risks at micro-administrative units in the coastal plains of Tamil Nadu, India. Geomat Nat Haz Risk 12(1):605–630. https://doi.org/10.1080/19475705.2021.1886183
3. Bhattacharya AK (2010) Artificial ground water recharge with a special reference to India. IJRRAS 4(2):214–221
4. Buckley JJ (1985) Fuzzy hierarchical analysis. Fuzzy Sets Syst 17(3):233–247. https://doi.org/10.1016/0165-0114(85)90090-9
5. Central Ground Water Board (2014) Dynamic ground water resources of India. Ministry of Water Resources, River Development and Ganga Rejuvenation, Government of India
6. Chowdhury A (2009) Integrated remote sensing and GIS-based approach for assessing groundwater potential in West Medinipur district West Bengal, India. Int J Remote Sens 30:231–250
7. Das S (2017) Delineation of groundwater potential zone in hard rock terrain in Gangajalghati block, Bankura district, India using remote sensing and GIS techniques. Model Earth Syst Environ 3(4):1589–1599
8. Deepika B et al (2013) Integration of hydrological factors and demarcation of groundwater prospect zones: insights from remote sensing and GIS techniques. Environ Earth Sci 70(3):1319–1338
9. Etikala B et al (2019) 'Deciphering groundwater potential zones using MIF technique and GIS: a study from Tirupati area Chittoor District, Andhra Pradesh, India. HydroResearch. https://doi.org/10.1016/j.hydres.2019.04.001
10. Ghosh PK et al (2016) Mapping of groundwater potential zones in hard rock terrain using geoinformatics: a case of Kumari watershed in western part of West Bengal. Model Earth Syst Environ 2(1):1
11. Gopinath S et al (2016) 'Hydrochemical characteristics and salinity of groundwater in parts of Nagapattinam district of Tamil Nadu and the Union Territory of Puducherry, India. Carbonates Evaporites. https://doi.org/10.1007/s13146-016-0300-y

12. Holden J (2014) Water resources: an integrated approach. Routledge, New York
13. Yeh H-F et al (2016) 'Mapping groundwater recharge potential zone using a GIS approach in Hualian River, Taiwan. Sustain Environ Res 26:33–43
14. Hutti B, Nijagunappa R (2011) Identification of groundwater potential zone using Geoinformatics in Ghataprabha basin, North Karnataka, India. Int J Geomat Geosci 2(1):91–109
15. Ibrahim-Bathis K, Ahmed SA (2016) 'Geospatial technology for delineating groundwater potential zones in Doddahalla watershed of Chitradurga district, Egypt India. J Remote Sens Space Sci. https://doi.org/10.1016/jejrs.2016.06.002
16. Jha MK et al (2007) Groundwater management and development by integrated remote sensing and geographic information systems: prospects and constraints. Water Resour Manag 21(2):427–467
17. Kantamaneni K at al (2022) A comprehensive assessment of climate change and coastal inundation through satellite-derived datasets: a case study of Sabang Island, Indonesia. Remote Sens 14(12):1–20. https://doi.org/10.3390/rs14122857
18. Machiwal D et al (2011) Assessment of groundwater potential in a semi-arid region of india using remote sensing, GIS and MCDM techniques. Water Resour Manag
19. Mallick J et al (2015) Geospatial and geostatistical approach for groundwater potential zone delineation. Hydrol Process
20. Muralitharan J, Palanivel K (2015) 'Groundwater targeting using remote sensing, geographical information system and analytical hierarchy process method in hard rock aquifer system, Karur district, Tamil Nadu, India. Earth Sci 8:827–842
21. Naghibi SA et al (2015) Groundwater qanat potential mapping using frequency ratio and Shannon's entropy models in the Moghan Watershed, Iran. Earth Sci Inform
22. Nitheshnirmal S et al (2019) Erosion risk assessment through morphometric indices for prioritisation of Arjuna watershed using ALOS-PALSAR DEM. Model Earth Syst Environ 5:907–924. https://doi.org/10.1007/s40808-019-00578-y
23. Kumar P et al (2016) Mapping of groundwater potential zones in Killinochi area, Sri Lanka, using GIS and remote sensing techniques. Sustain Water Resour Manag 2:419–430. https://doi.org/10.1007/s40899-016-00725
24. Patel P, Desai MD (2010) Artificial ground water recharge field study: site characterization and test results. Int J Adv Eng Tech v.I(II):150–164
25. Rahmati O et al (2015) Groundwater potential mapping at Kurdistan Region of Iran using analytic hierarchy process and GIS. Arab J Geosci
26. Rao NS (2006) Groundwater potential index in a crystalline terrain using remote sensing data. Environ Geol 50:1067–1076
27. Ravichandran et al (2022) Monitoring of multi-aspect drought severity and socio-economic status in the semi-arid Regions of Eastern Tamil Nadu, India. Water 14(13):1–20. https://doi.org/10.3390/w14132049
28. Saaty TL (1980) The analytical hierarchy process. McGraw Hill, New York
29. Sahoo S et al (2016) Grey analytic hierarchy process applied to effectiveness evaluation for groundwater potential zone delineation. Geocarto Int 1–18. https://doi.org/10.1080/10106049.2016.1195888
30. Senanayake IP et al (2016) An approach to delineate groundwater recharge potential sites in Ambalantota, Sri Lanka using GIS techniques. Geosci Front 7(1):115–124
31. Venkatesh R et al (2020) Eco-environmental vulnerability zonation in essence of environmental monitoring and management. In: The international archives of the photogrammetry, remote sensing and spatial information sciences, XLIII-B5-2020, pp 149–155. https://doi.org/10.5194/isprs-archives-XLIII-B5-2020-149-2020

Chapter 13
Morphometric Analysis Using Geospatial Techniques in the Pandameru River Basin, Anantapur District, Andhra Pradesh, India

Ravi Kumar Pappaka, Srinivasa Gowd Somagouni, Krupavathi Chinthala, and Anusha Boya Nakkala

Abstract Morphometry is a technique used to evaluate various landforms and drainage systems in order to search for available water resources. The drainage system is an essential requirement for analyzing the basin hydrology using morphometric characteristics. The morphometric properties of the basin influence the streaming system through which the water enters the streams. Using Strahler's method to classify the obtained drainage system, it was revealed that the basin has a dendritic drainage structure. A morphometric investigation of several hydrological characteristics of the streams in the Pandameru River basin, Anantapur district, South India, is the main purpose of this research. Among the hydrological parameters studied here are the drainage system, linear, areal, and relief factors, as well as the particular statistics of stream variables in the Pandameru river basin. Stream length (1261.7 km), stream order (1–6), stream length ratio (0.85), bifurcation ratio (3.4–4.0), and mean bifurcation ratio (4.084) are all linear morphometric results from the analyses. The areal morphometric parameters that were obtained include drainage density (1.88 km/km^2), basin perimeter (147 km), basin size (671 km^2), circularity ratio (0.39), and relief aspects, with the maximum value being 636 m and the lowest elevation being 63 m. As a result of the study, it is feasible to infer that satellite data (SRTM-DEM) combined with geospatial technology is a useful method in the morphometric investigation, and the results might be utilized for watershed conservation and other watershed research in the future.

R. K. Pappaka · S. G. Somagouni (✉) · K. Chinthala · A. B. Nakkala
Department of Geology, Yogi Vemana University, Kadapa, Andhra Pradesh 516005, India
e-mail: ssgowd@gmail.com

R. K. Pappaka
e-mail: ravigpcs75@gmail.com

K. Chinthala
e-mail: chinthalakrupavathi95@gmail.com

A. B. Nakkala
e-mail: anushabadapalli@gmail.com

E. Balaji et al. (eds.), *Emerging Technologies for Water Supply, Conservation and Management*, Springer Water, https://doi.org/10.1007/978-3-031-35279-9_13

Keywords Stream order · Pandameru · Structural · DEM · Remote sensing · GIS

13.1 Introduction

The physical environment's most dynamic and significant component is represented by rivers. Rivers often have well-defined geographical limits that serve as a conduit for the transfer of energy from one location to another in the surrounding environment. River basin planning investigations are crucial in semi-arid areas for protecting groundwater, as surface water supplies are scarce in some areas and absent in others. Groundwater recharge in these areas is purely reliant on rainfall. Knowing the morphometric parameters at the basin level is a prerequisite for comprehending the rainwater recharge process and calculating groundwater costs at the basin level. Morphometry comprises a numerical investigation of the region under consideration's size, elevation, length, slope, site conditions, and drainage features. Morphometric analysis is a critical tool for assessing and explaining the behavior of hydrological systems because it provides the quantitative needs of watershed morphology to understand the preliminary gradient and the lithological and geomorphological history of river basins [2, 16, 38]. The morphometric features of every river basin cover a variety of upstream factors, such as weather, geography, and landform, which together influence the aqueous regime.

The interaction of spatial variables is well studied in the current scientific and satellite era, including RS and GIS approaches [1, 4, 6]. GIS is a highly effective investigative technique because it can quickly generate, alter, store, process, and output high-quality spatial data. It also effectively integrates thematic maps for a geospatial approach [5, 8, 11] On the subject of river basin morphometric analysis, numerous studies were conducted and reported, particularly on the river system of the Penna River in South India [10]. In India, topographic sheet-based, descriptive, and classificatory methods have traditionally been used in river geomorphology. But over the past couple of decades, field-based investigations into the properties and behavior of river channels, along with advancements in computer technology, GIS, and remote sensing, have improved river landforms in India [7, 35].

The current research focuses on the investigation of the morphometric aspects of the Pandameru basin, a major tributary stream of the Penna River, where hydraulic activities have been dominant, highlighting the significance of river morphometric investigations. The findings of a morphometric examination of the Pandameru River basin are given in this study. In the current research work, we made an attempt to assess the soil erosion susceptibility along with the flooding areas of the river basin. The outcomes of the research can helpful for environmentalists, geologists, and many planners for the assessment of hydrological conditions and the making of action plans toward a sustainable and resilient society.

13.2 Study Region

The current study is focused on the Pandameru River, which originates in Kangalakunta village in Ramagiri Mandal and flows through Kanaganapalle, Rapthadu, and Anantapur Mandal before joining the Anantapur tank. The mean annual precipitation is 563 mm, and it is situated between the latitudes of 14°15′00″ to 14°40′00″N and the longitude of 77°25′00″ to 77°40′00″E in the south-western part of Anantapur, which is experiencing drought. The basin has a 671 km^2 area, and a perimeter of 147 km, and it is depicted on toposheets in scales of 1:50,000 for 57F/6, 57F/7, 57F/10, and 57F/11. The PGC of Archaean age, hornblende biotite gneiss, granites, and granodiorite are the primary rock types exposed in the region [4] (Fig. 13.2). Gneisses are the major component of the PGC, which is more widely distributed and they include biotite, hornblende, and hornblende-biotite gneisses. The landscape has various denudational ridges and hills, and it is undulating. Figure 13.1 illustrates a location map including the drainage of the research region.

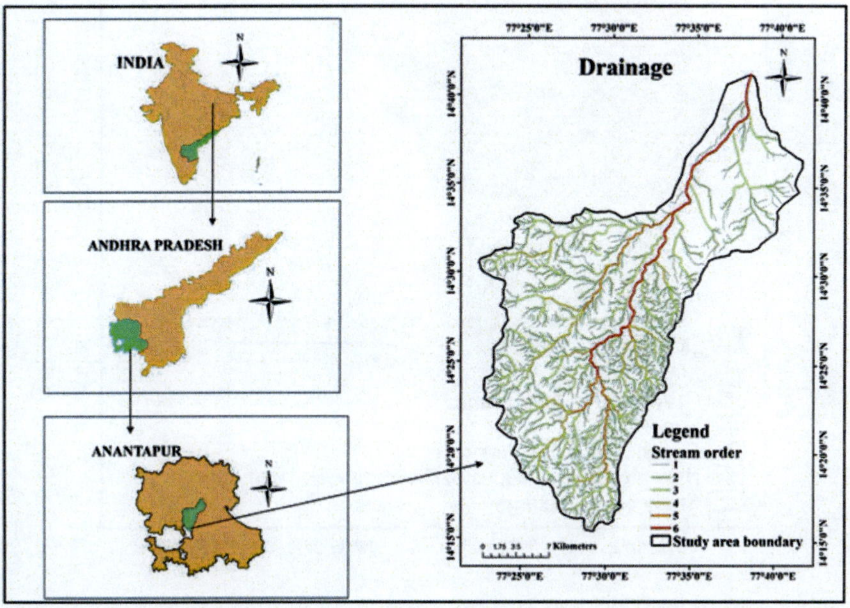

Fig. 13.1 Drainage map of the study area

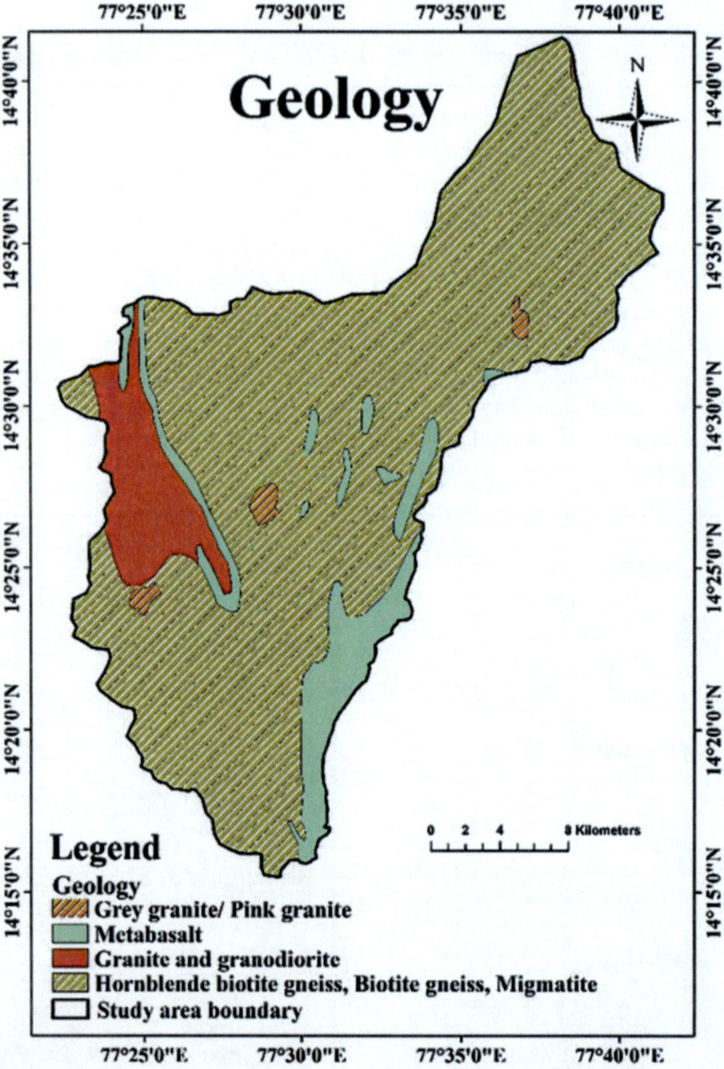

Fig. 13.2 Geology map of the research region

13.3 Methodology

The Pandameru River basin was morphometrically studied using 1:50,000 scale SOI topographical maps 57F/6, 57F/7, 57F/10, and 57F/11. The investigation takes into account linear, areal, and relief characteristics. Horton [13] and Strahler [38] methodologies were used to investigate the basin's drainage network. The WGS-84 datum and zone 44N of the Universal Transverse Mercator projection were used to georeference SOI topographic maps in ArcGIS Desktop 10.8. The current work included the creation of a database and the extraction of multiple drainage characteristics using SRTM-DEM, and SOI topographical data [17, 28]. Table 13.1 provides details about the information used. A global positioning system is used during fieldwork to record the altitude and additional parameters of the research area. This research outlined the Pandameru River and extracted its drainage network using SRTM-DEM (1-arc s (90 m)) and SOI toposheets. Only two more important steps are listed: fixing the digital elevation model and replenishing the sinks [34]. Following the completion of the digital elevation model, the flow path of each pixel was determined, and the flow accumulation was calculated using the flow path of each pixel to build a watershed (Fig. 13.4). The analysis and classification of morphometric data into three classes based on their spatial orientation. They are calculated based on the formulas presented in the following and include linear, aerial, and relief aspects (Table 13.1). Figure 13.3 Shows the methodology of the study region.

13.4 Results and Discussion

The characteristics of landforms and the arrangement of the subsurface are evaluated in morphometric analysis. This analysis is done to evaluate the drainage basin quantitatively and to plan and manage the use of water resources. Linear, areal, and relief are the three main factors that have been described in the study. The stream order, stream number, stream length, and ratio of the bifurcations all explain the linear aspect of morphometry. The catchment area's features are represented by the areal aspect, which also explains how the catchment region regulates and controls the hydrological behavior. The catchment's terrain arrangement and the characteristics of the terrain are defined by the relief aspect. The findings of the morphometry parameters of the research area are shown in Table 13.1.

13.4.1 Liner Characteristic

13.4.1.1 Stream Order (U)

The quantitative assessment of the drainage basin starts with stream order. The term "u" describes where a stream falls in the tributary hierarchy [23]. First-order streams, according to [36], are those without tributaries. I order channels are the only tributaries of second-order streams. When two II order channels combine, a section of the III order is created. IV order channel is created when two third-order segments combine, and so on. The Pandameru River Basin was determined to be a sixth-order drainage system, according to the ordering scheme. An essential part of effective

Table 13.1 Findings of morphometry parameters of the research area

S.no	Parameters	Formula	References	Stream order					
1	Stream order (u)	Hierarchical rank	[38]	I	II	III	IV	V	VI
2	Number of streams (N_u)	$N_u = N1 + N2 + ... + N6$	[13]	1069	309	74	14	4	1
3	Length of streams (L_u)	$L_u = L1 + L2... + L6$	[13]	619.53	319.97	187.4	58.87	34.87	41.04
4	Mean stream length (Lsm)	$L_{sm} = Lu/Nu$	[38]	0.58	1.03	2.53	4.21	8.72	41.04
5	Stream length ratio (R_l)	$R_l = Lu/Lu - 1$	[13]		II/I	III/II	IV/III	V/IV	VI /V
					1.78	2.44	1.66	2.07	4.71
6	Bifurcation ratio (R_b)	$R_b = N_u/(N_{u+1})$	[38]		I/II	II/III	III/IV	IV/V	V/VI
					3.46	4.17	5.28	3.5	4
7	Mean bifurcation ratio (R_{bm})		[38]	4.084					
8	Rho coefficient (ρ)	$(\rho) = R_l/R_b$	[13]	0.62					
9	Stream frequency (F_s)	$F_s = N_u/A$	[12]	2.19					
10	Perimeter (P)		[29]	147					
11	Length (Lb)		[29]	50.82					
12	Area (A)		[29]	671					
13	Drainage density (D_d)	$Dd = L_u/A$	[12]	1.88					
14	Drainage texture (T)	$T = D_d \times F_s$	[13]	10.0028					
15	Infiltration number (I_f)	$If = F_s \times D_d$	[9]	4.12					
16	Elongation ratio (R_e)	$R_e = D/L = 1.128\sqrt{}A/L$	[29]	0.8					
17	Circulatory ratio ($_{Rc}$)	$R_c = 4\pi A/P2$	[20]	0.39					
18	Time of concentration (T_c)		–	25.64					
19	Total relief (R)	$H = Z\text{-}z$	[38]	573					

(continued)

Table 13.1 (continued)

S.no	Parameters	Formula	References	Stream order					
1	Stream order (u)	Hierarchical rank	[38]	I	II	III	IV	V	VI
20	Relief ratio (R_r)	$R_r = H/Lb$	[29]	0.00112					
21	Shape factor (F_s)	$F_s = Lb^2/A$		1.89					
22	Form factor (F_f)	$F_f = A/Lb^2$	[12]	0.78					
23	Constant of channel maintenance (C)	$C = 1/Dd$		0.53					

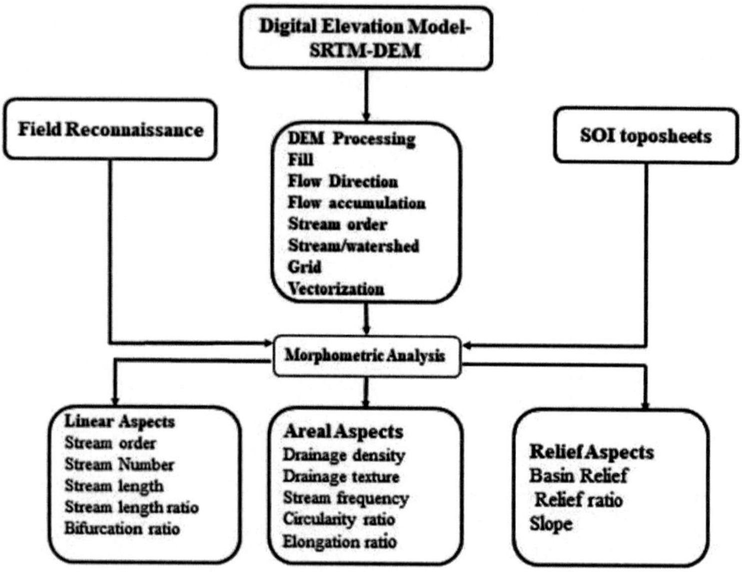

Fig. 13.3 Methodology

water management is the study of stream order. 1471 streams were identified in the river basin.

13.4.1.2 Stream Number (N_u)

The total number of stream segments in the order "u" is the stream number (N_u). Pandameru River Basin is counted and tabulated in this study for analysis. The table below lists the number of streams in the basin belonging to each order (N_u). In the study region, it is found that the overall number of streams reduces as the stream order rises (Table 13.1). In the study area, there are a total of 1471 stream segments, of which 72.67% (1069) fall under the I order stream, 21.00% (309) fall under the II order stream, 5.03% (74) fall under the III order, 0.95% (14) fall under the IV order stream, 0.27% (4) fall under the V order stream, and 0.06% (1) fall under the VI order stream, respectively.

13.4.1.3 Stream Length (L_u)

Stream length (L_u) is an important characteristic of every drainage basin because it exposes surface runoff characteristics of relatively shorter drainage lengths as well as characteristics of higher slope areas and finer textures [31]. The GIS platform was used to calculate the different "u" that make up the basin's total "L_u". The total

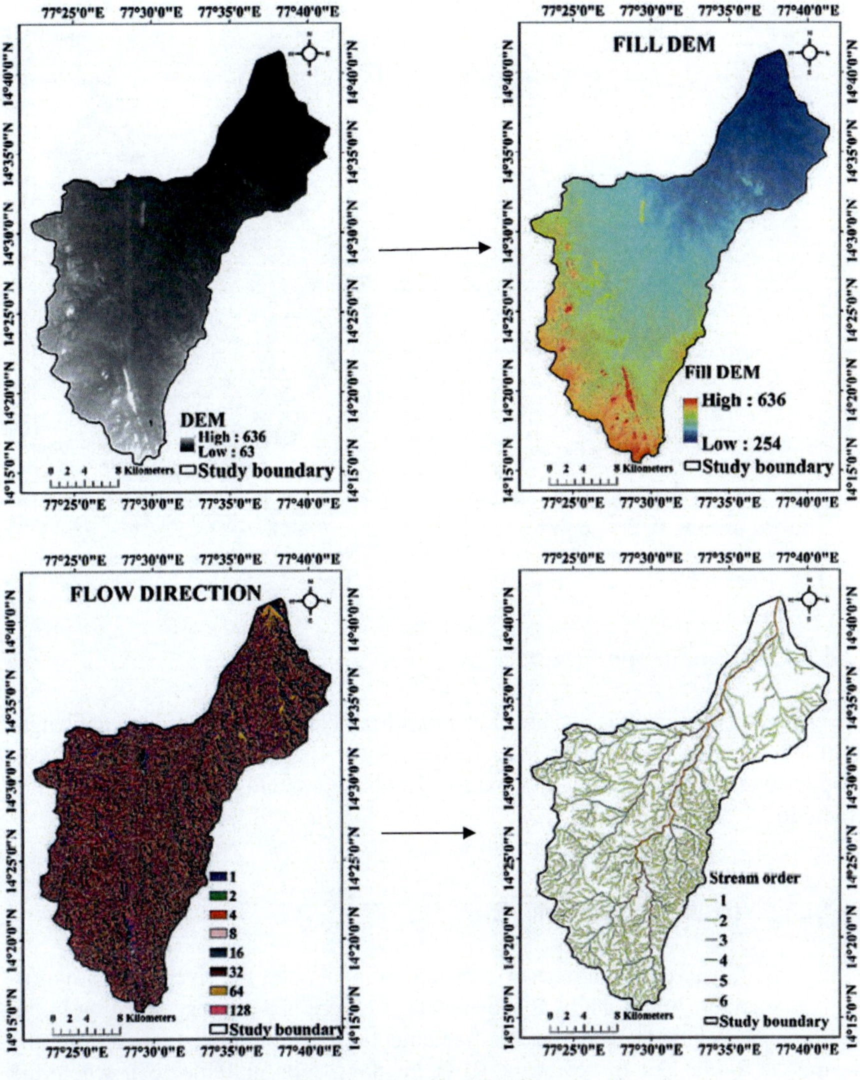

Fig. 13.4 Drainage networks are extracted from SRTM-DEM

length of stream segments often increases for first-order streams while decreasing for higher-order streams. The length of each stream for each order is then added together to yield the total length (L_u) of each stream for each order. The SRTM-DEM data was used to extract stream networks comprising 1261.73 km, with the I order being 619.53 km (49.1%), followed by 319.97 km (25.35%), 187.44 km (14.85%), 58.87 km (4.66%), 34.87 km (2.76%), and 41.05 km (3.25%) as the II, III, IV, V, and VI orders, respectively.

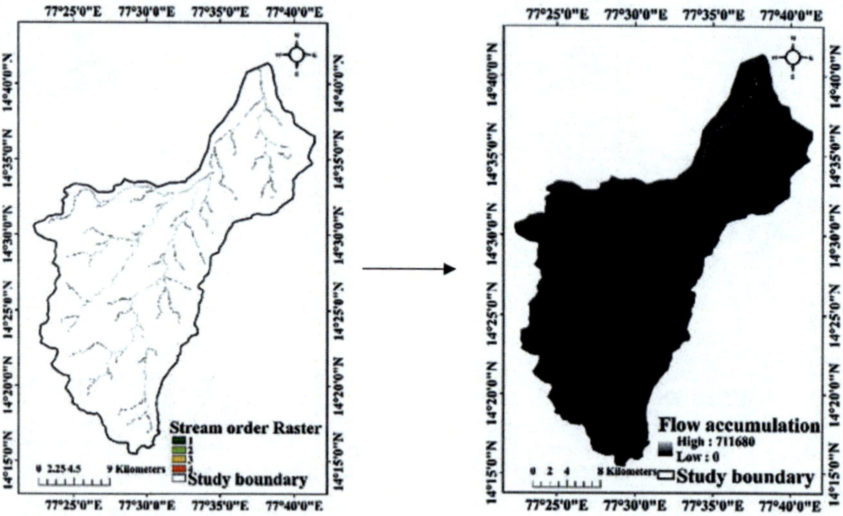

Fig. 13.4 (continued)

13.4.1.4 Mean Stream Length (L_{sm})

The mean stream length is estimated by measuring the order's overall stream length by the number of streams (L_{sm}). Pandameru River Basin's L_{sm} value of 0.85 km, the contributing surface has a fine texture, gentle slopes, and mild drainage network gradients.

13.4.1.5 Stream Length Ratio (R_l)

R_l is a significant characteristic to consider for analyzing the river basin's hydrological behavior. Additionally, it shows whether there is a primary variation in the hydrological behavior of the basin's fundamental rock formations [18]. Using the formula in Table 13.1 to determine R_l yields the results in Table 13.1, where the values of R_l range from 1.78 to 4.71.

13.4.1.6 Rho Coefficient (P)

Horton [13] defines it as the bifurcation ratio divided by the stream length ratio. The link between drainage density and basin geomorphology is critical. This facilitates the evaluation of the drainage network's storage capacity and, as a result, determines the final level of drainage growth in a particular basin [32]. It is also affected by human, geological, biological, biologic, and environmental factors. The Pandameru basin's ρ value is 0.62, which indicates more hydrologic storage during floods.

13.4.1.7 Stream Frequency (F_s)

The quantity of streams per unit area provides a measure of a basin's stream frequency [13]. A greater F_s denotes higher ground relief, steeper terrain, impermeable subsurface conditions, larger surface runoff, and sparser vegetation [19]. Low relief and high permeable geology are indicators of low stream frequency. The F_s of the Pandameru river basin is 2.19 km^2.

13.4.1.8 Bifurcation Ratio (R_b)

According to Horton [13], the R_b is the proportion of streams of one order (Nu) to streams of the next higher order (Nu + 1). The bifurcation ratios in the research area ranged from 3.46 to 4.00, with a mean R_b of 4.08. The R_b standards for channels growing on uniform rocks typically range from 3.0 to 5.0. The decreased bifurcation ratios signal fewer structural issues because the drainage pattern is unaltered [22, 38]. The basin features are dominated by structural control when the bifurcation ratio is greater than 10.00. According to the bifurcation ratio for the Pandameru basin, there are not many structural disturbances, and the basin is geologically uniform.

13.4.2 Areal Parameters

A basin's areal characteristics are its two-dimensional properties. Table 13.1 displays the findings of the estimates for the aerial characteristics of the Pandameru basin.

13.4.2.1 Drainage Density (D_d)

Drainage density is calculated as the sum of all stream lengths divided by the basin area [13]. Runoff potential and terrain fragmentation are both quantitatively indicated by D_d. The catchment's vegetation cover and infiltration capability are displayed. D_d affects erosion susceptibility as well as the catchment area's discharge of water and sediment. Climate, flora, relief, soil and rock composition, and landscape all have an impact on the D_d of the watershed [24]. The potential for groundwater in an area may be determined using drainage density because of its link to surface flow and permeability. Low relief, dense vegetation, and permeable subsoil are typical outcomes of low drainage density, while low vegetation, hilly terrain, and impermeable underlying material are the causes of high drainage density [15, 21]. The Pandameru River basin has a D_d value of 1.88 km/km^2, classifying it in the medium category, which indicates topography with mild to steep slopes, moderate flora, and reduced permeability with medium rainfall (Fig. 13.5 and Table 13.1).

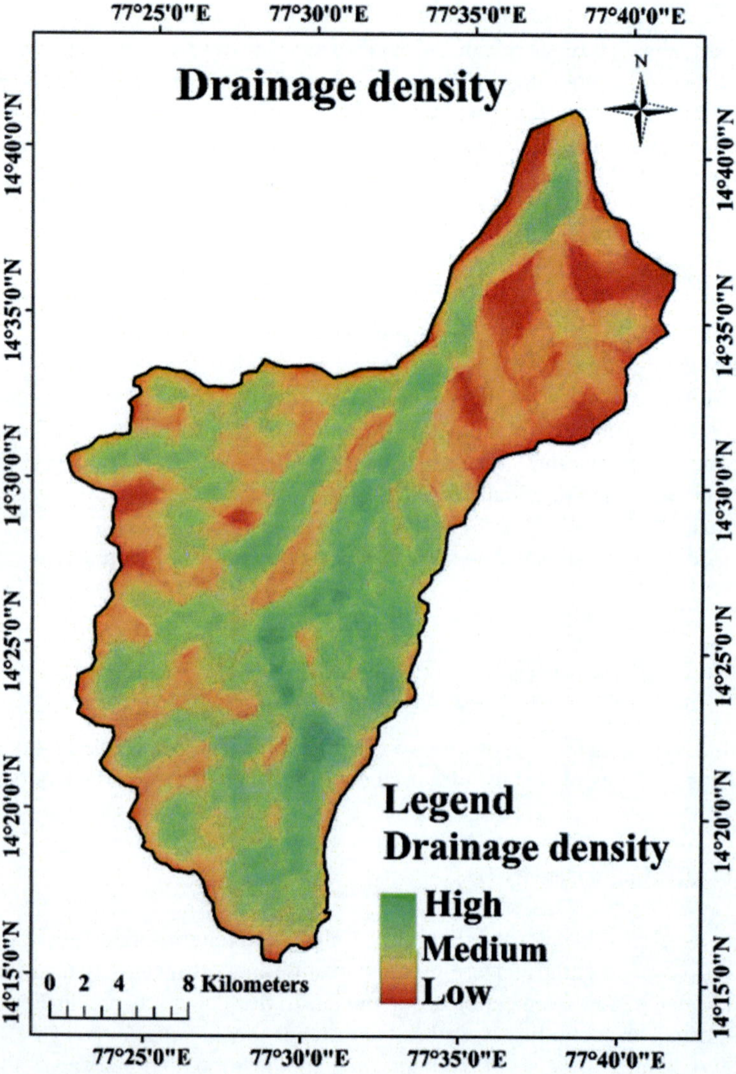

Fig. 13.5 Drainage density

13.4.2.2 Drainage Texture (T)

The drainage texture determines the channel spacing. The climate, amount of precipitation, kind of soil and rocks, rate of permeation, relief, and stage of drainage improvement all play a role [33]. While the texture of hard rocks becomes coarse, that of weak rocks becomes fine. The sparse flora of an arid climate leads to the formation of finer textures on identical rocks than a humid climate would produce. Plant life and weather have an impact on the texture of rocks [20]. The Pandameru basin has a drainage texture value of 10 km, with less intense rainfall, the intermediate texture is brought on by gentle slopes and nearly level topography (Table 13.1). The sub-humid climate and hard rock formations may be responsible for the intermediate texture.

13.4.2.3 Infiltration Number (I_f)

The ratio between the frequency of streams and the density of drainage is known as a drainage basin's infiltration number [9]. It provides information about the basin's infiltration features. Low infiltration and heavy runoff are indicated by the greater value. Such a basin's lithology is dense and impermeable. In the current investigation, a result for the infiltration number of 4.12 (Table 13.1) was obtained, showing limited infiltration and excessive runoff.

13.4.2.4 Elongation Ratio (R_e)

R_e is the proportion of a circle's width to its largest basin length [29]. R_e values for most of the basins typically range from 0.6 to 1.0. In areas with high relief, the value is between 0.6 and 0.8, whereas areas with extremely low relief and circular shapes have values near to 1.0. Because circular basins require less concentration time, they are more effective at discharging runoff than elongated basins. Pandameru basin's R_e value is 0.8, indicating considerable relief with a steep slope and an elongated form.

13.4.2.5 Circularity Ratio (R_c)

The area of a circle with a diameter equal to the diameter of the basin divided by its size is known as the circularity ratio [20]. Stream length and frequency, geological units, climate, terrain, and basin slope are only a few of the many factors that affect it and cause it to fluctuate. If the R_c number is close to 1, it means that the basin is circular, there is scope for homogeneous infiltration, and it will take longer for the extra water to flow out of the basin [27]. The level of the basin's evolution is reflected in the value of R_c. Low and high numbers, respectively, represent the basin's youth and maturity stages. The Pandameru River basin's estimated circularity ratio is 0.39. The R_c reveals the topography's high-yielding stage.

13.4.2.6 Shape Factor (B_s)

The ratio of the square to the watershed length is known as the shape factor (B_s). The Bs value for the research regions is 1.89 (Table 13.1), which reflects the basin's uneven form.

13.4.2.7 Constant of Channel Maintenance (C)

The proportional of drainage density is the constant of channel maintenance (C). It shows the size of the various landforms within a river system in relation to one another [37]. The constant denotes the quantity of basin square kilometers needed to create and sustain a stream channel that is one kilometer long. The basin "C" value is 0.53 km^2 (Table 13.1).

13.4.2.8 Form Factor (F_f)

Form Factor (F_f) is a metric that predicts the concentration of a basin in a specific location and is directly correlated with peak discharge [13]. F_f can be described as the proportion between a basin's area and the square of the basin length [14]. For a basin that is perfectly circular, the F_f value will never be less than 0.78. The shortest number indicates a longer F_f. Specifically, 0.54 (Table 13.1) shows the basin's elongated shape.

13.4.2.9 Time of Concentration (T_c)

The proportion between the length of the mainstream and the relief provided by the basin is used to define the time of concentration. Tc represents the amount of time it takes for a water molecule to move from the watershed's source, which is its location with the least amount of hydrological connectivity, to its collection point. From 0.77 to 8.09, the T_c values range. The total basin T_c is 25.64. (Table 13.1). The basin's most distant point and its outflow are separated by the basin with the highest T_c value, which signifies the largest distance traveled by water.

13.4.3 Relief Aspect

13.4.3.1 Total Relief (R)

Total relief (R), which is the variation in elevation between the basin's maximum and minimum points, is a vital parameter for comprehending the drainage analysis's denudation characteristic [30]. The relief map, created with the use of DEM, displays elevations as high as 636 m and as low as 63 m. The Pandameru River Basin has a 573 m total relief.

13.4.3.2 Relief Ratio (R_r)

R_r is measured by the difference between the highest relief and the horizontal distance along the longest dimension that is parallel to the primary drainage path of the basin. The relief ratio quantifies a drainage basin's overall steepness and is correlated with the rate at which erosion is occurring on the basin's slope [29]. The research area's relief ratio, which is 0.0112 (Table 13.1), indicates modest relief and a gentle slope.

13.4.3.3 Aspect

The aspect map often has a connection to the slope face direction (Fig. 13.6). The aspect map, a vital element, must be understood in order to comprehend how the sun's rays affect where the neighbourhood time occurs [26]. Most of the time, a sloped western side may be warmer than an eastern slope shielded from the sun's harshest rays. The distribution of the geometry of the field vegetation is significantly influenced by the aspect map. The direction of the compass is depicted on the aspect map created using SRTM-DEM: 0 is north with time, and 90 issues to the east. Slopes in the region face east, and as a result, have more flora and moisture than their western-facing equivalents.

13.4.3.4 Slope

The slope is a significant morphometric characteristic that is influenced by morpho-climatic processes in any region that is supported by rocks with variable degrees of surface resistance [3]. Understanding the nature of slope in every place is essential because slope controls the relationship between infiltration and runoff. The slope and infiltration rate are mutually exclusive. The southern portions of the Pandameru basin have a highly steep slope (>35°), which is a suggestion of minimal infiltration and high D_d and frequency (Fig. 13.7). Additionally, it shows the early phases of geomorphic evolution. But once one enters foothill plain areas, the slope rapidly diminishes. This distinctive slope characteristic resulted in both distinct geomorphic landforms, such as alluvial fans, and strong flood characteristics.

13.5 Evaluating the Impact of Morphometric Factors on Soil Erosion Susceptibility

Morphometry is extremely important in understanding soil quality, erosion features, and hydrological activity. Morphometry provides the most accurate relationship between land surface processes and many basin characteristics such as geomorphology, geology, land use, land cover, and hydrology. Groundwater recharge and

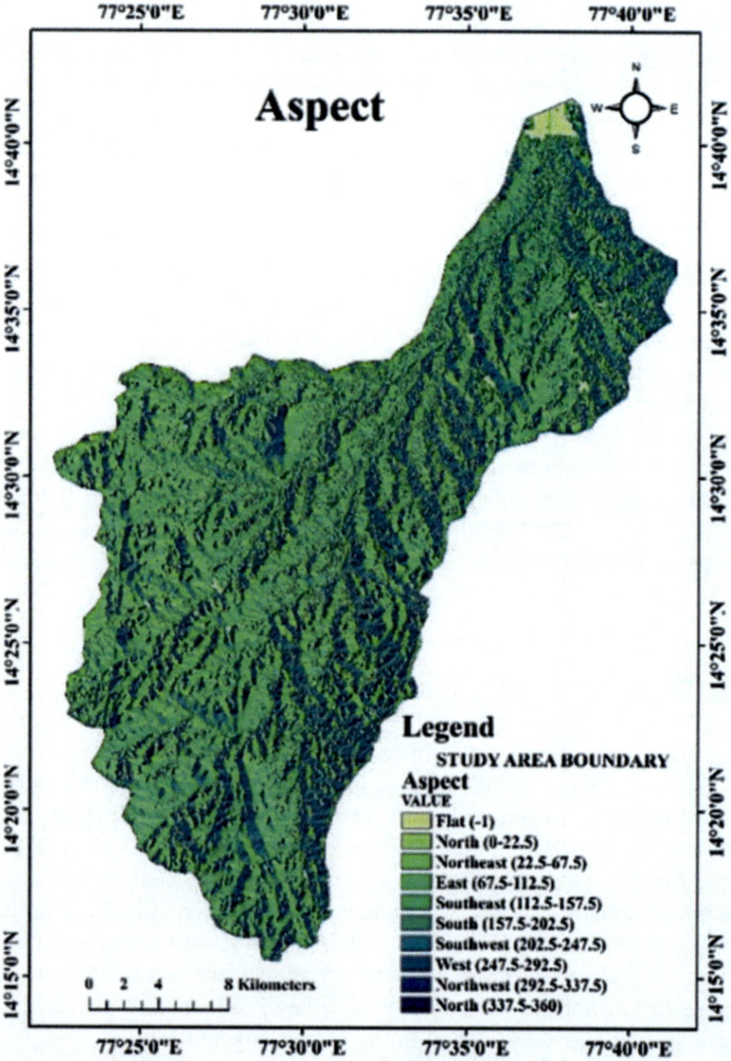

Fig. 13.6 Aspect map

outflow are influenced by a watershed's drainage system. Pandameru basin's south, southwest, and north-eastern regions have extremely steep erosion (>35°), which suggests little infiltration and a lot of runoffs (Fig. 13.8). The basin's northern region has a low erosion susceptibility (<20°), which increases the probability of flooding.

According to all the influencing characters of the slope, soil textures, drainage density, and morphometric studies of the basin, the Pandameru river basin erosion susceptibility has been categorised into four different categories viz., 'very high', 'high', 'moderate', and 'low'. Due to heavy rains, and the overflowing of stream

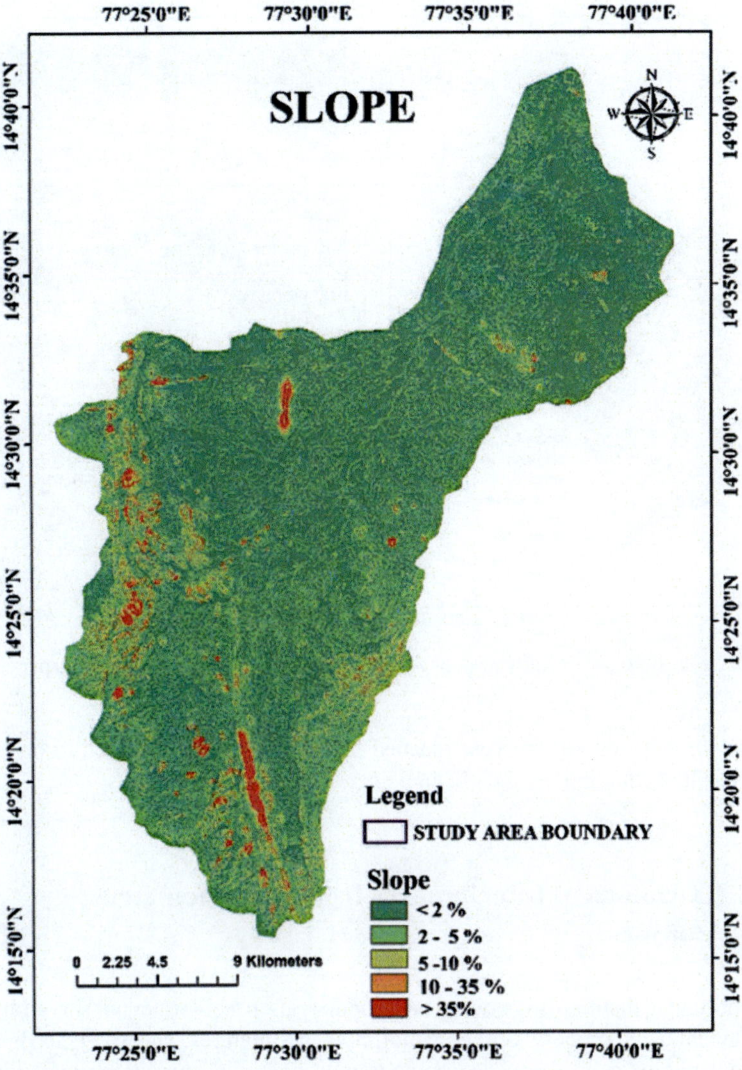

Fig. 13.7 Slope map

channels during 10th–12th October 2022, some parts of the Anantapur, and Rapthadu regions facing flood conditions. Flash floods pounded Anantapur from above due to heavy overnight rain in the district on November 11, several colonies in and around the city were cut off as roads were blocked and water entered the residences. Figure 13.8b–e show the flood conditions of the Pandameru river boundary. The Rayalaseema region receives 335 mm of rain annually. But even in Anantapur's rain

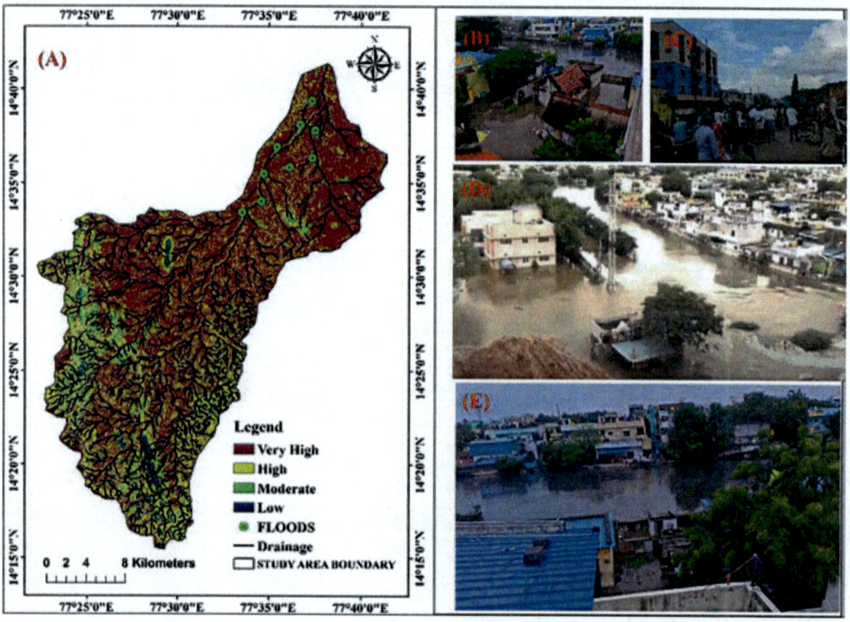

Fig. 13.8 **a** Erosion susceptibility map, **b–e** floods intensity in and around the Anantapur city

shadow, surplus rainfall this year reached 521 mm. This is attributed to breaches, seepage, silt accumulation, and channel encroachment.

13.6 Hydrological Interferences from Morphometric Analysis

It is important to distinguish between basin demarcation, soil-water conservation, and management using properly observed morphometric features. The watershed has low relief and an elongated shape, according to morphological studies conducted inside the Pandameru River Basin. Based on small-scale topographic maps, the research region mostly chooses artificial recharge and rainwater gathering for the advancement of groundwater management. Given the intention of creating artificial recharge structures, the application of RS and GIS tools in drainage evaluation is advantageous. The parameters have been investigated in order to create requirements for the permeability of rock formations in particular drainage basin components [25]. Groundwater development and support plans are increased when the mode of recharge technique and groundwater recharge strategies are integrated with other hydrological drainage basin characteristics [39]. The study area's dendritic watershed may be influenced by the almost uniform major lithology, high D_d regions over highly porous subsoils, and low relief parts that are advantageous for groundwater. Mountainous topography and impermeable hard rock are characteristics of high D_d regions. Slope and infiltration have an adverse correlation; when the slope is decreased, infiltration increases while runoff decreases.

13.7 Conclusion

The findings of this research have revealed that the geospatial approach employed in GIS and RS is an efficient way of calculating and analyzing different morphometric features of the basin and aids in understanding the basic terrain factors, such as the kind of geology, infiltration rate, runoff, etc. The Pandameru basin is naturally well-drained, and the order of the stream's ranges from I to VI. Lower-order streams predominate in the basin, whereas I-order streams have the longest overall distance of stream segments. According to morphometric evaluation, the Pandameru River basin contains a VI-order stream with a dendritic drainage system. The implication is that the rocks exhibit uniform horizontal erosion resistance due to their high homogeneity. But lithology and structure have a main role in determining the basin's drainage pattern. The logarithmic translation revealed that the number, arrangement, and length of the streams in the basin are geometrically related. The logarithmic translation revealed a simple geometric link between the number, configuration, and length of streams in the basin. This connection strongly supports observable facts, such as the basin's length being greater than its width. High runoff, low recharge, and low precipitation concentrations all have an impact on the region's main streams' ability to discharge water according to the basin's R_b, D_d, and R_r. The average bifurcation ratio suggests that rock formations influence the drainage pattern in some way. The circularity ratio reflects early to late mature topography due to variations in slope, elevation, and structural conditions. The basin features high relief and increased slopes based on the elongation ratio. The predominant slope is mild, S-SW-NE oriented, and loaded with high levels of water, and there is a reduced evaporation rate in the basin. According to the basin morphometric study, the southwest, south, and northeast Pandameru river basin requires emergency treatment for their protection. These areas represent a significant basin-wide perennial source of flowing water. The river basin originates on the elevated hillside area. The majority of streams are found in relatively flat locations with permeable underlying layers. As a result, the plains are an important place for creating a man-made recharge system for improving groundwater. The geospatial techniques utilized in this study will aid decision-making in basin development and management studies.

Acknowledgements The first author P. Ravi Kumar is greatly thankful to the Department of Science and Technology (DST), Government of India, for financial support through Inspire program (Sanction order No. DST/INSPIRE Fellowship/2018/IF180877) and also thankful to the Department of Geology, Yogi Vemana University, for necessary facilities for carrying out my research work.

References

1. Anusha BN, Kumar BP, Rajasekhar M, Babu KR (2022) Delineation of groundwater potential zones using geospatial and MCDM approaches in urban areas of Anantapur District, AP, India. Urban Clim 46:101341
2. Asode AN, Sreenivasa A, Lakkundi TK (2016) Quantitative morphometric analysis in the hard rock Hirehalla sub-basin, Bellary and Davanagere Districts, Karnataka, India using RS and GIS. Arab J Geosci 9(5):1–14
3. Badapalli PK, Kottala RB, Madiga R, Golla V (2022) An integrated approach for the assessment and monitoring of land degradation and desertification in semi-arid regions using physico-chemical and geospatial modeling techniques. Environ Sci Pollut Res 1–14
4. Badapalli PK, Kottala RB, Madiga R, Mesa R (2021) Land suitability analysis and water resources for agriculture in semi-arid regions of Andhra Pradesh, South India using remote sensing and GIS techniques. Int Journal of Energy Water Resour 1–16
5. Bera A, Mukhopadhyay BP, Das D (2018) Morphometric analysis of Adula River Basin in Maharashtra, India using GIS and remote sensing techniques. Geo-Spatial Data Nat Resour 13–35
6. Burrough PA, McDonnell RA, Lloyd CD (2015) Principles of geographical information systems. Oxford University Press
7. Das S, Pardeshi SD (2018) Morphometric analysis of Vaitarna and Ulhas river basins, Maharashtra, India: using geospatial techniques. Appl Water Sci 8(6):1–11
8. Etikala B, Golla V, Li P, Renati S (2019) Deciphering groundwater potential zones using MIF technique and GIS: a study from Tirupati area, Chittoor District, Andhra Pradesh, India. HydroResearch 1:1–7
9. Faniran A (1968) The index of drainage intensity: a provisional new drainage factor. Aust J Sci 31(9):326–330
10. Gowd SS (2004) Morphometric analysis of Peddavanka basin, Anantapur district, Andhra Pradesh, India. J Appl Hydrol 17(1):1–8
11. Golla V, Badapalli PK, Telkar SK (2022) Delineation of groundwater potential zones in semi-aridregion (Ananatapuram) using geospatial techniques. Mater Today: Proc 50:600–606
12. Horton RE (1932) Drainage-basin characteristics. Trans Am Geophys Union 13(1):350–361
13. Horton RE (1945) Erosional development of streams and their drainage basins; hydrophysical approach to quantitative morphology. Geol Soc Am Bull 56(3):275–370
14. Kaliraj S, Chandrasekar N, Magesh NS (2015) Morphometric analysis of the River Thamirabarani sub-basin in Kanyakumari District, South west coast of Tamil Nadu, India, using remote sensing and GIS. Environ Earth Sci 73(11):7375–7401
15. Kannan R, Venkateswaran S, Prabhu MV, Sankar K (2018) Drainage morphometric analysis of the Nagavathi watershed, Cauvery River basin in Dharmapuri district, Tamil Nadu, India using SRTM data and GIS. Data Brief 19:2420–2426
16. Kumar B, Venkatesh M, Triphati A (2018) A GIS-based approach in drainage morphometric analysis of Rihand River Basin, Central India. Sustain Water Resour Manag 4(1):45–54
17. Magesh NS, Jitheshlal KV, Chandrasekar N, Jini KV (2013) Geographical information system-based morphometric analysis of Bharathapuzha river basin, Kerala, India. Appl Water Sci 3(2):467–477
18. Mahala A (2020) The significance of morphometric analysis to understand the hydrological and morphological characteristics in two different morpho-climatic settings. Appl Water Sci 10(1):1–16
19. Mangan P, Haq MA, Baral P (2019) Morphometric analysis of watershed using remote sensing and GIS—A case study of Nanganji River Basin in Tamil Nadu, India. Arab J Geosci 12(6):1–14
20. Miller V (1953) A quantitative geomorphic study of drainage basin characteristics in the Clinch Mountain Area, Virginia and Tennessee. Project NR 389-402, Technical Report 3, Columbia University, Department of Geology, ONR, New York
21. Panda B, Venkatesh M, Kumar B (2019) A GIS-based approach in drainage and morphometric analysis of Ken River basin and sub-basins, Central India. J Geol Soc India 93(1):75–84

22. Patel DP, Dholakia MB, Naresh N, Srivastava PK (2012) Water harvesting structure positioning by using geo-visualization concept and prioritization of mini-watersheds through morphometric analysis in the lower Tapi basin. J Indian Soc Remote Sens 40(2):299–312
23. Prabhakaran A, Jawahar Raj N (2018) Drainage morphometric analysis for assessing form and processes of the watersheds of Pachamalai hills and its adjoinings, Central Tamil Nadu, India. Appl Water Sci 8(1):1–19
24. Prakash K, Rawat D, Singh S, Chaubey K, Kanhaiya S, Mohanty T (2019) Morphometric analysis using SRTM and GIS in synergy with depiction: a case study of the Karmanasa River basin, North central India. Appl Water Sci 9(1):1–10
25. Rai PK, Chandel RS, Mishra VN, Singh P (2018) Hydrological inferences through morphometric analysis of lower Kosi river basin of India for water resource management based on remote sensing data. Appl Water Sci 8(1):1–16
26. Rajasekhar M, Raju GS, Raju RS (2020) Morphometric analysis of the Jilledubanderu river basin, Anantapur District, Andhra Pradesh, India, using geospatial technologies. Groundw Sustain Dev 11:100434
27. Sahu N, Obi Reddy GP, Kumar N, Nagaraju MSS, Srivastava R, Singh SK (2017) Morphometric analysis in basaltic Terrain of Central India using GIS techniques: a case study. Appl Water Sci 7(5):2493–2499
28. Sarkar D, Mondal P, Sutradhar S, Sarkar P (2020) Morphometric analysis using SRTM-DEM and GIS of Nagar River basin, Indo-Bangladesh barind tract. J Indian Soc Remote Sens 48(4):597–614
29. Schumm SA (1956) Evolution of drainage systems and slopes in badlands at Perth Amboy, New Jersey. Geol Soc Am Bull 67(5):597–646
30. Siddi Raju R, Sudarsana Raju G, Rajasekhar M (2020) Morphometric analysis of Mandavi River Basin in Rayalaseema region of Andhra Pradesh (South India), using remote sensing and GIS. J Ind Geophys Union 24(1):54–67
31. Singh P, Thakur JK, Singh UC (2013) Morphometric analysis of Morar River Basin, Madhya Pradesh, India, using remote sensing and GIS techniques. Environ Earth Sci 68(7):1967–1977
32. Singh AP, Arya AK, Singh DS (2020) Morphometric analysis of Ghaghara River Basin, India, using SRTM data and GIS. J Geol Soc India 95(2):169–178
33. Smith KG (1950) Standards for grading texture of erosional topography. Am J Sci 248(9):655–668
34. Soni S (2017) Assessment of morphometric characteristics of Chakrar watershed in Madhya Pradesh India using geospatial technique. Appl Water Sci 7(5):2089–2102
35. Sreedevi PD, Sreekanth PD, Khan HH, Ahmed S (2013) Drainage morphometry and its influence on hydrology in an semi arid region: using SRTM data and GIS. Environ Earth Sci 70(2):839–848
36. Strahler AN (1952) Hypsometric (area-altitude) analysis of erosional topography. Geol Soc Am Bull 63(11):1117–1142
37. Strahler AN (1957) Quantitative analysis of watershed geomorphology. EOS Trans Am Geophys Union 38(6):913–920
38. Strahler AN (1964). Quantitative geomorphology of drainage basin and channel networks. In: Handbook of applied hydrology
39. Sujatha ER, Selvakumar R, Rajasimman UAB, Victor RG (2015) Morphometric analysis of sub-watershed in parts of Western Ghats, South India using ASTER DEM. Geomat Nat Haz Risk 6(4):326–341

Chapter 14
Ground Water Quality Assessment Using Water Quality Index and Geographical Information System of Mogamureru River Basin, Y.S.R. Kadapa District, Andhra Pradesh, India

Krupavathi Chinthala, Srinivasa Gowd Somagouni, Ravi Kumar Pappaka, and Harish Vijay Gudala

Abstract Human activities have a significant negative impact on the quality of water, which is a global issue. The quality of drinking and irrigation water varies depending on how groundwater is collected and used, the intensity of rainfall, and the subsequent aquifer recharge in different countries, nations, and places. The current research intends to use a water quality index to estimate the groundwater quality in the Mogamureru river basin using a geographical information system. The WQI will analyze groundwater sustainability for irrigation and drinking in this assessment. A total of 46 pre-monsoon samples were obtained from the Mogamureru river basin. In order to use them in evaluating water quality, 11 parameters (Ca^{2+}, Mg^{2+}, $CO_3{}^{2-}$, K^+, $HCO_3{}^-$, $SO_4{}^{2-}$, Na^+, Cl^-, pH, EC, and TDS) were evaluated for each groundwater sample. Numerous water parameters are combined into a single indicator number that represents the degree of water quality using a mathematical model known as the water quality index (WQI). WQI's findings revealed that 95.7% of groundwater samples are unfit for consumption, with 4.3% of samples having very low quality. Irrigation indices include the hazards of salinity and alkalinity (SAH), the sodium adsorption ratio (SAR), the sodium percentage (Na %), the residual sodium carbonate (RSC), the magnesium adsorption ratio (MAR), the permeability index (PI), the Kelley's ratio (KR), and the synthetic hazard coefficient (K) are used to determine the 100%

K. Chinthala · S. G. Somagouni (✉) · R. K. Pappaka · H. V. Gudala
Department of Geology, Yogi Vemana University, Kadapa, AP 516005, India
e-mail: ssgowd@gmail.com

K. Chinthala
e-mail: chinthalakrupavathi95@gmail.com

R. K. Pappaka
e-mail: ravigpcs75@gmail.com

H. V. Gudala
e-mail: harishvijay@outlook.com

E. Balaji et al. (eds.), *Emerging Technologies for Water Supply, Conservation and Management*, Springer Water, https://doi.org/10.1007/978-3-031-35279-9_14

of samples are suitable for irrigation. The analysis demonstrated that the most of the samples are appropriate for irrigation. Geospatial tools are utilized for the preparation of spatial distribution maps of each parameter as well as WQI, SAR, SSP, and PI indices.

Keywords Water quality index (WQI) · Hazards of salinity and alkalinity (SAH) · Sodium percentage (Na %) · Geographic information system · Mogamureru River Basin

14.1 Introduction

About one-third of freshwater use worldwide comes from groundwater, which is crucial for home, industrial, and agricultural usage, especially in arid as well as semi-arid regions, in which sources of water are scarce and unevenly distributed [1, 24, 46]. The constantly growing demographic, environmental conditions, as a result of additional emissions of atmospheric phenomenon gases (Greenhouse gases), and escalations of cultivation, place to acute extreme strain on the global major non-renewable assets are water and soil, and as a consequence affects huge mission to produce enough food to satisfy food demand in the present-days. The existing global population of 7.3 billion people is predicted to grow to more than nine billion by 2050, with the majority of such as population raise happening in emergent nations that already face the problem of food scarcity [3].

Forty-six groundwater samples from boreholes and dug wells have been examined in this study for physicochemical characteristics, and results were compared to BIS standards for potable and residential usage requirements based on the Water Quality Index (WQI). Horton [20] created the WQI for the earliest time using weighted arithmetic calculations. Brown et al. [9], Shankar and Kawo [37] Bawoke and Anteneh [7], and Ram et al. [32] devised a variety of WQI models based on the weighing and rating of various water quality parameters that are obtained by the weighted arithmetic method. WQI's importance: WQI, in general, is the comparison of the amount with a predetermined base, a predetermined standard, or both [35]. As a result, the WQI tracked environmental changes and status and reported on them statistically. By eliminating an excessive amount of data to show water quality, the WQI approach effectively informs on the level of purity and pollution of water [4, 15, 40, 45]. A faultless quality monitoring system is also made accessible by the WQI tool. To achieve this, the monitoring data should be used to create executive management indices that are clear and simple to understand.

A 70% growth in modern cultivation yield can be necessary to fabricate enough food if this population rises forecast demonstrates to be accurate. In these circumstances, collaborative activities are being assembled worldwide to enhance the water capability that will be utilized to raise the crop yield. The imprudent utilized brackish or saline water is excessively frequently connected with the evolution of soil salinity, sodium hazard, ionic ratio, and groundwater contamination. Due to the poor consequences, it is crucial for good understanding of precisely in what way water quality control of the administration of irrigated agriculture, for the most part in arid and semiarid regions. Sodicity, ion toxicity, and salinity are major problems in irrigation waters. In dry regions, if rainfall does not penetrate salts from the soil, an accretion of salts will come about in the root zone of the crop. Consequently, a regular try-out of soils and waters was needed for the detection some modifications in salt concentration. Sodium hazard: the appearance of sodium (Na^+) will cause in a worsening of the soil composition, thereby decreasing water infiltration that can enter and pass through the soil. The ionic ratio means the censoriousattentiveness of several salts such as chloride, sodium, and some trace elements, at which these salts are harmful to plant growth. Some characteristics of irrigation water quality and standards to find out the quality. It intends to cover social control problems and soil properties through the use of irrigation water of diverse quality. Remote sensing (RS) and geographic information systems (GIS) are both extensively used to evaluate water quality around the world. GIS development makes it easier for us to include laboratory data.

14.2 The Study Area's Description

The Mogamureru River begins in Gurujala, Simhadripuram Mandal, flows from west to east, and meets the Papaghni River in Animela, Veerapunayunipalle Mandal. It is 174.82 km long and covers an area of 1332.47 km^2. This location (latitude 14°40′30″ N; longitude 78°30′0″ E) is covered by India toposheets 57J/2, 57J/3, 57J/6, 57J/7, 57J/11, 57F/14, and 57 F/15 (Fig. 14.1).

The study area consists of quartzite with conglomerate of Gulcheruquartzite and dolomites, volcanic flows, and quartzite of the Vempalle Formation of the Papaghni group. Quartzite with conglomerate of Pulivendula quartzite and shale, dolomite of the Tadipatri Formation of the ChitravatiGroup. In the study region is predominatedby denudational, structural landforms and small amounts of anthropogenic landforms. Clay, loamy and loamy to clay soils are presented in the study area. The average rainfall of the study area is 143 mm.

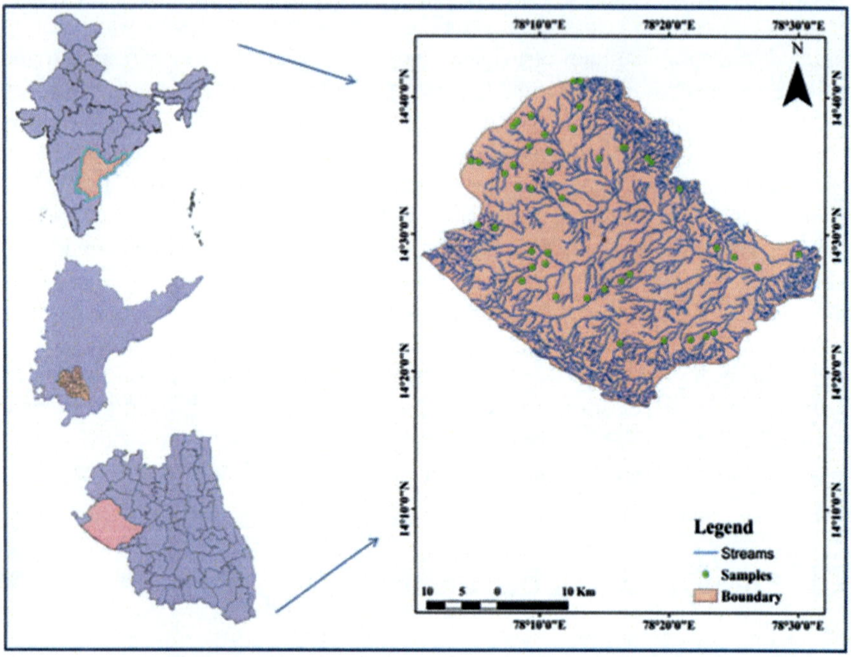

Fig. 14.1 Location map

14.3 Materials and Methods

14.3.1 Analysis and Sampling of Water

Water samples were collected from 30 shallow and 16 deep locations in the river basin. Borewell samples were collected from Kamassamudram, Diddekunta, Simhadripuram, Vemula, Pulivendula, and Vempalli villages. Well samples were collected from JangamReddypalle, Bhadrampalle, Akkuvaripalle, Lingala, and Veerapunayunipalle villages. The upper section of the river basin has ten shallow wells each in Simhadripuram, Thondur, and Veerapunayunipalle mandals, while the bottom part of the river basin has sixteen deep wells in Lingala, Pulivendula, Vemula, and Vempalle. Sampling was planned to take place throughout both the rainy and dry seasons. In May 2021, pre-monsoon season samples were collected.

Standard Methods for the examination of water and wastewater and the Global Environmental Monitoring Systems/Water Operational Guide were used for sampling and analysis [41]. One-liter polyethylene bottles were used to collect and clean the samples. Sulfuric acid is used to acidify groundwater samples in order to preserve them.

14.3.2 Water Quality Index

Potential hydrogen, electrical conductivity, Cl^-, dissolved particles; hardness, SO_4^{2-}, Na^{2+}, Mn^{2+}, Ca^{2+}, K^+, and Mg^{2+} have all been taken into account when computing the Water Quality Index (WQI). The weighted computation by the water quality index method has been used in many kinds of research for the estimate [29, 38].

The weighted arithmetic water quality index method clarifies the water quality according to the degree of purity by using the most commonly measured water quality factors. Many scientists have used the technique [4], and the following equations were used to calculate WQI [9]:

$$WQI = \frac{Wi\,Qi}{Wi} \tag{14.1}$$

where: The water quality index, or WQI, does have a value ranging from 0 and 100;

qi stands for a relative water quality rating for each parameter;

i represent the range of parameters.

The sub-index (Qi) is calculated by the way of the using the parameter

$$Qi = 100\left[\frac{Vi - V0}{Si - V0}\right] \tag{14.2}$$

where Vi is the parameter's concentration while evaluating water,

Si = Standard acceptable value of parameter,

Vo = Real value of this parameter in uncontaminated water (for natural water Vo = 0 for hardness).

The sub-index (Wi) turned into calculated via using the ideal value of the parameter

$$Wi = \frac{K}{Si} \tag{14.3}$$

where k = proportionality steady,

Si = Standards of parameter.

Where K is a constant which could result from making use of the formulation

$$K = \frac{1}{\sum(\frac{1}{Si})} \tag{14.4}$$

Based on the value acquired for the Weighted Arithmetic WQI method, the water ecology.

Popularity may be decided.

14.3.3 Irrigation Water

In the Mogamureru river basin, groundwater is frequently used for irrigation. The mineral makeup of irrigation water and how it affects plants and soil are both reflected in its quality. Plants and agricultural soils are directly impacted by the chemical makeup of irrigation water, which lowers productivity. Consequently, water quality monitoring for cultivation had becomes crucial for the Mogamureru river basin's growing agricultural production. In this study, the indices of the SH, SAR, Na%, RSC, MH, KR, PI, and K [23] were used to estimated the quality of irrigation water.

14.3.3.1 Total Hardness (TH)

Total Hardness is an important parameter that has been used in irrigation. In order to understanding water hardness, ions of calcium and magnesium (Ca^{2+} + Mg^{2+}) are higher in a few samples due to the existenceof alkaline earth metals. $CaCO_3{}^{2-}$has been formed when calcium combines with water in the limestone and chalk formations.

14.3.3.2 Hazards of Salinity and Alkalinity (SAH)

When there is a lot of salt in the groundwater, it becomes saline. The assessment of salinity risk is a significant characteristic of agricultural water, because excessive sodium concentration in irrigation water due to the soil to become saline and has a negative impact on plant salt consumption performance.

14.3.3.3 Sodium Percentage (Na%)

The salt danger index is expressed as a percentage. Soil structure, aeration, and penetration will all be harmed by a high sodium content [14, 17].

14.3.3.4 Sodium Adsorption Ratio (SAR)

The SAR is also a sodium threat since it can decrease the permeability of the soil, obstructing agricultural absorption of water. The groundwater is then classified using SAR: excellent, good, dubious, and unsuitable [5, 26]. All of the samples in this category are considered excellent.

14.3.3.5 Residual Sodium Carbonate (RSC)

Bicarbonate is a significant aspect for assessing irrigation water quality [23, 47]. RSC is a useful tool for determining the importance of irrigation water by calculating the carbonate and bicarbonate sum, as well as the calcium and magnesium sum connection [8, 36]. Because of sodium carbonate deposition, irrigation soil with excessive RSC containing water might become sterile [33, 43].

14.3.3.6 Magnesium Adsorption Ratio (MAR)

The Magnesium Adsorption Ratio (MAR) describes the property of soil composition modification due to the magnesium in irrigation water [19]. Excess magnesium in groundwater causes soil alkalinity; additionally, magnesium and clay sediments absorb a large amount of water, reducing infiltration soil potentiality and crop yield [18, 28].

14.3.3.7 Permeability Index (PI)

The permeability index (PI) is a factor in ascertain whether or not the water is suitable for irrigation. Continuous irrigation water use regulates soil water motion capacity (porosity and permeability), which is impacted by Na^{2+}, Ca^{2+}, Mg^{2+}, and $HCO_3{}^{2-}$ ions in the soil [25, 30, 31].

14.3.3.8 Kelley's Ratio (KR) and Synthetic Harmful Coefficient (K)

Kelley's ratio is utilized to estimate the fitting of groundwater cultivation [6, 11–13, 16, 27]. The alkali and salt risks could be masked by the synthetic hazardous coefficient.

The following formulas used to derive these indicators:

$$\mathbf{SSP} = \frac{Na^{+}}{Ca^{2+} + Mg^{2+} + Na^{+} + K^{+}} \times 100 \tag{14.5}$$

$$\mathbf{SAR} = \frac{\mathbf{Na}^{+}}{\frac{\sqrt{\mathbf{Ca}^{2+} + \mathbf{Mg}^{2+}}}{2}} \tag{14.6}$$

$$\mathbf{RSC} = \left(\mathbf{HCO}_3^{2-} + \mathbf{CO}_3^{2-}\right) - \left(\mathbf{Ca}^{2+} + \mathbf{Mg}^{2+}\right) \tag{14.7}$$

$$MAR = \frac{Mg^{2+}}{Ca^{2+} + Mg^{2+}} \times 100 \tag{14.8}$$

$$\mathbf{PI} = \frac{\mathbf{Na} + \sqrt{\mathbf{HCO_3}^-}}{\mathbf{Ca^{2+}} + \mathbf{Mg^{2+}} + \mathbf{Na^+}} \times 100 \tag{14.9}$$

$$KR = \frac{\mathbf{Na^+}}{\mathbf{Ca^{2+}} + \mathbf{Mg^{2+}}} \tag{14.10}$$

$$K = 12.4M + SAR \tag{14.11}$$

14.4 Results and Discussion

14.4.1 Evaluation of the Drinking Water Quality

The analytical results examined whether the groundwater in the study region was suitable for drinking and agricultural use. Table 14.1 shows the research findings, including all parameters for the groundwater samples in the research area.

14.4.1.1 The Water Quality Index

The Water Quality Index is an indicator that recognizes the underlying reasons for contrasting water quality metrics. The component suitable for people, ingestion of

Table 14.1 Water quality rating as in the step with weight arithmetic water quality index method

Parameters	S_n	$\frac{1}{S_n}$	Σ 1/ Sn	$K = \frac{1}{\Sigma \frac{1}{S_n}}$	$Wi = \frac{K}{S_n}$	V0	V_n	$\frac{V_n}{S_n}$	$Q_n = \frac{V_n}{S_n} \times 100$	WnQn
pH	8.5	0.12	0.24	4.16	0.49	7	8.1	0.73	73.33	35.9
EC	300	0.0033	0.24	4.16	0.014	0	2.11	−3.26	−326	−4.52
TDS	500	0.002	0.24	4.16	0.008	0	1025	678.66	67,867	564.82
TH	300	0.0033	0.24	4.16	0.014	0	50.03	28.68	2868.6	39.78
Na^+	50	0.02	0.24	4.16	0.08	0	3.39	−2.4	−240.5	−20.01
Ca^{2+}	75	0.013	0.24	4.16	0.056	0	1.27	−3.82	−381.8	−21.18
Mg^{2+}	30	0.033	0.24	4.16	0.14	0	11.37	2.917	291.77	40.47
$CO_3{}^{2-}$	200	0.005	0.24	4.16	0.02	0	**0**	−4.66	−466.7	−9.7
$HCO_3{}^-$	30	0.033	0.24	4.16	0.14	0	0.09	−4.6	−460.1	−63.82
Cl^-	250	0.004	0.24	4.16	0.017	0	0.12	−5.25	−525.3	−8.74
$SO_4{}^{2-}$	200	0.005	0.24	4.16	0.021	0	49.44	26.96	2696.2	56.09
Total					1					609.09

Table 14.2 Using the WQI rating to classify water quality

Water quality class	Range	No. of samples
Excellent	<50	–
Good	50–100	2
Poor	100–200	4
Very poor	200–300	5
Unsuitable	>300	35

groundwater, is used to calculate the WQI. The current activity's verification goal is to deal with the fittest of people, groundwater consumption, and reliance on the numerical quantity of the water quality index. Horton [20] coined the term "water quality index," which has since evolved into an economic tool for water quality reflection and social awareness.

Water quality indexes differ depending on which of the five classifications they belong to. The forty-four samples are all considered excellent. Two more samples (Table 14.2) fall into the good and extremely poor categories and representing in spatial distribution Map (Fig. 14.2).

14.4.2 Irrigation Purposes

Groundwater Suitability for irrigation system is heavily on water quality estimation [10, 28, 44]. Groundwater estimation in the Mogamureru river basin was carried out utilizing metrics such as EC, TH, SAR, SSP, RSC, MAR, PI, KR, and K, which are important parameters for irrigation (Tables 14.3 and 14.4).

14.4.2.1 Total Hardness (TH)

Four samples had moderate hardness (9%), three samples had hardness (6%), and thirty-nine samples had extremely hardness (85%), which is favorable for ingesting [34]. TH is represented in the spatial distribution map (Fig. 14.3).

14.4.2.2 Hazards of Salinity and Alkalinity (SAH)

Salinity concerns in irrigation water are estimated using EC and TDS ranges. In the current research area, the EC ranges from 720 to 4000 S–1 cm^{1} and TDS from 300 to 2000 mg L1 in the current research location; the average is 38.3 S–1 cm^{1} and 918.9 mg L1; standard deviations are 824.7 S–1 cm1 and 410.5 mg L1 correspondingly (Table 14.1).

According to the Total Dissolved Solids ranges, 26% of the samples fell into the good category, 36% into the fair category, 19% into the poor category, and 19% into

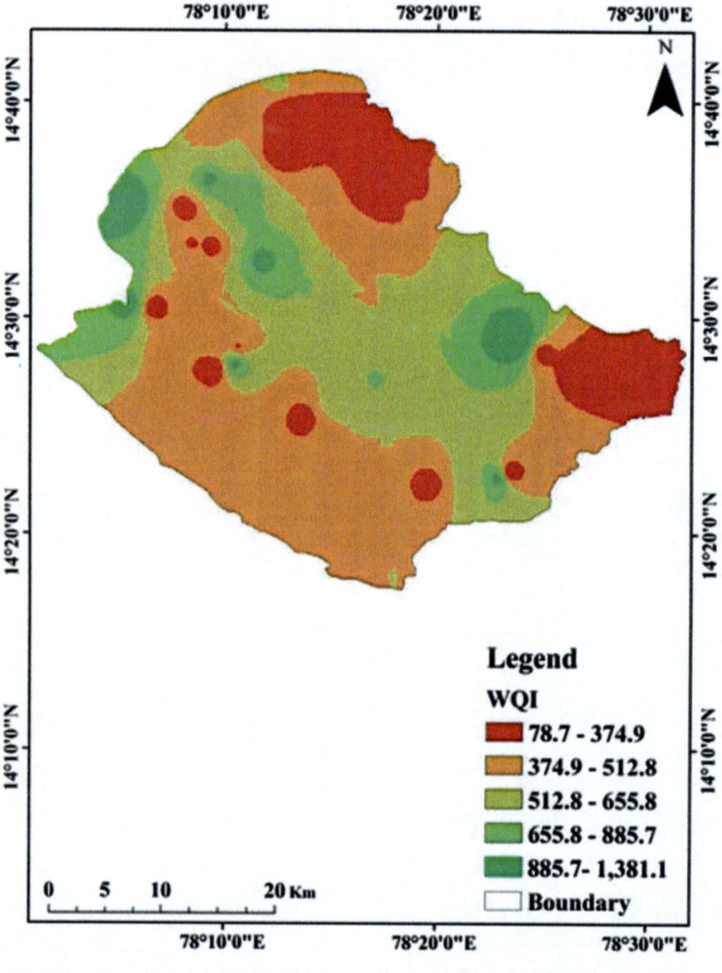

Fig. 14.2 Spatial distribution map of WQI

the unacceptable category. The electrical conductivity of water samples was divided into four categories depending on salinity, with the excellent category indicating that no sample has fallen.1% (one) of the samples came into the good category, 70% (36) of the samples fell into the permissible category, 25% (9) of the samples went into the questionable category, and 4% (6) of the samples fell into the unsafe category. According to the salinity, it was determined that none of the samples were considered unsuitable for irrigation and are representing in the spatial distribution map (Fig. 14.4).

Table 14.3 Statistical analysis for irrigation

Samples references id	EC	TH	TDS	SAR	MAR	SSP	RSC	KR	PI	K
M1	2110	632.5	1025	1.35	0.7	21.15	−12.6	0.3	0.003	1.7
M2	2210	820.2	1101	1.73	0.6	23.21	−16.3	0.3	0.003	1.8
M3	1630	571.9	816	1.75	0.5	26.75	−11.3	0.4	0.01	2.5
M4	720	226.3	374	2.46	0.6	44.96	−4.35	0.8	0.003	1.9
M5	820	505.6	411	1.88	0.4	29.44	−9.9	0.42	0.002	1.3
M6	1700	895.3	862	1.26	0.4	17.45	−17.8	0.21	0.001	1.1
M7	3930	1961.9	2000	1.04	0.6	10.52	−39.1	0.12	0.002	1.5
M8	3340	1127	1675	1.45	0.5	17.77	−22.4	0.22	0.004	2.7
M9	910	402.8	465	2.69	0.8	40.1	−7.94	0.67	0.004	2.4
M12	2360	538.2	1192	1.07	0.6	11.83	−31.7	0.13	0.002	1.5
M13	2630	1258.7	1315	1.51	0.4	19.61	−18.9	0.24	0.001	1.1
M14	2950	1588	1469	1.41	0.7	16.85	−23.9	0.2	0.002	1.52
M15	3050	954	1523	2.41	0.5	31.3	−13.9	0.45	0.002	1.4
M16	2810	1202	1414	1.79	0.5	22.71	−18.4	0.29	0.003	2.4
M17	1030	701.8	547	1.45	0.5	18.26	−20.9	0.22	0.003	1.8
M18	1410	925.4	711	1.6	0.4	20.91	−18.1	0.26	0.002	1.5
M19	1730	1051.6	872	2.68	0.5	47.51	−4.2	0.9	0.002	1.6
M20	1480	913.4	746	1.77	0.7	28.76	−9.4	0.4	0.005	2.7
M21	910	218.6	459	2.64	0.6	41.4	−6.9	0.71	0.003	1.8
M22	1320	479.5	654	1.27	0.8	15.04	−25.4	0.18	0.004	2.6
M23	1790	347.9	871	3.62	0.5	61.78	−2.4	1.62	0.002	1.3
M24	3380	1278.1	1678	2.3	0.8	43.45	−4.4	0.77	0.01	3.6
M25	1720	125.1	885	1.23	0.6	16.39	−19.4	0.19	0.005	2.3
M26	1780	223.6	807	1.6	0.4	23.35	−13.6	0.3	0.002	1.2
M27	1670	976.4	847	1.37	0.6	20.72	−13.6	0.26	0.003	1.6
M28	1460	685.24	743	1.38	0.4	19.59	−15.9	0.24	0.002	1.4
M29	1060	682.34	510	1.78	0.4	26.24	−12.3	0.35	0.003	1.4
M30	1790	803.34	895	2.03	0.5	27.9	−13.7	0.39	0.003	1.8
M31	1110	623.64	552	1.48	0.8	18.89	−20.0	0.23	0.003	2
M32	1350	691.4	676	1.58	0.7	22.06	−15.6	0.28	0.002	1.5
M33	1790	1008.5	873	1.7	0.7	25.12	−12.7	0.33	0.003	1.6
M34	1590	783.9	790	1.35	0.7	19.88	−14.6	0.25	0.003	1.7
M35	2040	641.9	1002	0.85	0.5	9.912	−29.8	0.11	0.002	1.4
M36	1120	736.5	606	1.63	0.6	27.19	−9.4	0.37	0.001	0.9
M37	3070	1502.3	1514	5.06	0.7	73.72	−1.5	2.8	0.003	1.6

(continued)

Table 14.3 (continued)

Samples references id	EC	TH	TDS	SAR	MAR	SSP	RSC	KR	PI	K
M38	1320	477.5	671	0.99	0.8	12.94	−22.1	0.15	0.01	5.1
M39	1250	81.4	628	2.11	0.6	27.47	−15.4	0.38	0.001	1.01
M40	2320	1112.1	1131	2.34	0.5	33.04	−11.2	0.49	0.003	2.1
M41	1700	775.1	863	2.51	0.6	34.73	−10.9	0.5	0.004	2.4
M42	1200	564.1	596	2.46	0.6	30.68	−15.3	0.44	0.004	2.5
M43	1530	555.9	764	3.93	0.7	60.66	−3.05	1.54	0.003	2.5
M44	2020	771.4	997	1.82	0.7	26.4	−12.7	0.36	0.01	3.9
M45	1150	162.2	569	1.06	0.8	11.79	−31.4	0.13	0.003	1.8
M46	1550	642.8	780	3.9	0.4	58.55	−3.7	1.41	0.001	1.1
Max	720	81.4	300	0.9	9.9	−39.1	0.4	0.001	0.11	0.9
Min	4000	1962	2000	5.1	74	−1.5	0.8	0.01	2.8	5.1
Avg	1838	760.8	919	1.9	28	−14.9	0.6	0.003	0.5	1.9
STDV	824.7	416.5	411	0.9	15	8.4	0.13	0.002	0.5	0.8

14.4.2.3 Sodium Percentage (Na%)

The concentration of sodium in groundwater ranges from 9.9 to 73.7 meq/L (Table 14.1), with a mean of 28.1 meq/L and a standard deviation of 14.7. The groundwater samples have an allowed water quality category (Table 14.2) and represented in the spatial distribution map (Fig. 14.5).

The Wilcox diagram [2, 21, 22, 42, 48] is utilized to determine whether groundwater is suitable for irrigation. No sample fell into the excellent to good category, the maximum (thirty) samples fell into the good to permissible category (65%); the two samples (5%) fell into the permissible to doubtful category; the two samples (15%) fell into the doubtful category; and the two samples (15%) fell into the doubtful category (Fig. 14.6). This indicates that the groundwater is suitable for irrigation. The vigorous interplay between groundwater and lithology, which results in cation interchange in particular, is the cause of excess sodium (Na^+) in most samples. Agricultural practices, like the utilization of fertilizers and pesticides, also raise sodium levels.

14.4.2.4 Sodium Adsorption Ratio (SAR)

All of the samples in this category are considered excellent. The SAR of groundwater samples ranged from 0.9 to 5.1 meq/L, with a mean of 1.5 meq/L and a standard deviation of 0.9. A penetration through the fitting for irrigation water could be obtained by plotting the sodium absorption ratio and electrical conductivity using the US Salinity

Table 14.4 Water quality for irrigation

Parameter	Range	Water class	Sample no	Sample no (percentage)
EC	<250	Excellent	Nil	Nil
	250–750	Good	46	1%
	750–2000	Permissible	3–6, 9, 15–21, 23–32, 35, 36, 37, 39–44	70%
	2000–3000	Doubtful	1, 2, 10–12, 14, 33, 38, 42	25%
	>3000	Unsafe	7, 8,13, 22, 34, and 44	4%
TDS	<300	Excellent	Nil	–
	300–600	Good	4, 5, 9, 15, 19, 20, 27, 29, 34, 40, 43 and 46	26%
	600–900	Fair	3, 6, 16–18, 21, 23–26, 28, 30–32, 36, 37, 39, 40, 43, and 460	36%
	900–1200	Poor	1, 2, 7, 8, 10, 33, 38, 41 and 44	19%
	>1200	Unacceptable	7, 8, 11, 12, 13, 14, 22, 35, and 45	19%
TH	<75	Soft	Nil	–
	75–150	Moderate	23, 37, 43, and 46	9%
	150–300	Hard	4, 19, and 24	6%
	>300	Very Hard	1–3, 5–18, 20–22, 25–42, 44, 45	85%
SAR	<20	Excellent	All	100%
	20–40	Good	Nil	–
	40–60	Permissible	Nil	–
	60–80	Doubtful	Nil	–
	>80	Unsuitable	Nil	–
MAR	<50	Suitable	All	100%
	>50	Unsuitable	Nil	
Na%	<200	Suitable	All	100%
	>200	Unsuitable	Nil	–
RSC	1.25	Safe	All	100%
	1.25–2.50	Marginally Safe	Nil	–
	>2.50	Unsafe	Nil	–
KR	<1.0	Suitable	All	100%
	>1.0	Unsuitable	Nil	–
PI	<80	Good	All	100%
	80–100	Moderate	Nil	–
	100–120	Poor	Nil	–

(continued)

Table 14.4 (continued)

Parameter	Range	Water class	Sample no	Sample no (percentage)
K	<25	Excellent	All	100%
	25–36	Good	Nil	–
	36–44	Injurious	Nil	–
	>44	Unsuitable	Nil	–

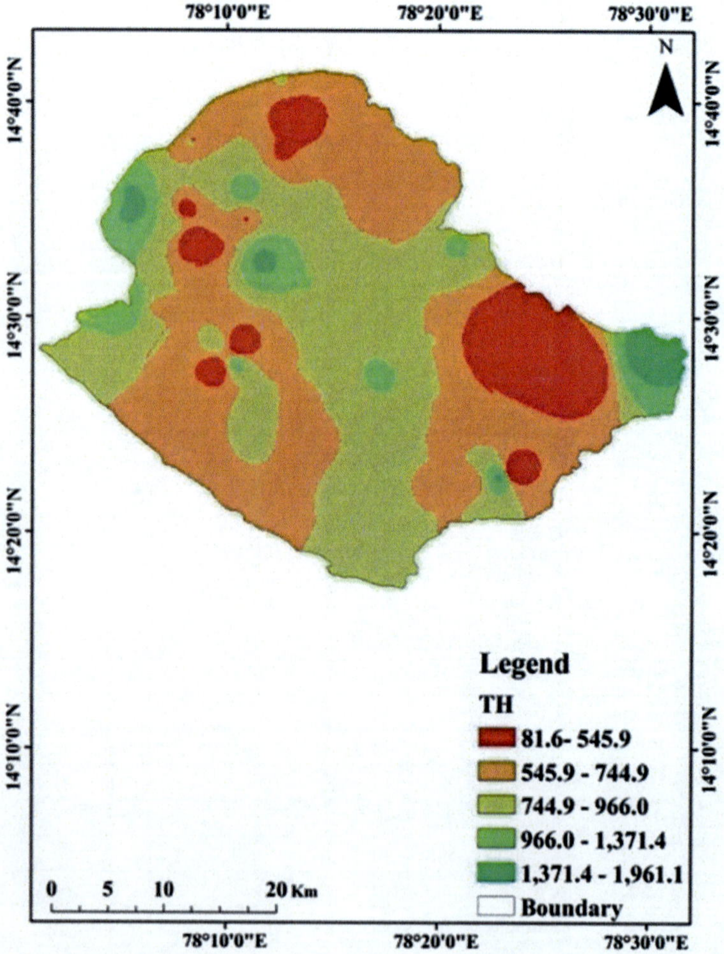

Fig. 14.3 Spatial distribution map of total hardness

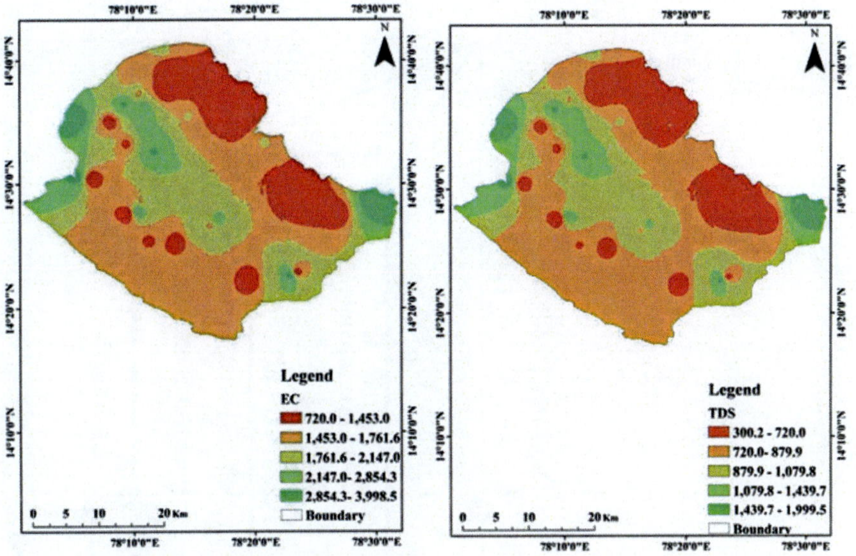

Fig. 14.4 Spatial distribution map of EC and TDS

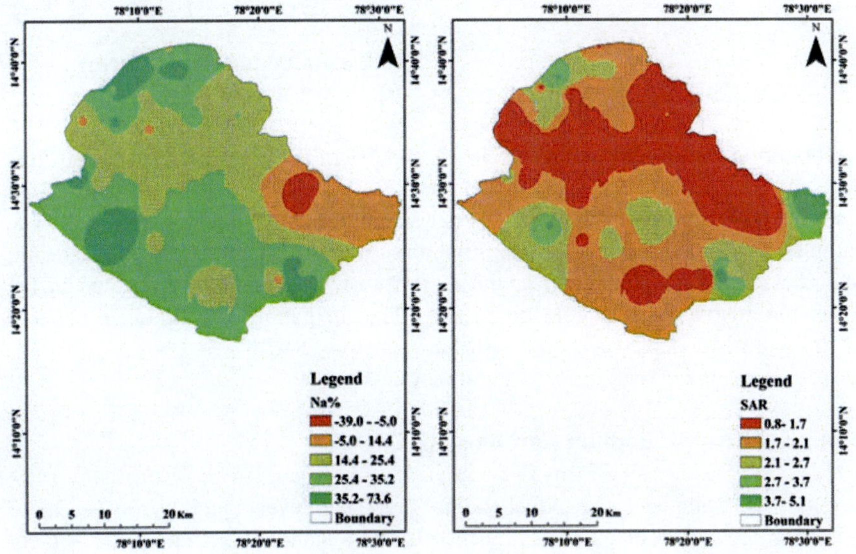

Fig. 14.5 Spatial distribution map of Na% and SAR

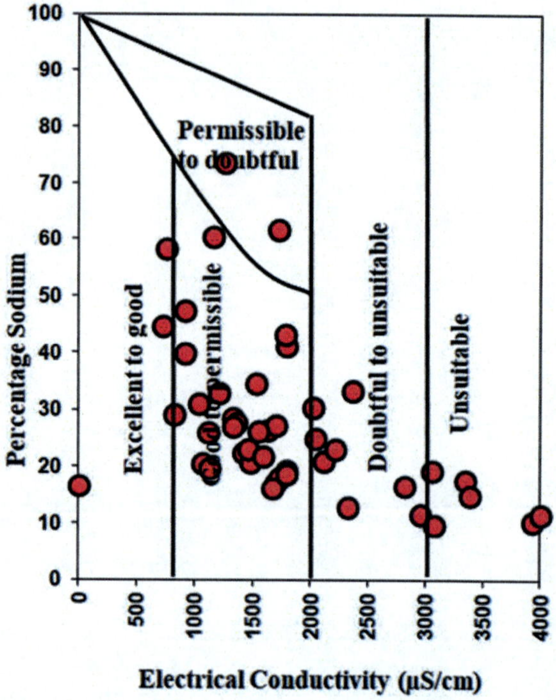

Fig. 14.6 Wilcox diagram for assessing the quality of irrigation water

Laboratory diagram [39]. About 9, 70, 20, and 1% of the samples, respectively, had been categorized as C2S1, C3S1, C4S1, and C2S2, denoting high to extremely high salinity and little sodium (Fig. 14.8). As a result, it can be used for irrigation. The samples were mostly divided into high salinity groups (C3S1 and C3S2). As a result, regardless of salinity dominance, the majority of water samples collected in the Mogamureru river basin could be used for land irrigation with drainage patterns [23]. Figure 14.5 depicts a spatial distribution map of SAR.

14.4.2.5 Residual Sodium Carbonate (RSC)

Groundwater samples had residual sodium carbonate levels ranging from −39.1 to −1.5 meq/L, with a mean of −14.9 meq/L and a standard deviation of 8.4 meq/L. (Table 14.1). The residual sodium carbonate within the river basin is used to categorize groundwater. There were three zones (67) in the samples: safe (1.25), somewhat safe (1.25–2.50), and unsafe (>2.50), with all samples falling into the safe zone (100%). RSC is representing in the spatial distribution map (Fig. 14.7).This could be owing to the Cuddapah basin's extensive dispersion in dolomite-rich areas in the current region.

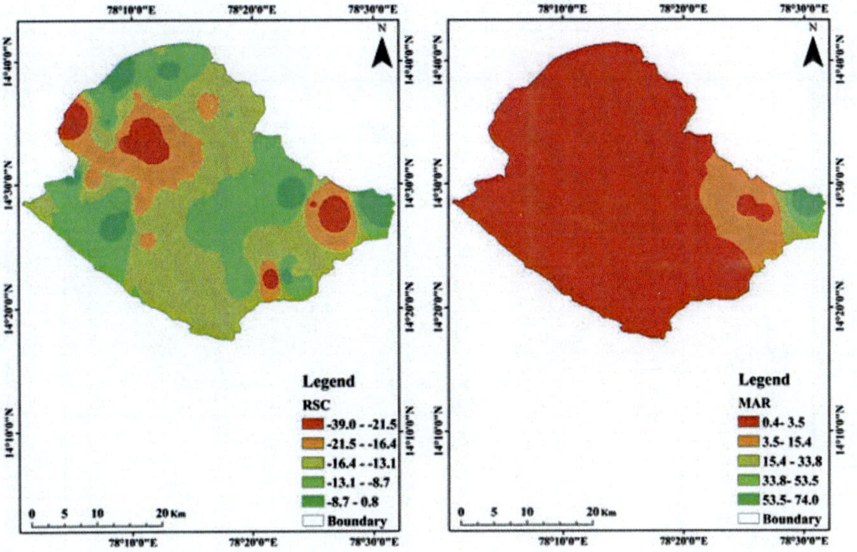

Fig. 14.7 Spatial distribution map of RSC and MAR

14.4.2.6 Magnesium Adsorption Ratio (MAR)

Based on magnesium adsorption ratio ranges, Sawye and McCarty [34] classified them as appropriate (50) or unsuitable (>50). The samples range in concentration from 0.4 to 0.8 meq/L, with a mean concentration of 0.6 meq/L and a standard deviation of 0.13 (Table 14.1). Furthermore, all of the samples (100%) were appropriate for cultivationand representing in spatial distribution Map (Fig. 14.7).

14.4.2.7 Permeability Index (PI)

The permeability index is characterized as into three categories: good (less than 80), moderate (80–100), and poor (more than 100). In all of the categories, all of the samples are good (100%). The sample permeability index ranged from 0.001 to 0.01 meq/L, with a mean of 0.003 meq/L and a standard deviation of 0.002 meq/L.

The permeability index formula calculates water mobility capacity in the soil while the fitting of whichever variety origin for irrigation water, Class I is >75% appropriate, class II is 25–75% good, while class III is less than 25% unfit. Every sample (100%) of the study fell into class I (Fig. 14.9), which is recommended for irrigation purposes [30, 31].

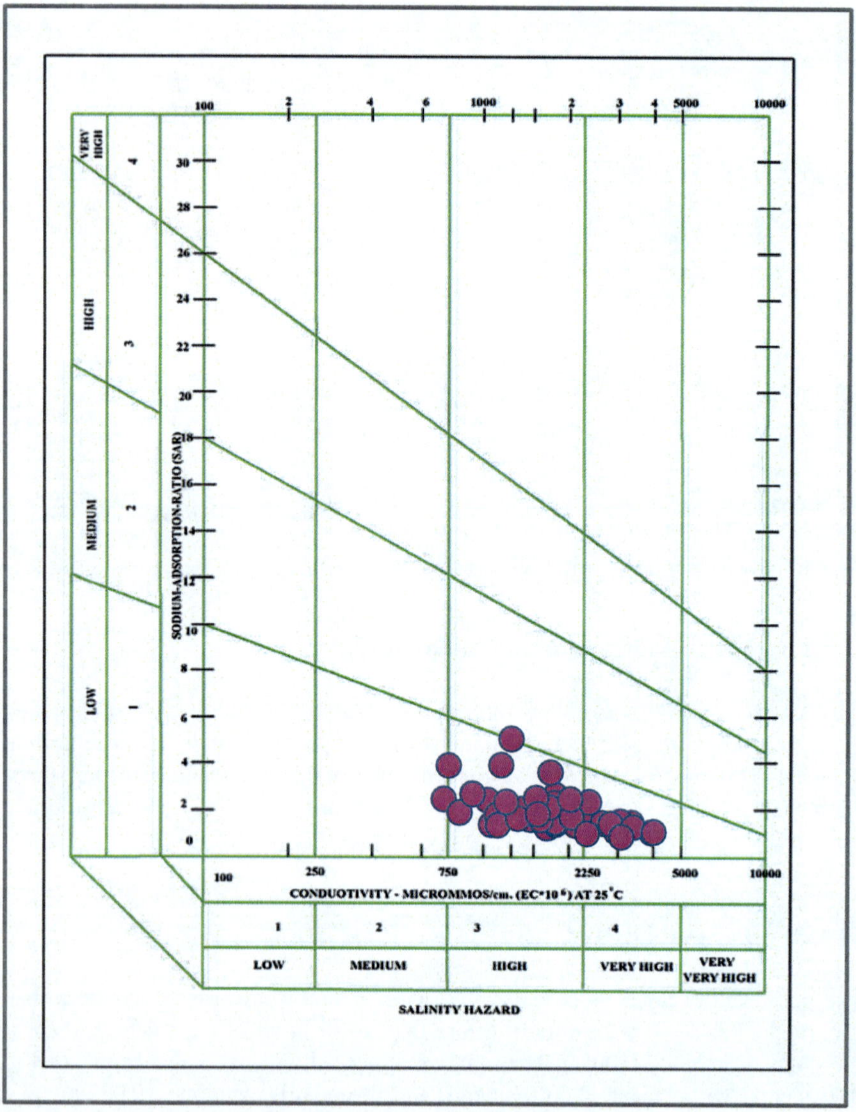

Fig. 14.8 USSL diagram for determining the quality of irrigation water

14.4.2.8 Kelley's Ratio (KR)

The samples' Kelley's ratios varied from 0.11 to 2.8, had a mean of 0.5 and a standard deviation of 0.5.According to Kelly's ratio, which categorizes samples as appropriate 1or unsuitable (>1), groundwater samples fall into the suitable group, and hence are used for irrigation purposes. KR representing in spatial distribution Map (Fig. 14.10).

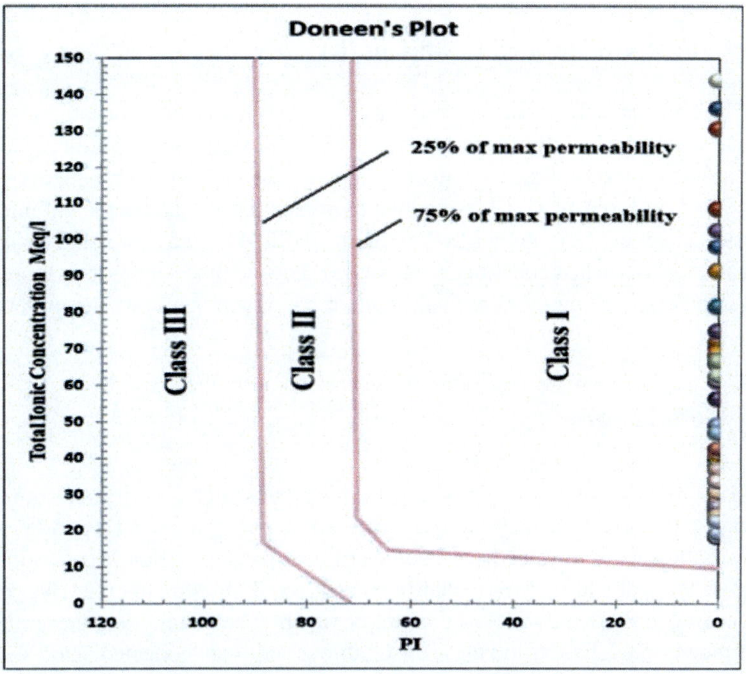

Fig. 14.9 Doneen diagram for assessing irrigation water quality

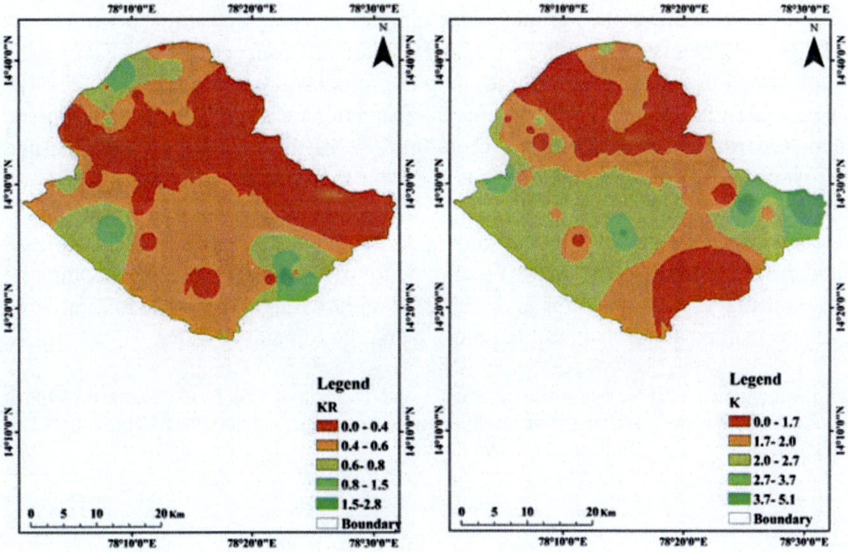

Fig. 14.10 Spatial distribution maps of KR and K

14.4.2.9 Synthetic Harmful Coefficient (K)

The alkali and salt risks could be totally masked by the synthetic hazardous coefficient. The average groundwater sample was 1.9, with a standard deviation of 0.8. The synthetic hazardous coefficient concentrations of the groundwater samples varied from 0.9 to 5.1 (Table 14.1). Classified the Synthetic hazardous coefficient into four categories: excellent (25), good (25–36), harmful (36–44), and unsuitable (>44). All of the samples are in the good range (Fig. 14.10). This method's evaluation resolution is conservative, resulting in water quality under good light in a short amount of time.

14.5 Conclusions

Groundwater is essential for improving farming particularly in dry and semi-arid locations; therefore, its quality and suitability for irrigation are significant. The chief ions of shallow groundwater have been examined for a progressed description of hydrogeochemical and irrigation quality estimation of the Mogamureru river basin.

According to WQI, only 24% of the samples are fit for consumption, the remaining 76% of the samples are not, and these unsuitable samples are collected in the Vemula, Vempalle and Pulivendula Mandals. In the water samples, sodium, calcium, magnesium, potassium, carbonates, and bicarbonates are present in sufficient amounts. Therefore, the irrigation parameters like theindices SAR, MAR, SSP, RSC, PI, and KR indicate that all (100%) groundwater samples are suitable for irrigation.

Other parameters such as the EC and TDS, the EC concentration of 71% of the samples are suitable, and 29% of the samples are not suitable for irrigation. The EC concentration is exceeded because the water contains a large amount of dissolved salts. According to the TDS concentration, 66% of the samples are suitable and the rest of the samples are not suitable. Seawater intrusion can cause EC, while salt leaching from soil, improper sewage disposal, and mineral breakdown can cause TDS concentration. Groundwater in the study area is very useful for irrigation purposes, but in the future, this water will not be suitable for irrigation. Therefore, I recommend to use fertilizers and pesticides in moderation and treating wastewater to remove as much suspended solids as possible before using the remaining water.

Acknowledgements The Department of Science and Technology (DST), Government of India, New Delhi, funded the first author's research (Sanction order No. DST/INSPIRE Fellowship/2017/IF170787), which is gratefully acknowledged.

References

1. Adeaga O, Eslamian S (2021) Assessment of freshwater and conservation in Africa. In: Handbook of water harvesting and conservation: case studies and application examples, pp 119–139
2. Adimalla N, Venkatayogi S (2018) Geochemical characterization and evaluation of groundwater suitability for domestic and agricultural utility in semi-arid region of Basara, Telangana State, South India. Appl Water Sci 8(1):1–14
3. Ajibola AF, Raimi MO, Steve-Awogbami OC, Adeniji AO, Adekunle AP (2020) Policy responses to addressing the issues of environmental health impacts of charcoal factory in Nigeria: necessity today; essentiality tomorrow. Commun Soc Media 2576–5388
4. Akhtar N, Ishak MIS, Ahmad MI, Umar K, Md Yusuff MS, Anees MT, Qadir A, Ali YK (2021) Modification of the water quality index (WQI) process for simple calculation using the multi-criteria decision-making (MCDM) method: a review. Water 13(7):905
5. Akakuru OC, Eze CU, Okeke OC, Opara AI, Usman AO, Iheme O, Ibeneme SI, Iwuoha PO (2022) Hydrogeochemical evolution, water quality indices, irrigation suitability and pollution index of groundwater (PIG) around eastern Niger Delta, Nigeria. Int J Energy Water Resour 1–23
6. Barkat A, Bouaicha F, Bouteraa O, Mester T, Ata B, Balla D, Rahal Z, Szabó G (2021) Assessment of complex terminal groundwater aquifer for different use of Oued Souf Valley (Algeria) using multivariate statistical methods, geostatistical modeling, and water quality index. Water 13(11):1609
7. Bawoke GT, Anteneh ZL (2020) Spatial assessment and appraisal of groundwater suitability for drinking consumption in Andasa watershed using water quality index (WQI) and GIS techniques: Blue Nile Basin. Northwestern Ethiopia Cogent Eng 7(1):1748950
8. Bishwakarma K, Wang GX, Zhang F, Adhikari S, Karki K, Ghimire A (2022) Hydrochemical characterization and irrigation suitability of the Ganges Brahmaputra River system: review and assessment. J Mt Sci 19(2):388–402
9. Brown RM, Mccleiland NJ, Deiniger RA, O'Connor MF (1972) Water quality index-crossing the physical barrier. Res Jerusalem 6:787–797
10. Devi G, Goswami L, Kushwaha A, Sathe SS, Sen B, Sarma HP (2021) Fluoride distribution and groundwater hydrogeochemistry for drinking, domestic and irrigation in an area interfaced near Brahmaputra floodplain of North-Eastern India. Environ Nanotechnol Monitor Manag 16:100500
11. Erukubami OF, Adelakun AT, Asiwaju-Bello YA, Olabode OF (2021) Conceptualization of pollution of groundwater resources in Ilaje Area, Nigeria, West Africa: contribution from hydrogeochemical and numerical index techniques
12. Etikala B, Arveti N, Yenamala S, Arveti T, Sharifi ZY (2017) Hydrochemical characterization of groundwater in around Tirupati Area, Chittoor District, Andhra Pradesh, South India. Appl Water Sci 7:1203–1212. https://doi.org/10.1007/s13201-016-0448-6
13. Etikala B, Golla V, Adimalla N, Surasura R, Marapatla S (2021) Hydrogeochemical characterization of groundwater using conventional graphical, geospatial and multivariate statistical techniques. In Geostatistics and geospatial technologies for groundwater resources in India. Springer, Cham, pp 81–96
14. Farahani E, Emami H, Fotovat A, Khorassani R, Keller T (2020) Soil available water and plant growth in relation to K: Na ratio. Geoderma 363:114173
15. Gad M, Saleh AH, Hussein H, Farouk M, Elsayed S (2022) Appraisal of surface water quality of NileRiver using water quality indices, spectral signature and multivariate modeling. Water 14(7):1131
16. Gaikwad S, Meshram D, Wagh V, Kandekar A, Kadam A (2020) Geochemical mobility of ions in groundwater from the tropical western coast of Maharashtra, India: implication to groundwater quality. Environ Dev Sustain 22(3):2591–2624

17. Garcia-Franco N, Wiesmeier M, Hurtarte LCC, Fella F, Martínez-Mena M, Almagro M, Martínez EG, Kögel-Knabner I (2021) Pruning residues incorporation and reduced tillage improve soil organic matter stabilization and structure of salt-affected soils in a semi-arid Citrus tree orchard. Soil and Tillage Res 213:105129
18. Gebrehiwet M, Tafesse NT, Habtu S, Alemaw BF, Laletsang K, Lasarwe R (2021) The contribution of groundwater to the salinization of reservoir-based irrigation systems. Agronomy 11(3):512
19. Gowd SS (2005) Assessment of groundwater quality for drinking and irrigation purposes: a case study of Peddavanka watershed, Anantapur District, Andhra Pradesh, India. Environ Geol 48(6):702–712
20. Horton RK (1965) An index number system for rating water quality. J Water Pollut Control Federation 373:303–306
21. Iqbal J, Nazzal Y, Howari F, Xavier C, Yousef A (2018) Hydrochemical processes determining the groundwater quality for irrigation use in an arid environment: The case of Liwa Aquifer, Abu Dhabi, United Arab Emirates. Groundw Sustain Dev 7:212–219
22. Ismail AH, Hassan G, Sarhan AH (2020) Hydrochemistry of shallow groundwater and its assessment for drinking and irrigation purposes in Tarmiah district, Baghdad governorate, Iraq. Groundw Sustain Dev 10:100300
23. Kawo NS, Karuppannan S (2018) Groundwater quality assessment using water quality index and GIS technique in Modjo River Basin, central Ethiopia. J Afr Earth Sc 147:300–311
24. Li P, He S, Yang N, Xiang G (2018) Groundwater quality assessment for domestic and agricultural purposes in Yan'an City, northwest China: implications to sustainable groundwater quality management on the Loess Plateau. Environ Earth Sci 77(23):775
25. Mandal SK, Dutta SK, Pramanik S, Kole RK (2019) Assessment of river water quality for agricultural irrigation. Int J Environ Sci Technol 16(1):451–462
26. Mal U, Adhikari K (2021) Groundwater quality and hydrological stress induced by Lower Gondwana open cast coal mine. J Earth Syst Sci 130(1):1–24
27. Mao M, Guo P, Zhu X, Yang J, Pan D (2021) Hydrochemical characterization and quality assessment of groundwater in the hilly area of the Taihang Mountains in Henan Province, China. Environ Sci Pollut Res 28(32):43853–43871
28. Markus UI, Ilori OP, Wada IM, Musa ST, Peter JE (2022) Hydrogeochemical evaluation and geospatial distribution modeling of the major ion chemistry of groundwater and their suitability for drinking and irrigation in Lagelu, Southwestern Nigeria. Appl Water Sci 12(7):1–19
29. Mohammed I, Al-Khalaf SKH, Alwan HH, Naje AS (2022) Environmental assessment of Karbala water treatment plant using water quality index (WQI). Mater Today: Proc 60:1554–1560
30. Muniz DH, Malaquias JV, Lima JE, Oliveira-Filho EC (2020) Proposal of an irrigation water quality index (IWQI) for regional use in the Federal District, Brazil. Environ Monitor Assess 192(9):1–15
31. Qadir M, Sposito G, Smith CJ, Oster JD (2021) Reassessing irrigation water quality guidelines for sodicity hazard. Agric Water Manag 255:107054
32. Ram A, Tiwari SK, Pandey HK (2021) Groundwater quality assessment using water quality index (WQI) under GIS framework. Appl Water Sci 11:46
33. Rawat K, Pradhan S, Tripathi V, Jeyakumar L, Singh SK (2019) Statistical approach to evaluate groundwater contamination for drinking and irrigation suitability. Groundw Sustain Dev 9:100251
34. Sawyer CN, McCarty PL (1967) Chemistry for sanitary engineers
35. Şener E, Şener Ş, Varol S (2022) Appraisal of groundwater quality with WQI and human health risk assessment in karamuk wetland and surrounding (Afyonkarahisar/Turkey). Environ Geochem Health 1–25
36. Setia R, Lamba S, Chander S, Kumar V, Dhir N, Sharma M, Singh RP, Pateriya B (2021) Hydrochemical evaluation of surface water quality of Sutlej River using multi-indices, multivariate statistics and GIS. Environ Earth Sci 80(17):1–17

37. Shankar K, Kawo NS (2019) Groundwater quality assessment using geospatial techniques and WQI in North East of Adama Town, Oromia Region. Ethiopia Hydrospatial Anal 3(1):22–36
38. Tripathi M, Singal SK (2019) Allocation of weights using factor analysis for development of a novel water quality index. Ecotoxicol Environ Saf 183:109510
39. United States Salinity Laboratory Staff (2019). Diagnosis and improvement of Saline and Alkali Soils. https://www.ars.usda.gov/ARSUserFiles/20360500/hb60_pdf/hb60complete.pdf
40. Vaseashta A, Gevorgyan G, Kavaz D, Ivanov O, Jawaid M, Vasović D (2021) Exposome, biomonitoring, assessment and data analytics to quantify universal water quality. In: Water safety, security and sustainability. Springer, Cham, pp 67–114
41. World Health Organization (2015) World health statistics 2015. World Health Organization
42. Wilcox L (1955) Classification and use of irrigation waters, no 969. US Department of Agriculture
43. Xu P, Zhang Q, Qian H, Zheng L (2021) Spatial distribution characteristics of irrigation water quality assessment in the Central-Western Guanzhong Basin, China. IOP Conf Ser: Earth Environ Sci 647(1):012143
44. Yu H, Gui H, Zhao H, Wang M, Li J, Fang H, Jiang Y, Zhang Y (2020) Hydrochemical characteristics and water quality evaluation of shallow groundwater in Suxian mining area, Huaibei coalfield, China. Int J Coal Sci Technol 7(4):825–835
45. Yuan H, Yang S, Wang B (2022) Hydrochemistry characteristics of groundwater with the influence of spatial variability and water flow in Hetao Irrigation District, China. Environ Sci Pollut Res 1–15
46. Zhang Q, Qian H, Xu P, Hou K, Yang F (2021) Groundwater quality assessment using a new integrated-weight water quality index (IWQI) and driver analysis in the Jiaokou Irrigation District, China. Ecotoxicol Environ Saf 212:111992
47. Zemunac R, Savic R, Blagojevic B, Benka P, Bezdan A, Salvai A (2021) Assessment of surface and groundwater quality for irrigation purposes in the Danube-Tisa-Danube hydrosystem area (Serbia). Environ Monit Assess 193(8):1–19
48. Zhou Y, Li P, Xue L, Dong Z, Li D (2020) Solute geochemistry and groundwater quality for drinking and irrigation purposes: a case study in Xinle City, North China. Geochemistry 80(4):125609

Chapter 15
Using Geo-Spatial Technologies for Land and Water Resource Development Planning: A Case Study of Tirora Tehsil, India

Nanabhau Kudnar and M. Rajashekhar

Abstract In this study presents the using geo-spatial technologies for land and water resource development planning in Tirora tehsil, India. Using the SOI topographic maps for physical parameters and water quality analysis parameter analysis for inverse distance weighting (IDW) interpolation method in the in study area. The study area is small in terms of land; therefore, no variation in the physiography, soils, climatic conditions and many more results single crop pattern. As per the study of total crop productivity and irrigated area productivity; crop productivity in irrigated area is high than total cropped area. The percentage of the total geographical area under crop is 56.78% which means that agriculture is found only in the river belt and plains while the rest of the hills and other areas are not cultivated. The main crop is paddy in the region; therefore another crops are produces in very much less amount in the area. The average pH value of groundwater here is 8.01 with the highest value being 8.90 and the lowest value being 7.21 which means that the nature and quality of water here is good. The highest TDS pattern of water you find is 2011 and the lowest is 310 in which average value is 604.75 which mean TDS of water is good but TDS of water you see increased in river side and in hilly region it is good. This study can be used for future planning of farm management, crop growth and sustainable development.

Keywords Geospatial modeling · Environmental resources · RS · GIS

N. Kudnar
Department of Geography, C. J. Patel College, Tirora, Gondia, Maharashtra 441911, India
e-mail: nanabhaukudnar@gmail.com

M. Rajashekhar (✉)
Department of Geology, Yogi Vemana University, Kadapa, AP 516005, India
e-mail: sekhar.raja042@gmail.com

E. Balaji et al. (eds.), *Emerging Technologies for Water Supply, Conservation and Management*, Springer Water, https://doi.org/10.1007/978-3-031-35279-9_15

Abbreviations

AHP	Analytical hierarchical process
APML	Adani Power Maharashtra Limited
CWC	Central Water Commission
IMD	Indian Meteorological Department
MWRRA	Maharashtra Water Resources Regulatory Authority
NEERI	National Environmental Engineering Research Institute
SOI	Survey of India
WGS	World Geodetic System
WQI	Water quality index

15.1 Introduction

If sustainable development is to be achieved, soil and water must be protected. Soil can be said to be a gift created by nature for living things. Soil is made up of various rocks. This type of soil is formed like a rock. Degraded rock fragments, partially or completely decomposed organic matter and numerous micro-organisms are present in the soil. There is a constant interaction between organic and inorganic elements in the soil [1, 5, 40, 42, 49, 53].

They get the nutrients they need for plant growth from the soil. Soil is a perfect ecosystem. Soil is a mixture of different minerals, organic matter, gases, different substances and innumerable micro-organisms. Which together contribute to life on the Earth, naturally occurring soils have four major functions. It is a medium for plant growth and development. It holds supplies and purifies water. It changes the earth's atmosphere. It is the habitat of living beings. All of these types ultimately change the soil. Soil types are regur soil, red soil, black soil etc. It is also a solid, air and water retention system [14, 15, 20, 38]. Preventing soil erosion as well as conserving water is a major challenge in multiple types of conditions such as areas with heavy rainfall, drought prone areas, land under oligopolies, land in mining areas, and areas prone to landslides. Therefore, there is a huge responsibility to do research on the geographical and scientific study of the area, to find alternatives to it, to determine its method, to verify which cropping system can be effective, etc. [9–11, 28, 29, 33, 51, 56].

There is currently no area where water is not used. Secondly, water is essential for drinking but due to lack of water, the importance of water is increasing in other areas. Currently, the amount of rainfall has decreased a lot. River water is polluted due to industry and other reasons [2–4, 19, 26, 32, 36, 39]. Decreased availability of water means that the farmer becomes weak. If it is to be improved, water availability must be increased. India is an agricultural country. Farmers are the backbone of the country. Water is the blood of farmers, then if there is no water, farmers will not survive [18, 37, 41, 43, 48, 55]. The work plan of watershed management can be prepared with the help of GIS and RS techniques. Many environmental disasters have been created during the last century because of human activities. Man has affected water quality [8, 12, 16, 17, 30, 35, 54] every-where by pursuing fast economic development and by adopting a physical life and a materialistic culture. Even though water is a renewable resource it is possible to develop sustainable practices for its use provided that the quality of water is not compromised. The most important and critical principle stated by [13, 23, 24, 31, 50, 52] is that environmental flows must be maintained in all perennial rivers, and if there are impounding structures like dams, then enough water must be saved for releasing it into the main river.

In the study, geomorphology, soil types, geology, slope, relief, landforms, climate and rainfall, hydrology, water availability analysis, water inflow, water balance, water quality analysis, measurement of water discharge, demographics, water resource management, resource units, and an inventory of existing medium/minor/lift irrigation schemes were used to build the areas for sustainable development in Tirora tehsil, central India.

15.2 Study Area

Tirora tehsil, north latitudes 21°13′05″ to 21°33′30″ and East longitudes 79°47′50″ to 80°05′00″ and is located in Gondia district, Maharashtra (Fig. 15.1). Tirora tehsil situated western part of Gondia district of Maharashtra. Tirora tehsil is selected as a study region because it is having variation in physiography, culture, population and other aspects of physical geography. Total geographical area of the tehsil is 51616 hector (617.1 km^2); out of which 33,588 ha area is under agriculture, 2974 ha area is barren land and 35,347 ha area is not useful for agriculture. Out of total area under agriculture 20,650 ha area irrigated. Total 2065 ha irrigated area 12,750 ha area irrigated by tanks. For administrative purpose Tirora tehsil divided in 5 Revenue divisions. In this tehsil total no of tanks is 152; out of which 2 medium projects, 27 small irrigation projects and 123 Malgujari tanks in Tirora tehsil. During these consultations and subsequent visits to some village tanks on the ground, local people said that "Maji Malgujari" tanks, which have a unique history of being run by the community, are in a state of rapid deterioration. The second reason to choose this area for the study is that it has the most village tanks. On average, each village in the Gondia district has 1.5 tanks. This is the highest after population density, so the area is called the "district of tanks." The focus of this tank study is on Malgujari tanks, which are more important for medium irrigation projects and small irrigation projects. According to the official classification, there are two types of village tanks: those built by the government and those built by the community hundreds of years ago. Development of agriculture and fishery occupation is happened due to more number of tanks in this study area. In the study reason majority of the tanks having less amount of water capacity due to ignorance of Government and locale community water capacity of tanks reducing day by day. Water capacity of tanks determined by physical environment of the region. Irrigation facility in a particular region more

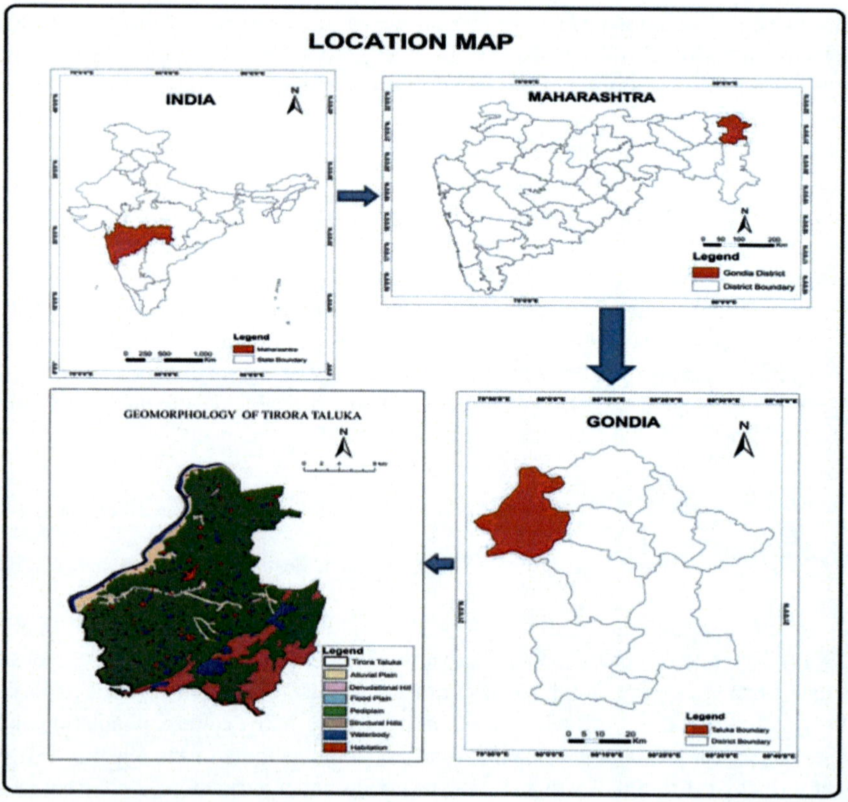

Fig. 15.1 Study area, Tirora Tehsil

important input as compare to other input. For the development of agriculture sector irrigation is most important tool. As agriculture sector develop it automatically affect economical level of farmers in particular area. In India more than 70% population living in rural sector and agriculture is main occupation in rural peoples. tanks of the study area are not equally distributed in more number of tanks situated in south and south east direction of the study area. In Tirora tehsil tanks plays important role in agriculture rather than other means of irrigation.

15.3 Data Collection and Methodology

The SOI toposheets are used to look at topographical maps with a scale of 1:50,000 for this study. Analysis of the geomorphic unit, geology, land use, soil, and water, The collection of data, maps, and reports about rainfall and runoff. Long-term rainfall data for the whole Wainganga basin and small tributaries, especially in the tehsil area, has been collected by IMD [46]. Stream flow and sediment load data at Rajegaon, as well as water availability and discharge data near Dhapewada dam, has also been collected and is being watched by CWC and MWRRA. Samples of water have been taken from the tehsil area and the river Wainganga and analysed for their physiochemical properties. Multispectral images have been taken so that different hydrological and geomorphological features of the study area can be looked at over time. The following steps were taken to look at the tehsil's environmental network for the study. Using SOI topographic maps and UTM zone 44N projection, WGS 84 datum was used to georeference the data in ArcGIS desktop 9.3. Water samples have been taken from both upstream and downstream of the river Wainganga [25], and the environment as well as geomorphology, soil types, geology, slope, relief, landforms, climate and rainfall, hydrology, water availability analysis, water inflow water balance, measurement of water discharge, demographics, and prioritisation of land resource management have been studied. To do this Water Quality analysis (WQA), water samples were analysed for correlation, ranking, prioritisation, and priority mappings for parameters like TSS, TDS, pH, Total Alkalinity, COD, BOD, Total Coliform, Chloride, Iron, Fecal, Mg, and Nitrate using the AHP method and GIS techniques. Using the SOI topographic maps for physical parameters and water quality analysis parameter analysis for inverse distance weighting (IDW) interpolation method in the study area.

15.4 Results and Discussion

15.4.1 Physical Resources

The Wainganga is the district's primary river, which is fed by five tributaries. Geologically, the tehsil is part of the Deccan plateau, which is mostly composed of volcanic basalt rock (Fig. 15.2a). However, differentiates from the remainder of the Deccan plateau area. Large amounts of Gneiss rocks that are older than the basalt and impermeable may be found in the region. The Kali, Kanhar, Morand, Khardi, Sihar, and Bardi soils (Fig. 15.2b), which cover the whole region, are utilised to cultivate rice, the district's most important crop. The area is part of the Wainganga sub-basin and has undulating topography with altitudes between 263 and 312 m above mean sea level. Physiographically, may be roughly split into two physiographic units: the one controlled by structural characteristics, also known as the Structural Origin, and the one controlled by differential weathering, also known as the Denudation Origin. Denudational characteristics, such as pediments and pediplains, are more prevalent in the north-central, west-central, and south-western regions. Structural hills and ridges are more prevalent in the east and south-east. The widely diverse soils of the taluka result from the weathering of crystalline metamorphic and igneous rocks. With 14.3% of the forest land designated as reserve forest, 8.3% as preserve forest, and 4.7% as unclassified forest, the total forest area is 27.3%. In this forest, essential trees like as Bija, Halda, Tivas, Sisam, Moh, Khair, and Garadi are used for various

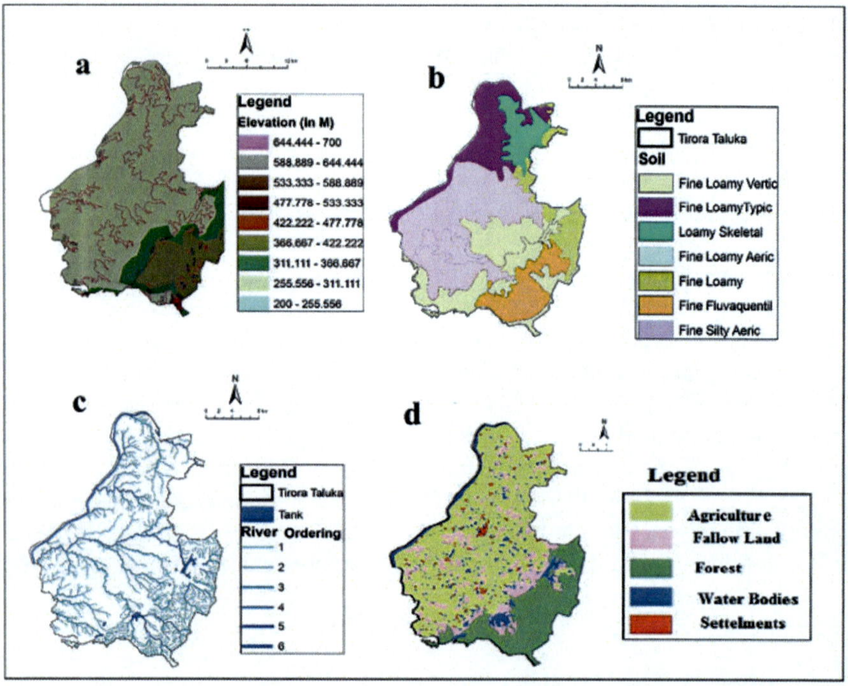

Fig. 15.2 **a** Relief, **b** soil, **c** drainage & tank and **d** land use

purposes. The average height from sea level is 300 m, although the physiography of the region is plateau-like; thus, the majority of the taluka's surface is undulating, resulting in natural tanks. Tehsil is a portion of the Deccan plateau, which stretches throughout Central India. The average annual precipitation recorded for the region is 1200 mm. In the months of June, July, August, and September, the south western monsoon receives over 90% of the annual precipitation. Normal annual precipitation in the region varies from 1300 to 1500 mm. It is at its lowest in the north-west to south-east direction and grows in the north-east. Tirora gets the least amount of precipitation, 1357.8 mm [21, 22, 27, 44, 45].

15.4.2 Water Resources

In this study area, ponds of water holding capacity are more common towards hilly areas and the area below them is known as agriculturally rich area. How to change this, Bodhalkasa Dam, Chorkhamara, Risala, Nandalpar etc. all dams are useful for agricultural production. Rainfall in the study area is decreasing from south-east to north-west. From the surface showing the slope, it can be seen that the slope decreases from the south-east direction to the north-west direction. There are 22 ponds spread over 250 ha. These ponds are used for agribusiness, fishing business, shingada and vegetable farming. The number of these ponds may be two to three in a village, some of which have canals built on them, while others are pumped to supply water to agriculture. The total number of conventional ponds is 123. The percentage of the total geographical area under crop is 56.78% which means that agriculture is found only in the river belt and plains while the rest of the hills and other areas are not cultivated. Products like rice, wheat, gram, tur etc. are grown here. Among the various crops, rice is the most important and 91% of the total area under crop is under rice crop. Irrigation from various tanks used for rabbi and kharip seasons. Tank irrigation is more important than Different means of irrigation. Bodalkasa and Chorkhamara surrounded by hilly region. Relationship between tanks physiography is more important in these regions. For the classification of tanks some parameters are used by Government like water capacity, occupied area of tank and ownership of tank. Out of these three parameters capacity of water is more important. Above 500 ha area of tank categorized as medium irrigation tank, 250–500 ha categorized small irrigation tank, 100–250 ha area called minor irrigation and 0–100 ha area called Malgujari tank in the area. Those tanks having irrigation capacity is above 500 ha called minor irrigation projects. Some places have heavy rainfall and some places have drought conditions. This has an adverse effect on agricultural yields, requiring artificial irrigation of crops, hence the increase in agricultural activity due to irrigation, and this increase is seen in the Wainganga River basin. In Mandikota and Wadegaon areas, the land is found in dry and semi-arid form, this land is in dire need of water. Due to irrigation, we can grow crops like paddy in that place, artificial irrigation is required here. Along with the increasing population, the demand for food grains is also increasing, so even though the population has increased tremendously, there has not been much increase in the agricultural sector. Although Green Revolution and Agricultural Biotechnology are being used all over India, the main reason why it is used very little here is that the farmers here are poor and have financial problems and due to this the production per hectare in India is very low [6, 7].

The stored water of the tank is not useful for only agriculture but also it is used for other uses. For the purpose of irrigation 100% water is not used for irrigation; some amount of water kept for other uses. This water used by fishermen to cultivate fish throughout the year. Water capacity of medium irrigation tanks (above 500 ha) in Tirora tehsil is 38.444 m^3, out of which 37.254 m^3 water can be used for irrigation purpose and remaining 1.19 m^3 dead water will be available for Fishery. Water capacity of small irrigation tanks (above 250 ha irrigation capacity) 11.169 m^3, out of which 10.6278 m^3 water can be used for irrigation purpose and remaining 0.5412 m^3 dead water will be available for Fishery. Water capacity of small irrigation tanks (100–250 ha irrigation capacity) in is 4.11 m^3, out of which 3.78 m^3 water can be used for irrigation purpose and remaining 0.32 m^3 dead water will be available for Fishery. A small irrigation tank (0–100 ha irrigation capacity) in having Water capacity (Fig. 15.2c) of is 11.26 m^3 and it is completely used for irrigation purpose. Rainfall is Main source of water in all tanks of the taluka. Bodalkasa and Chorkhamara tank water source is number of small seasonal flow of rainy water. These tanks surrounded by hilly region as well as forest, therefore sufficient water catchment area available for the tanks. Medium irrigation project having the control of Government; therefore, the status of these tanks is good as compare the community based tanks. As it is explained that the importance of tanks in irrigation facilities in the tehsil; it is necessary to explain the water capacity and actual available water in tanks. Bodalkasa is the medium irrigation project in the taluka having water capacity is 17.392 m^3 but actual available water in the tank is less than the capacity. Chorkhamara is another medium irrigation project in the taluka having water capacity is 20.80 m^3 but actual available water in the tank is less than the capacity. Rest of the crops are secondary and some are grown in Kharif season and some in Rabi season. Wheat is the second important crop here and the total area under this crop averages 322 ha, which means that the production of wheat is very low compared to that of rice. After this gram is grown in 375 hectares in this place [34, 40]. In this study area, the productivity of different crops is found in different forms in which for almost total crops 91% of the area is under rice crop while only nine percent is found under other crops. The productivity of wheat crop is 599 kg/ha and the highest wheat crop production is observed in the river belt. The productivity of gram crop is 650 kg/ha and the highest productivity is also found in Thanegaon. There the productivity of tur crop is 550 kg/ha and the highest productivity of tur crop is 605 kg/ha in Thanegaon division and the lowest productivity is 509 kg/ha in Mundikota division. Crop productivity depends on various geographical and human factors, but availability of irrigation facilities is equally important in modern agricultural system. Crops such as rice and wheat require a lot of irrigation while gram and gram do not require much irrigation. The productivity of rice here is 1605 kg/ha. Whereas the productivity of wheat crop is 613 kg/ha. The main ornament of the is 152 tanks. In these tanks number of fishermen involved in fishery occupation as a secondary activity. In Dhiwar community involved in this occupation; group of people constitute a co-operative society to get ownership of tank. Dhiwar community people involved in this occupation after the primary activity (agriculture). The most significant auxiliary activity is fishing. This endeavour is particularly significant because Tirora is regarded as the tehsil of lakes and ponds. In addition to Malgujari tanks, medium and small irrigation tanks are used for fishing. More emphasis is placed on production than on employment-oriented activities. All of the tehsil's ponds and tanks will be utilised for fish farming. For year-round fish farming, it is suggested to implement the RKVY scheme by deepening Malgujari tanks below sea level. Also suggested for the tehsil is the construction of a fish seed rearing pond. 'Singada Occupation' is a marginal activity dependent on the Tanks in the tehsil. This is a seasonal employment that supports the financial needs of the impoverished. 16 tanks out of 152 are involved in the occupation of Singada.

15.4.3 Land Recourses

The soil here is also very suitable for plants. Black soil is more common in Wainganga River basin. Along with this, there is also abundance of red point sand mixed soil, purple soil, all these soils are very suitable for the growth and development of the plant, so the maximum number of plants is found in this study area. Out of the total forest area in this study area, reserved forest area is 14.3%, protected forest area is 8.3% and unclassified forest area is 4.7%. A forest of good quality teak trees is found here along with forest trees like Bija Halda Divas Shisam Khair Garade etc. Useful trees such as Tendu leaves, Mocha flowers, Lakhakara, Palas for gum, Karavi Dhavada and Katara Khair are found in the forest. There are pastures in many areas for grazing and bamboo trees are cultivated here on a large scale for paper mills. A large amount of social forestry activities is being carried out by the Department of Social Forestry on the flat land on the hill slopes and along the roadside.

The agricultural productivity and production capacity is increasing day by day with changes in rice, wheat, pea, gram and sugarcane crops in the study area. This change is seen to gradually increase during the period 1971–1980 in which the effect of the drought of 1972 on the agriculture here and at the same times in 1972 the productivity and production capacity of the agriculture decreased. In1975, due to heavy rainfall, the agricultural production was found to be low, with a productivity as low as 405.5 kg/ha in 1973 as compared to 394 kg/ha. So much is found. A similar change is observed in wheat, pea, gram and sugar production. Drought in 1974 across India is still visible after 1972. At that time in this study area, the productivity of agriculture is reduced due to drought, in which the production of rice crop is found as low as 1056 and production capacity is 394. The quantity of this is found to be less than 1973. The quantity of wheat, peas, gram and sugar is also found to be reduced. The main crop is paddy in the region; therefore, another crop are produces in very much less amount in the area. Kharif crops completely dependent on monsoon rainfall; though the importance of irrigation could not be neglected. Gram and Tur are the crops need not irrigation; it produced with the help of soil moisture. 73.37% area irrigated to the total cropped area in the out of which 70% irrigated area under rice. Difference in the crop productivity of among all five revenue divisions. Rice productivity of is 1554 kg/ha; whereas Thanegaon division standing on top as per hector productivity of Rice. Rice productivity in the irrigated area is quite high with 1605 kg/ha production. Other crops also resulting high crop productivity in irrigated area rather than non-irrigated area. In Vadegaon revenue division having highest 1825 kg/ha rice productivity in the study area. Rice is the dominant crop in the therefore cropping pattern of the results single crop in all five-revenue division. In all five-revenue division having above 90% area under Rice crop. The study area is small in terms of land; therefore, no variation in the physiography, soils, climatic conditions and many more results single crop pattern. As per the study of total crop productivity and irrigated area productivity; crop productivity in irrigated area is high than total cropped area. Agriculture is the main occupation of about 70% of the people here. Due to the vagaries of the monsoon, crops are damaged, sometimes there is too much rain and sometimes there is too little rain, which leads to a scarcity situation when the agricultural yield declines. This requires artificial irrigation. Irrigation serves a dual role. Losses due to uncertainty of rainfall can be avoided. Increase in yield per hectare of crops. The distribution of rainfall in this place is very variable in its consistency (Fig. 15.2d).

15.4.4 Demographic Characteristics

As per the Census 2011, the population in the tehsil is reported to be 176,254. Out of this, 85.7% of the population resides in the rural part and 14.3% resides in urban area. Average population density is 333 km^2. Male to female ration in the district is reported to be 1000:981. Scheduled Castes and Scheduled Tribes constitute 22,254 and 14,381 of the total population. Average literacy rate of in 2011 were 87.78% in which, male and female literacy were 94.34% and 81.12% respectively. Total literate in were 138,188 of which male and female were 74,760 and 63,428 respectively. In out of total population, 87,828 were engaged in work activities (Fig. 15.3a). 68.5% of workers describe their work as Main work, while 31.5% were involved in Marginal activity providing livelihood for less than 6 months. Of 87,828 workers engaged in main work, 21,421 were cultivators (owner or co- owner) while 22,385 were Agricultural labourer. In study area another occupation is fishing due to total more number of tanks and water capacity. 27 tanks under custody of Zila parishad Gondia and fishing activity done by fishery co-operative societies. In the study area situated on the broad-gauge railway track Mumbai-howrah useful for transportation of agriculture production. To analysis the farmers demographic characteristics, tanks details, fishery occupation, singada occupation etc. primary survey conducted on the basis of random sampling information. Total number of villages in is 114; out of which 20% villages selected for random sampling to collect the information as required for research problem (Fig. 15.3b).

After the analysis of collected information, it is clear that the having rural background. Rural population fully dependant on agriculture for their basic needs. Maximum population of 15–59 age group involved in agriculture. Out of total population 51.5% are male and 48.5% are female, of Gondiya district has total population of 176,254 as per the Census 2011. Out of which 88,955 are males while 87,299 are females. In 2011 there were total 39,286 families residing in Tirora. The Average Sex Ratio of is 981. The average literacy rate in is 66.96% and the working population of the surveyed respondent is 51.48% out of which 21.39% cultivators, 19.30% agriculture labour, 7.78% other worker and 3.30% population involved in service sector. In the 32-cultivator having their own agriculture land. 28.46% cultivators having 0–2 hector land, and average annual income of the people of the tehsil is one lakh. 57% cultivators are used tanks as means of irrigation, is the very much important with the irrigation point of view; total 152 tanks control the whole economy of the area. Agriculture as a main occupation (Fig. 15.3c). depending upon irrigation because the uncertainty of the monsoon. Therefore, people says that Agriculture on monsoon is the 'big gamble' in the tehsil. Out of 152 total tanks 123 Maji Malgujari Tank are more important than medium irrigation projects. These Maji Malgujari tanks distributed in the fishery is the additional occupation distributed in the area. 78% production of Fish sold by local market; fishing is the seasonal occupation.

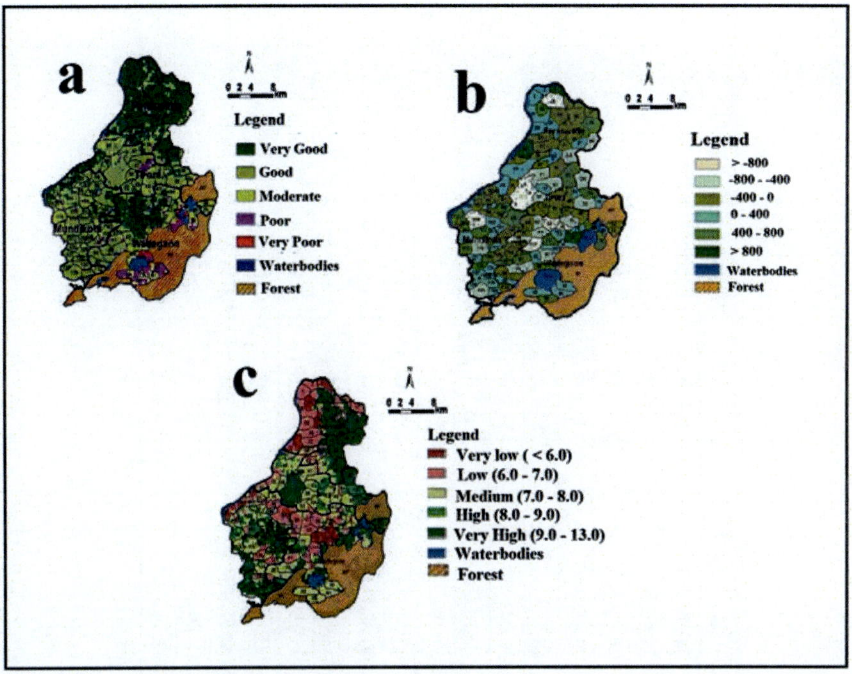

Fig. 15.3 **a** Biophysical resource, **b** status of human carrying capacity and **c** prioritized areas for land resources management

15.4.5 Water Quality Analysis (WQA)

Applied data 2018 Pre-Monsoon (mg/L) calculate correlations (Table 15.1) and AHP method, weight was found to be: EC < TDC < TH < Cl^- < SO_2^- 4 < pH < NO_3^- < F^-. Chemical oxygen demand is an accurate measurement of the amount of oxygen present in the wastewater process. Its range is 16.70, the lowest is 1.40 and the highest is 18.10. Which the shows Figs. 15.4a–g and 15.5a–g that the correlation is in plus and the relativity of electronic conductivity is in minus. The average value of TA is 404 with a minimum of 107 and a maximum of 511. EC has an average of 1091.51 with a low of 540 and a high of 3139 [47, 48]. The average pH value of groundwater here is 8.01 with the highest value being 8.90 and the lowest value being 7.21 which means that the nature and quality of water here is good. The highest TDS pattern of water you find is 2011 and the lowest is 310 in which average value is 604.75 which means TDS of water is good but TDS of water you see increased in river side and in hilly region it is good.

Table 15.1 Statistical analysis of ground water quality in Tirora Tehsil

Parameter	Range	Minimum	Maximum	Mean	Std. deviation	Variance	Skewness	Kurtosis
COD	16.70	1.40	18.10	11.8500	4.60486	21.205	−1.428	2.128
TA	404	107	511	174.44	133.679	17,869.996	2.295	4.116
EC	2599.0	540.0	3139.0	1091.513	594.1086	352,965.083	3.030	10.473
pH	1.69	7.21	8.90	8.0113	0.41635	0.173	0.545	1.518
TH	547	133	680	244.44	158.372	25,081.729	1.691	2.569
TDS	1701	310	2011	604.75	436.041	190,131.933	2.451	7.254
F^-	1.71	0.06	1.77	0.5556	0.60928	0.371	0.995	−0.756
NO_3^-	278.89	0.11	279.00	19.3100	69.28936	4801.015	3.992	15.957
Ca^{2+}	219.960	20.040	240.000	86.47163	66.449737	4415.568	1.086	0.149
HCO_3	412	239	651	320.69	90.080	8114.363	3.682	14.380
Cl^-	3001.80	13.00	3014.80	690.4219	882.47737	778,766.316	1.541	1.978
K^+	399	12	411	48.63	98.780	9757.450	3.733	14.354
Na^+	490	190	680	377.69	160.968	25,910.763	0.517	−0.667
Do	2.58	6.32	8.90	7.2613	0.74218	0.551	0.828	0.165
$SO_4{}^{2-}$	469.00	12.00	481.00	52.7981	118.89455	14,135.913	3.565	13.109
Mg^{+2}	101.00	10.00	111.00	44.1844	33.36655	1113.327	0.760	−0.746

Source Central Ground Water Board (CGWB) and Environmental Laboratory, APML, 2011

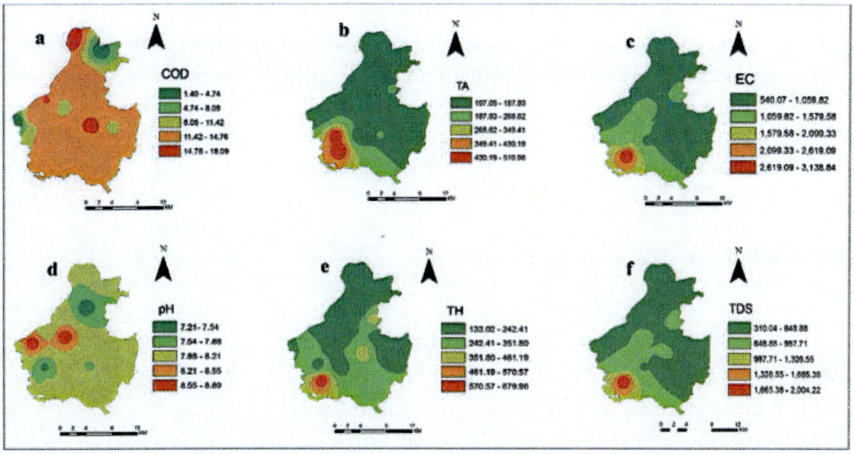

Fig. 15.4 Water quality analysis. **a** Chemical Oxygen Demand (COD). **b** Tantalum (TA). **c** Electric Conductivity (EC), **d** potential of hydrogen (pH). **e** Total Hardness (TH), **f** Total Dissolved Solids (TDS)

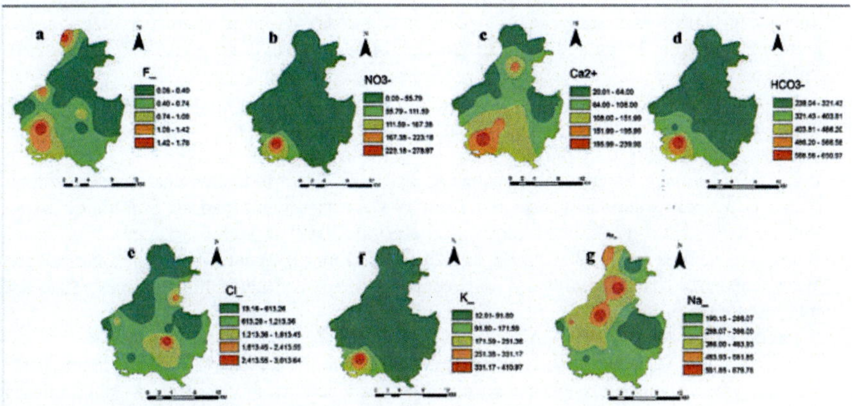

Fig. 15.5 Water quality analysis, **a** Fluoride (F^-), **b** Nitrate ($NO_3{}^-$), **c** Calcium ion, **d** Bicarbonate. **e** Chlorine (Cl^-), **f** Potassium (K^+), **g** Sodium ion (Na^+)

15.5 Conclusions

This study area explains how water and soil are used as resources. Since many of the lakes here are created by nature, their construction cost is not much. There are also some traditional lakes privately owned by farmers. These lakes are used to increase agricultural production and fishing is also done here on a large scale. But the usefulness of some lakes here has decreased. Some of these lakes are very ancient and are not maintained or modernized. The quality of water here is deteriorating day by day so you can see its adverse effects in the coming time. This study can be used for future planning of farm management and crop growth. From this research you can see that this can happen in the present and in the future to use and plan a region using geographical technology. It is necessary to take measures for the tribal's and Nasalizes in this region and we can take these measures in the future.

Conflict of Interest The authors declare no conflict of interest.

References

1. Aslam B, Ismail S, Ali I (2020) A GIS-based DRASTIC model for assessing aquifer susceptibility of Safdarabad Tehsil, Sheikhupura District, Punjab Province, Pakistan. Model Earth Syst Environ 6:995–1005. https://doi.org/10.1007/s40808-020-00735-8
2. Ahmad N, Pandey P (2018) Assessment and monitoring of land degradation using geospatial technology in Bathinda district, Punjab, India. Solid Earth 9(1):75–90. https://doi.org/10.5194/se-9-75-2018
3. Ahamed T, Noguchi R, Muhsin N et al (2020) Sustainable agricultural development: a micro-level GIS-based study on women's perceptions of environmental protection and entrepreneurship in Japan and Bangladesh. GeoJournal. https://doi.org/10.1007/s10708-020-10169-5
4. Andualem TG, Hagos YG, Kefale A et al (2020) Soil erosion-prone area identification using multi-criteria decision analysis in Ethiopian highlands. Model Earth Syst Environ. https://doi.org/10.1007/s40808-020-00757-2
5. Balaji E, Nagaraju A, Sreedhar Y, Thejaswi A, Sharifi Z (2016) Hydrochemical characterization of groundwater in around Tirupati Area, Chittoor District, Andhra Pradesh, South India. Appl Water Sci 7(3):1203–1212. https://doi.org/10.1007/s13201-016-0448-6
6. Bisen DK, Kudnar NS (2013) A sustainable use and management of water resource of the Wainganga river basin: a traditional management systems. J Contrib. https://doi.org/10.6084/m9.figshare.663573.v1
7. Bisen DK, Kudnar NS (2019) Climatology. SaiJyoti Publication, Nagpur, pp 1–211
8. Bouaicha F, Dib H, Belkhiri L, Bouaicha F, Manchar N, Chabour N (2017) Hydrogeochemistryand geothermometry of thermal springs from the Guelma region Algeria. J GeolSoc India 90:226. https://doi.org/10.1007/s12594-017-0703-y
9. Bouteraa O, Mebarki A, Bouaicha F, Nouaceur Z, Laignel B (2019) Groundwater quality assessment using multivariate analysis, geostatistical modeling, and water quality index (WQI): a case of study in the Boumerzoug-El Khroub valley of Northeast Algeria. Acta Geochim 38:796. https://doi.org/10.1007/s11631-019-00329-x
10. Cannata M, Neumann J, Rossetto R (2018) Open source GIS platform for water resource modelling: FREEWAT approach in the Lugano Lake. Spat Inf Res 26:241–251. https://doi.org/10.1007/s41324-017-0140-4
11. Chattaraj S, Srivastava R et al (2017) Semi-automated object-based landform classification modelling in a part of the Deccan Plateau of Central India. Int J Remote Sens 38(17):4855–4867. https://doi.org/10.1080/01431161.2017.1333652
12. Cieszynska M, Wesolowski M, Bartoszewicz M et al (2012) Application of physicochemical data for water-quality assessment of watercourses in the Gdansk Municipality (South Baltic coast). Environ Monit Assess 184:2017–2029. https://doi.org/10.1007/s10661-011-2096-5
13. Datta A, Gaikwad H, Kadam A et al (2020) Evaluation of groundwater prolific zones in the unconfined basaltic aquifers of Western India using geospatial modeling and MIF technique. Model Earth Syst Environ. https://doi.org/10.1007/s40808-020-00791-0

14. Dongare VT, Reddy GPO, Maji AK et al (2013) Characterization of landforms and soils in complex geological formations-a remote sensing and GIS approach. J Indian Soc Remote Sens 41:91–104. https://doi.org/10.1007/s12524-011-0195-y
15. Etikala B, Golla V, Adimalla N, Surasura R, Marapatla S (2021) Hydrogeochemical characterization of groundwater using conventional graphical, geospatial and multivariate statistical techniques. In: Adhikary PP, Shit PK, Santra P, Bhunia GS, Tiwari AK, Chaudhary BS (eds) Geostatistics and geospatial technologies for groundwater resources in India. Springer Hydrogeology, pp 81–96. https://doi.org/10.1007/978-3-030-62397-5_5
16. Guettaf M, Maoui A, Ihdene Z (2017) Assessment of water quality: a case study of the Seybouse River (North East of Algeria). Appl Water Sci 7:295–307. https://doi.org/10.1007/s13201-014-0245-z
17. Hamad A, Baali F, Hadji R, Zerrouki H, Besser H, Mokadem N, Legrioui R, Hamed Y (2018) Hydrogeochemical characterization of water mineralization in Tebessa-Kasserine karst system (Tuniso-Algerian Transboundry basin). Eur-Mediterr J Environ Integr 3:7. https://doi.org/10.1007/s41207-017-0045-6
18. Kadam AK, Umrikar BN, Sankhua RN (2020) Assessment of recharge potential zones for groundwater development and management using geospatial and MCDA technologies in semiarid region of Western India. SN Appl Sci 2:312. https://doi.org/10.1007/s42452-020-2079-7
19. Karande UB, Kadam A, Umrikar BN et al (2020). Environmental modelling of soil quality, heavy-metal enrichment and human health risk in sub-urbanized semiarid watershed of western India. Model Earth Syst Environ 6:545–556. https://doi.org/10.1007/s40808-019-00701-z
20. Khan MA, Gupta VP, Moharana PC (2001) Watershed prioritization using remote sensing and geographical information system: a case study from Guhiya, India. J Arid Environ 49(3):465–547. https://doi.org/10.1006/jare.2001.0797
21. Kudnar NS (2017) Morphometric analysis of the Wainganga river basin using traditional & GIS techniques. Ph.D. thesis. Rashtrasant Tukadoji Maharaj Nagpur University, Nagpur, pp 40–90
22. Kudnar NS (2018) Water pollution a major issue in urban areas: a case study of the Wainganga river basin. Vidyawarta Int Multidiscip Res J 2:78–84
23. Kudnar NS (2020) GIS-based assessment of morphological and hydrological parameters of Wainganga river basin, Central India. Model Earth Syst Environ 6:1933–1950. https://doi.org/10.1007/s40808-02000804-y
24. Kudnar NS (2020b) GIS-based investigation of topography, watershed, and hydrological parameters of Wainganga River Basin, Central India. In: Sustainable development practices using geoinformatics. Scrivener Publishing LLC, pp 301–318. https://doi.org/10.1002/9781119687160.ch19
25. Kudnar NS, Rajasekhar M (2020) A study of the morphometric analysis and cycle of erosion in Waingangā Basin, India. Model Earth Syst Environ 6:311–327. https://doi.org/10.1007/s40808-019-00680-1
26. Kudnar NS, Diwate P, Mishra VN et al (2022) Spatio-temporal variability and trend analysis of rainfall in Wainganga river basin, Central India, and forecasting using state-space models. Theor Appl Climatol. https://doi.org/10.1007/s00704-022-04168-4
27. Kudnar NS, Padole MS et al (2021) Traditional crop diversity and its conservation on-farm for sustainable agricultural production in Bhandara District, India. Int J Sci Res Sci Eng Technol 2394–4099(8-1):35–43. https://doi.org/10.32628/IJSRSET207650
28. Kushwaha N, Elbeltagi A, Mehan S et al (2022) (2022) Comparative study on morphometric analysis and RUSLE-based approaches for micro-watershed prioritization using remote sensing and GIS. Arab J Geosci 15:564. https://doi.org/10.1007/s12517-022-09837-2
29. Laurent R, Anker W, Graillot D (1998) Spatial modeling with geographic information system for determination of water resources vulnerability application to an area in Massif Central (France). J Am Water ResourAssoc 34(1):123–134. https://doi.org/10.1111/j.1752-1688.1998.tb05965.x

30. Lkr A, Singh MR, Puro N (2020) Assessment of water quality status of Doyang River, Nagaland, India, using water quality index. Appl Water Sci 10:46. https://doi.org/10.1007/s13201-019-1133-3
31. Ksiazek L, Wos A, Florek J et al (2019) Combined use of the hydraulic and hydrological methods to calculate the environmental flow: Wislokariver, Poland: case study. Environ Monit Assess 191:254. https://doi.org/10.1007/s10661-019-7402-7
32. Madhav S, Ahamad A, Singh AK, Kushawaha J, Chauhan JS, Sharma S, Singh P, (2020) Water pollutants: sources and impact on the environment and human health. In: Pooja D, Kumar P, Singh P, Patil S (eds) Sensors in water pollutants monitoring: role of material. advanced functional materials and sensors. Springer, Singapore, pp 43–62. https://doi.org/10.1007/978-981-15-0671-0_4
33. Mahadevan H, Krishnan KA, Pillai RR et al (2020) Assessment of urban river water quality and developing strategies for phosphate removal from water and wastewaters: integrated monitoring and mitigation studies. SN Appl Sci 2:772. https://doi.org/10.1007/s42452-020-2571-0
34. Mishra VN, Rai PK (2016) A remote sensing aided multi-layer perceptron-Markov chain analysis for land use and land cover change prediction in Patna district (Bihar), India. Arab J Geosci 9:249. https://doi.org/10.1007/s12517-015-2138-3
35. Mishra VN, Rai PK, Prasad R, Punia M, Nistor MM (2018) Prediction of spatiotemporal land use/land cover dynamics in rapidly developing Varanasi district of Uttar Pradesh, India using Geospatial approach: a comparison of hybrid models. Appl Geomat 10(3):257–276
36. Mogaji KA, Lim HS (2020) A GIS-based linear regression modeling approach to assess the impact of geologic rock types on groundwater recharge and its hydrological implication. Model Earth Syst Environ 6:183–199. https://doi.org/10.1007/s40808-019-00670-3
37. Rahaman KM, Ahmed FRS, Nazrul Islam M (2016) Modeling on climate induced drought of north-western region, Bangladesh. Model Earth Syst Environ 2:45. https://doi.org/10.1007/s40808-016-0089-7
38. Rajasekhar M, Gadhiraju SR, Kadam A et al (2020) Identification of groundwater recharge-based potential rainwater harvesting sites for sustainable development of a semiarid region of southern India using geospatial, AHP, and SCS-CN approach. Arab J Geosci 13–24. https://doi.org/10.1007/s12517-019-4996-6
39. Rajasekhar M, SudarsanaRaju G, SiddiRaju R (2019). Assessment of groundwater potential zones in parts of the semi-arid region of Anantapur District, Andhra Pradesh, India using GIS and AHP approach. Model Earth Syst Environ 5:1303–1317. https://doi.org/10.1007/s40808-019-00657-0
40. Rajasekhar M, Sudarsana R et al (2021) Multi-criteria land suitability analysis for agriculture in SemiArid Region of Kadapa District, Southern India: geospatial approaches. Remote Sens Land 5(2):59–72. https://doi.org/10.21523/gcj1.2021050201
41. Ramadan EM, Fahmy MR, Nosair AM et al (2019) Using geographic information system (GIS) modeling in evaluation of canals water quality in Sharkia Governorate, East Nile Delta, Egypt. Model Earth Syst Environ 5:1925–1939. https://doi.org/10.1007/s40808-019-00618-7
42. Reddy GP, Sarkar D, Prasad J, Ramamurthy V (2013) Geospatial modeling in assessment of biophysical resources for sustainable land resource management. Trop Ecol 54(2):227–238
43. Salami ES, Ehteshami M (2016) Application of neural networks modeling to environmentally global climate change at San Joaquin Old River Station. Model Earth Syst Environ 2:38. https://doi.org/10.1007/s40808-016-0094-x
44. Salunke VS, Lagad SJ et al (2021) A geospatial approach to enhance point of the interest and tourism potential Centers in Parner Tehsil in Maharashtra, India. Int J Sci Res Sci Eng Technol (IJSRSET) 8(1):186–196. Online ISSN:2394-4099, Print ISSN:2395-1990. https://doi.org/10.32628/IJSRSET218136
45. Salunke VS, Bhagat RS et al (2020) Geography of Maharashtra, Prashant Publication, Jalgaon, pp 1–229
46. Shailaja G, Kadam AK, Gupta G et al (2019) Integrated geophysical, geospatial and multiple-criteria decision analysis techniques for delineation of groundwater potential zones in a semi-arid hard-rock aquifer in Maharashtra, India. Hydrogeol J 27:639–654. https://doi.org/10.1007/s10040-018-1883-2

47. Sharip Z, Saman JM, Noordin N et al (2016) Assessing the spatial water quality dynamics in PutrajayaLake: a modelling approach. Model Earth Syst Environ 2:46. https://doi.org/10.1007/s40808-016-0104-z
48. SiddiRaju R, SudarsanaRaju G, Rajsekhar M (2018) Estimation of rainfall-runoff using SCS-CN method with RS and GIS techniques for Mandavi Basin in YSR Kadapa District of Andhra Pradesh, India. Hydrospatial Anal 2(1):1–15. https://doi.org/10.21523/gcj3.18020101
49. Soutter M, Alexandrescu M, Schenk C et al (2009) Adapting a geographical information system-based water resource management to the needs of the Romanian water authorities. Environ Sci Pollut Res 16:33–41. https://doi.org/10.1007/s11356-008-0065-5
50. Teshome A, Halefom A (2020) Potential land suitability identification for surface irrigation: in case of Gumara watershed, Blue Nile basin, Ethiopia. Model Earth Syst Environ 6:929–942. https://doi.org/10.1007/s40808-020-00729-6
51. Thakur JK, Singh SK, Ekanthalu VS (2017) Integrating remote sensing, geographic information systems and global positioning system techniques with hydrological modeling. Appl Water Sci 7:1595–1608. https://doi.org/10.1007/s13201-016-0384-5
52. Theodoropoulos C, Georgalas S, Mamassis N, Stamou A, Rutschmann P, Skoulikidis N (2018) Comparing environmental flow scenarios from hydrological methods, legislation guidelines, and hydrodynamic habitat models downstream of the Marathon Dam (Attica, Greece). Ecohydrology 11:1–11. https://doi.org/10.1002/eco.2019
53. Tsihrintzis VA (2017) Integrated water resources management, efficient and sustainable water systems, protection and restoration of the environment. Environ Process 4:1–7. https://doi.org/10.1007/s40710-017-0271-6
54. Walega A, Młyński D, Bogdał A, Kowalik T (2016) Analysis of the course and frequency of high water stages in selected catchments of the upper Vistula Basin in the south of Poland. Water 8:394. https://doi.org/10.3390/w8090394
55. Xu H, Cai C, Du H et al (2020) Responses of water quality to land use in riparian buffers: a case study of Huangpu River, China. GeoJournal. https://doi.org/10.1007/s10708-020-10150-2
56. Zhang M, Yang F, Wu JX, Fan ZW, Wang YY (2016) Application of minimum reward risk model in reservoir generation scheduling. Water Resour Manag 30(4):1345–1355. https://doi.org/10.1007/s11269-015-1218-1

Chapter 16
Delineation of Seawater Intrusion into Freshwater Aquifers by Using VES & GIS in the Kakinada Region, East Godavari District, Andhra Pradesh, India

Vijayakumar Gundala and Vinoda Rao Tanikonda

Abstract Freshwater resources are severely harmed by seawater intrusion, especially in coastal regions where nearby fresh groundwater can be quickly altered by saltwater. A total of eighteen vertical electrical soundings (VES) to understand the resistivity distribution of the study area. The majority of the interpreted master curves are three-layered curves of H, K and Q types. A dominant Q-type curve indicates that resistivity values gradually decrease with depth ($1 < 2 < 3$). Successful delineation of iso-apparent resistivity and layer thickness, and spatial distribution maps are prepared by using the geographical information system (GIS). Because of the presence of topsoil, the initial layer has a high resistance. The leaching of salts from the older sediments is what causes the shallow aquifer's low resistivity. The influx of saline water is what causes the deeper aquifers to have low resistivity. The low iso-apparent resistivity regions below 10 m are found along the southern coastal tract and extend up to the centre, which amply supports the existence of salt water at various depths below the surface. Where the electrical conductivity (EC) exceeded the permissible limit. The analysis of groundwater quality values revealed that they were above the permissible limits of WHO drinking water standards. The IDW technique helped us identify the seawater intrusion areas requiring control, which pointed the way for further fragility management along the coastal areas of the Kakinada region.

Keywords Electrical resistivity · VES · Seawater intrusion · Groundwater quality · GIS · Kakinada

V. Gundala (✉)
St. Ann's College of Engineering & Technology, Vetapalem, Chirala, AP 523187, India
e-mail: vijaigk@gmail.com

V. R. Tanikonda
Department of Geology, Andhra University, Visakhapatnam, AP 530003, India
e-mail: vinodarao2004@yahoo.com

E. Balaji et al. (eds.), *Emerging Technologies for Water Supply, Conservation and Management*, Springer Water, https://doi.org/10.1007/978-3-031-35279-9_16

16.1 Introduction

The global world is currently confronting a significant problem in the form of securing water for sustainable development. Currently, emerging nations experience environmental strain brought on by a large population, fast urbanisation, and a lack of water sector services, which has an impact on how water resources are managed [18, 19]. Monitoring and determining the source of salt are essential for both water management and restoration as many coastal aquifers across the world are experiencing varying levels of salinization caused by both natural and anthropogenic processes [15]. The distribution, chemical make-up, temperature, and electrical resistance of water-bearing rocks are all highly influenced by water content. Using electrical resistivity methods and the proper electrode arrangement, one may distinguish between fresh water, brackish water, and pure saline water in the shallow coastal aquifers [10, 20]. Electrical resistivity provides distinct benefits over other hydrogeophysical techniques, including ease of use, low costs, and efficacy even when overburden and bedrock are worn [3, 7]. In order to obtain information on the stratification of the ground, vertical electrical sounding (VES) uses an electrical resistivity survey to measure potentials between two electrode pairs while transmitting direct current (DC) between additional electrodes in the homogeneous ground while varying the electrode separation [2]. Geographic Information Systems (GIS) are becoming more widely acknowledged as potent tools for managing spatial data holistically and retaining spatio-temporal variability in water quality assessment [11]. Geographical information system (GIS) techniques adds fundamental database operations with statistical analysis employing the primary mapping and geographical analysis features offered by maps and geographical databases [14]. This enables us to study visual representation of the cause-and-effect relationship. It also has the ability to model and study this complicated process, making it a useful tool for academics and managers of natural resources. The study area's coastal belt, Kakinada, presents a very different image. The resistivity studies in coastal areas have been used by writers from throughout the world, in order to draw attention to the traits of shallow aquifers and the impact of anthropogenic stress on these aquifers, many of which emerge as sweater invasion [6]. Since the clay layer was discovered to affect the resistivity survey, only shallow aquifers in the coastal belt were taken into account in the current investigation. More stress has been placed on this limited resource, which has resulted in reduced groundwater levels, as a result of increased urbanisation, rising population demands for freshwater, and bad management practices [1, 5, 13] the groundwater salinity from several wells in the study region is a significant problem.

16.2 Study Area

The region in the state of Andhra Pradesh, Kakinada, which is the capital of the East Godavari district, is the fifth-largest city located between 82°E to 82°20′ E longitude and its16°45′ N–17°5′ N latitude (Fig. 16.1). The delineated area falls in the toposheets nos. 65L, 65L1, published by the Survey of India (SOI), whereas the Kakinada's elevation is generally 2 m (6 feet) above sea level, some areas of the study area are below sea level, and its aerial extent is nearly 85 km^2. Two regions that are connected by bridges make up the research area. Because to the region's tropical environment, and the numerous cyclonic storms and depressions in the Bay of Bengal, rainfall is occasionally brought to Kakinada throughout the summer [12]. The city experiences extreme warmth from April to June, when the maximum mean average temperature is 32 °C, and extreme cold from December to January, when the minimum mean temperature is 25 °C.

16.2.1 *Geology and Hydrogeology*

The determination of the region's potential for groundwater is significantly influenced by the geology of the research area. The region is represented by Quaternary-aged

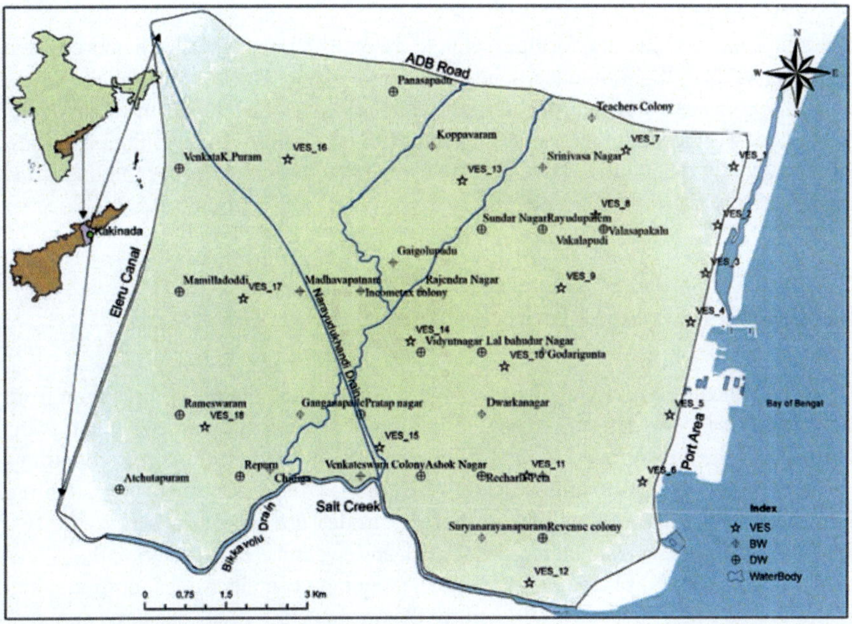

Fig. 16.1 Location map, VES and well locations

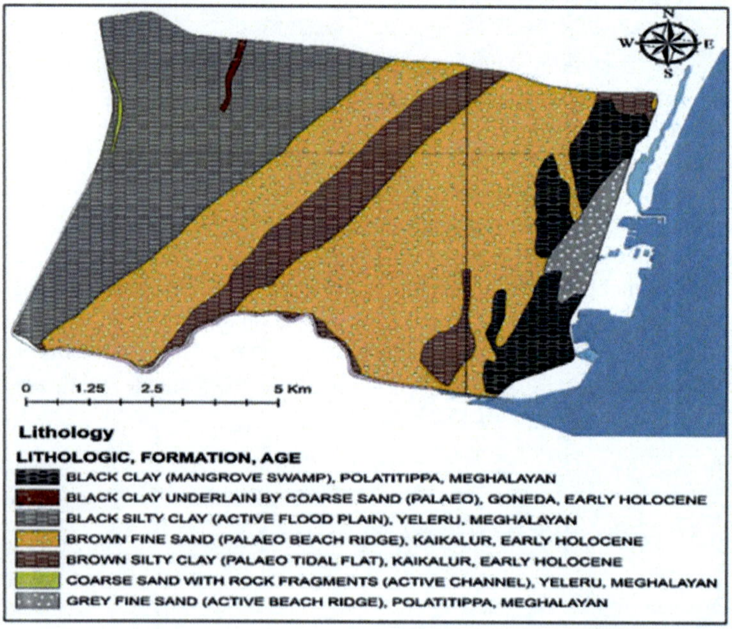

Fig. 16.2 Geology of study area

clay and sand sedimentary formations that are exposed near the shore and covered by Tertiary-aged Rajahmundry sandstone (Fig. 16.2). The research area, which primarily consists of an aquifer and an aquiclude, was developed throughout the Pleistocene-Holocene under the influence of fluvial, marine, and aeolian processes. The coastal alluvial aquifers have a relatively narrow saturated aquifer thickness and have relatively shallow aquifer depths.

16.3 Materials and Methods

The geophysical resistivity technique involves using apparent resistivity taken from the ground surface to determine the distribution of a physical parameter that is characteristic of the subsurface [16, 17]. Eighteen (18) in the research area, soundings (VES) were performed. using SSRMP-ATS 18.6.11 resistivity meter by adopting Schlumberger configuration. Where four electrodes are placed in a row, the two outer electrodes are current electrodes (A, B), and two inner electrodes are the potential electrodes (M, N) placed nearly. The spacing between electrodes minimum and maximum 1.5 and 100 m were adopted for the present study. The measured apparent resistivity for each location is obtained and plotted by using IPI2WIN 3.0, the resulted curves were examined for understanding the nature of layers and thickness present.

The resistivity of different layers and the corresponding thickness are reproduced by a number of inversions until the model parameters of all the VES curves are totally resolved with the fitting error (Gopinath 2017). The obtained results data has been used into Geographical Information System 10.3 (GIS), for preparation of spatial distribution maps to understand the extent layers in the study area.

16.4 Results and Discussion

16.4.1 Geographical Information System (GIS) Interpolation

The IDW interpolation is a method that is usually utilized in variable mapping. This interpolation technique, which is accurate and convex, is only appropriate for spatial variation in a continuous model. This method, which was developed in mining and geological engineering, uses the value acquired from the known place to assess a variable's value in some new locations [8, 9]. The fundamental idea behind IDW interpolation is to construct surfaces and predict unmeasured points using a sample point set with weighted linear combination and two statistical and mathematical techniques [9].

$$\breve{z}(x_{0)} = \sum_{i=1}^{n} z(x_i) \cdot d_{ij} - \dot{p} / \sum_{i=1}^{n} d_{ij-p} \tag{16.1}$$

where, Z is grid nodes interpolated value, Zi are data points for neighboring, dij measure the separations between grid nodes and data points.

To assist in the spatial organization of the study area's hazardous and contaminated areas and to determine the risk of contamination at various riverbank locations, IDW assumptions along the river were made, in addition to the individual sampling locations [21]. The targeting of eligibility in its main areas for interventions utilization are made possible by the spatial interpolation of its dangers. Other authors have also evaluated the surface water quality using GIS and IDW interpolation methods [4, 14].

16.4.2 A Geoelectrical Resistivity Survey

In the present study, The IPI2win software which is developed by Moscow State University is being interpreted to assess the resistivity of various strata [16]. A geoelectrical section of the VES location can be built using the data after the resistivity and thicknesses of various strata are determined. The root mean square (RMS) error varies 1.9–4.9%. Most of the VES curves are H, QH and KH curve types, which are interpreted as being the utmost appropriate curve, a horizontal current flow is one of the types used to evaluate the hydraulic parameters. Interpreted true resistivity curved data represented as well as four layer curves at a few sites, as well as three layer curve types in several locations. Aside from that, Madhavapatnam identified five different layer curve kinds at one place (VES 17). Overall, the layers' resistivity ranged from 0.185 to 200 m (Table 16.1). The first layer resistivity map (Fig. 16.3a) has brought out the surface distribution of the different lithounits, depending upon the resistivities. The first layer's thickness ranges from 1 to 6 m. The resistivities in the study vary between 1.33 and 138 Ω m. The low resistivities are mostly recorded along the coast (<10 Ω m) observed. The resistivities 10–20 Ω m in the study the clayey silt zone recorded between Hanuman Nagar, Gudarigunta, and Vakalapudi. A high resistivity zone is delineated along Achutapuram, Rameswaram, Madavapatnam south and extending towards north east. These resistivities are mostly above 50 Ω m and increases to more than 200 Ω m. The zone is mostly recorded over sands, between the sand zone and clayey silt zone, they spread as isolated patches along east of Kakinada, Valasapakala and west of Vakalapudi. These zones are mostly over sandy silt formations.

The thickness of the map and the second layer resistivity (Fig. 16.3b) were showed in spatial map. The resistivity less than 10 Ω m was observed as tiny patches along the coast of the study area. The resistivity values from 10 to 20 Ω m was noticed next to 10 Ω m, these two zones are clayey or brackish sand zones, because of the

Table 16.1 Quantitative resistivity and layer thickness analysis

Soundings location	p1	p2	p3	p4	p5	h1	h2	h3	h4	Total thickness
1	4.53	1.99	1.27	–	–	1.1	6.99	–	–	8.09
2	6.49	10.4	0.409	2.12	–	0.854	3.73	5.74	–	10.324
3	1.73	15.6	0.83	2.67	–	0.75	1.44	17	–	19.19
4	4.32	1.73	2.28	–	–	2.99	47	–	–	49.99
5	8.04	2.73	2.73	1.26	–	1.79	0.93	12.4	–	15.12
6	1.58	0.93	1.78	–	–	1.25	20.8	–	–	22.05
7	8.26	19.6	4.6	–	–	0.75	6.4	–	–	7.15
8	23.7	11.2	8.3	–	–	0.904	2.99	–	–	3.894
9	9.7	200	17	1.53	–	0.75	1.28	108	–	110.03
10	16.9	50.9	13.9	–	–	2.52	10.9	–	–	13.42
11	1.84	3.04	3.92	–	–	0.75	1.37	–	–	2.12
12	27	44.6	5.04	–	–	1.96	3.11		–	5.07
13	58.04	23.8	27.2	–	–	1.15	11.5	–	–	12.65
14	132	5.45	132	–	–	1.89	8.35	–	–	10.24
15	138	6.27	0.185	–	–	5	98.5	–	–	103.5
16	39.8	74.5	23.7	–	–	0.95	2.41	–	–	3.36
17	1.33	14.4	2.71	6.8	30.1	0.75	1.28	3.46	104	109.49
18	115	30.5	20	96.1	–	0.75	2.88	156	–	159.63

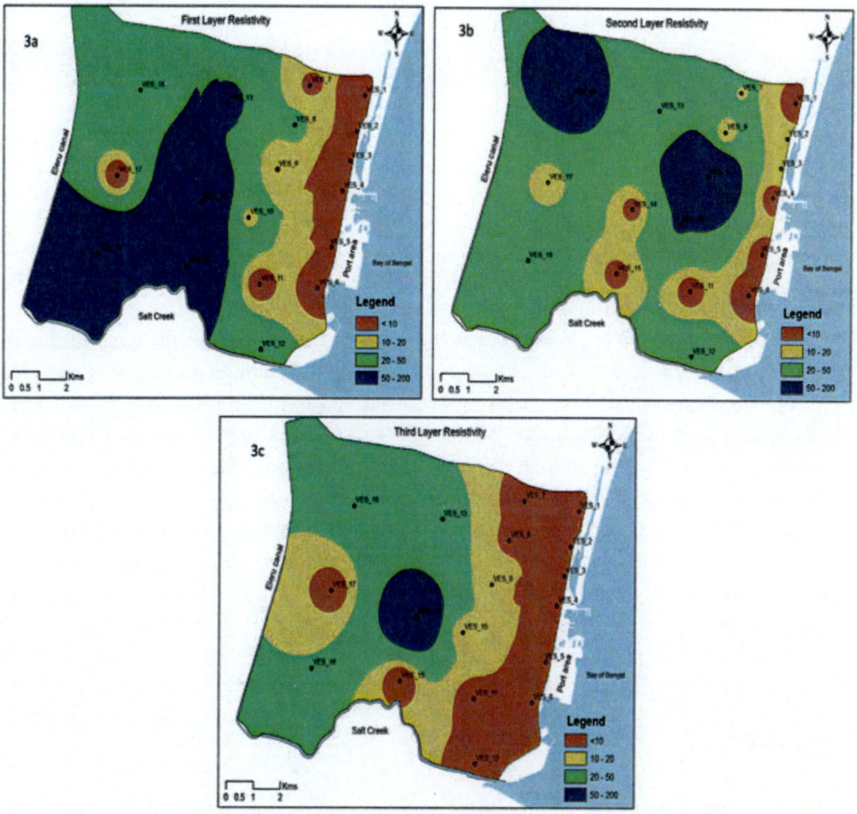

Fig. 16.3 **a–c** Resistivity interpolation spatial distribution maps

low resistivity range. The deeper zones with low resistivity along the coast may be useful as sites for industrial waste and those with high resistivity may be useful for development of ground water resources. Most of the study area resistivity ranges from 20 to 50 Ω m was observed, these zones can be interpreted as mostly sandy zones. The water may be potable in these zones. The small zone with the resistivity varying between 50 and 200 Ω m is recorded as patches at Gudarigunta and Panasapadu, these zones may be due to sandy sill and may yield good potable water.

The third layer resistivity of spatial map (Fig. 16.3c) has shown four types of resistivity variations, (i) <10 Ω m, (ii) 10–20 Ω m, and (iii) 20–50, and 50–200 Ω m. The distribution of less than 10 Ω m range covered Kakinada anchorage port, north of Valasapakala, the low resistivity zones may be due to clay or brackish sands, and the other zones with low resistivities along the coast may be useful sites for disposal of industrial waste. The resistivity 10–20 Ω m observed at places, these zones may be due to clayey sands and the quality of water may be good for human consumption. The resistivity rages from 50 to 200 Ω m observed as the patch, these deeper zones are useful for ground water development where ever the resistivities are higher than 50 Ω m.

16.5 Conclusion

Inverse Distance Weighted Interpolation (IDW) of the GIS method were used to analyse the temporal-spatial distribution of saltwater intrusion. Conducted resistivity surveys in the Kakinada region revealed that sounding results, a multilayer structure with low resistivity layers oriented eastward toward the shore and high resistivity layers oriented westward. These findings could help identify the freshwater and saltwater regions along Gudarigunta-Vakalapudi. In addition, they alternately followed the pockets of fresh and brackish water that pointed west. These investigations, in an interesting finding, showed that low resistivity layers were below high resistivity zones, and vice versa, and the analysis of the data indicating that the second and third layers of the study from southeastern to northern and along the shoreline side have more stress on coastal aquifers compared to central and northwestern parts of the study area.

References

1. Cosgrove WJ, Loucks DP (2015) Water management: current and future challenges and research directions. Water Resour Res 51(6):4823–4839. https://doi.org/10.1002/2014wr016869
2. Dahlin T (2001) The development of DC resistivity imaging techniques. Comput Geosci 27(9):1019–1029. https://doi.org/10.1016/s0098-3004(00)00160-6
3. Devaraj N, Chidambaram S, Panda B, Thivya C, Thilagavathi R, Ganesh N (2018) Geo-electrical approach to determine the lithological contact and groundwater quality along the KT boundary of Tamilnadu, India. Model Earth Syst Environ 4:269. https://doi.org/10.1007/s40808-018-0424-2
4. Dheenan PS, Jha DK, Vinithkumar NV, Ponmalar AA, Venkateshwaran P, Kirubagaran R (2014) Spatial variation of physicochemical and bacteriological parameters elucidation with GIS in Rangat Bay, Middle Andaman, India. J Sea Res 85:534–541. https://doi.org/10.1016/j.seares.2013.09.001
5. Etikala B, Adimalla N, Madhav S, Somagouni SG, Keshava Kiran Kumar PL (2021) Salinity problems in groundwater and management strategies in arid and semi-arid regions. Groundw Geochem 42–56. https://doi.org/10.1002/9781119709732.ch3
6. Gnanasundar D, Elango L (1999) Groundwater quality assessment of a coastal aquifer using geoelectrical techniques. Int J Environ Hydrol 7(2):21–33
7. Gopinath S, Srinivasamoorthy K, Saravanan K, Suma CS, Prakash R, Senthinathan D, Sarma VS (2017) Vertical electrical sounding for mapping saline water intrusion in coastal aquifers of Nagapattinam and Karaikal, South India. Sustain Water Resour Manag. https://doi.org/10.1007/s40899-017-178-4
8. Jones NL, Davis RJ, Sabbah W (2003) A comparison of three-dimensional interpolation techniques for plume characterization. Groundwater 41:411–419. https://doi.org/10.1111/j.1745-6584.2003.tb02375.x

9. Khouni I, Louhichi G, Ghrabi A (2021) Use of GIS based inverse distance weighted interpolation to assess surface water quality: case of Wadi El Bey, Tunisia. Environ Technol Innov 24:101892. https://doi.org/10.1016/j.eti.2021.101892
10. Lee S, Kim K, Ko I, Lee S, Hwang H (2002) Geochemical and geophysical monitoring of saline water intrusion in Korean paddy fields. Environ Geochem Health 24:277–291
11. Mtetwa S, Kusangaya S, Schutte CF (2003) The application of geographic information systems (GIS) in the analysis of nutrient loadings from an agro-rural catchment. Water Sa 29:189–194. https://doi.org/10.4314/wsa.v29i2.4855
12. NIH (2016) Groundwater table and hydrochemistry of Kakinada coastal Aquifer_NIH_1996-97
13. Nagaraju A, Balaji E, Dash P (2016) Land use/land cover analysis based on various comprehensive geospatial data sets: a case study from Tirupati Area, South India. Adv Remote Sens 5:73–82. https://doi.org/10.4236/ars.2016.52006
14. Oke A, Sangodoyin A, Ogedengbe K, Omodele T (2013) Mapping of river waterquality using inverse distance weighted interpolation in Ogun-Osun river basin, Nigeria. Acta Geogr Debrecina Landsc Environ 7:48–62
15. Prasanna MV. Geoelectrical investigation along Miri coast, East Malaysia: evaluate the vulnerability of coastal aquifer. https://doi.org/10.1088/1757-899X/495/1/012042
16. Senthilkumar S (2019) Integrated seawater intrusion study of coastal region of Thiruvallur district, Tamil Nadu, South India. Appl Water Sci 9:124. https://doi.org/10.1007/s13201-019-1005-x
17. Telford WM, Geldart LP, Sheriff RE (1990) Resistivity methods. In: Applied geophysics, 2nd Edition, Cambridge University Press, Cambridge, UK, pp 353–358. https://doi.org/10.1017/cbo9781139167932.012
18. UN-United Nations (2003) Water for people, water for life. UN World Development Report (WWDR)
19. WHO-World Health Organization (2006) Guidelines for drinking-water quality, 1st–3rd edns., vol 1. Geneva, Switzerland
20. Wilson SR, Ingham M, McConchie JA (2006) The applicability of earth resistivity methods for saline interface definition. J Hydrol 316(1–4):301–312
21. Yan CA, Zhang W, Zhang Z, Liu Y, Deng C, Nie N (2015) Assessment of water quality and identification of polluted risky regions based on field observations & GIS in the honghe river watershed, China. PLoS One 10:e0119130. https://doi.org/10.1371/journal.pone.0119130

Chapter 17
Systematic Approach of Groundwater Resources Assessment Using Remote Sensing and Multi-influence Factor (MIF) Techniques in Medchal Mandal, Telangana State, India

D. Naresh Kumar, Thumati Venkateswara Rao, Vamsi Kalyan Veerla, Balaji Etikala, Y. Mohana Prasada Rao, and Anvesh Jujjuvarapu

Abstract Integrated analysis of hydro-geomorphology is a vital aspect of locating groundwater perspective zones. It also provides a systematic approach to groundwater evaluation and promotes the sustainable development of resources. This study employed geospatial techniques to obtain the hydro-geomorphology of Medchal Mandal, located in the North-East direction of Hyderabad. The following analysis has been summarized from the study; (1) pediplain (5184.59 ha), which is moderately weathered, found the groundwater prospects in this zone ranging from poor to moderate (2) Pediplain shallow weathered (6066.10 ha), and noticed the groundwater condition varying from moderate to poor noticed (3) pediment (1554.75 ha) to hold moderate groundwater (4) pediment inselberg complex (5614.06 ha), was located in the study implied poor groundwater conditions (5) denudation hills (161.82 ha) has a groundwater potential ranging from poor to negligible and noticed that the cause

D. Naresh Kumar (✉) · V. K. Veerla · A. Jujjuvarapu
Civil Engineering Department, St. Martin's Engineering College, Secunderabad 500100, India
e-mail: naresh.geology@gmail.com

V. K. Veerla
e-mail: vamsikalyanveerla@gmail.com

A. Jujjuvarapu
e-mail: anveshce@smec.ac.in

T. Venkateswara Rao
Civil Engineering Department, VVIT, Nambur, AP 522508, India
e-mail: venkateswararaothumati@gmail.com

B. Etikala · Y. Mohana Prasada Rao
Department of Geology, Sri Venkateswara University, Tirupati, AP 517502, India
e-mail: balajiyvu@gmail.com

Y. Mohana Prasada Rao
e-mail: madhuteliki@gmail.com

E. Balaji et al. (eds.), *Emerging Technologies for Water Supply, Conservation and Management*, Springer Water, https://doi.org/10.1007/978-3-031-35279-9_17

for poor groundwater conditions was significant runoff produced and mild infiltration. Lineaments were identified in the following villages, Gaudavalli, Gosaiguda, Ghanpur, Sreerangavaram, and Nuthankal, indicating major orientations along with N-S and NE, SW, and SE directions. In the study area, the observed groundwater resource was found acceptably satisfactory. Furthermore, the thematic maps of hydro geomorphology and geomorphology showed possible artificial recharge areas that can be employed alternatively to manage groundwater resources in the study area about complex rock terrain localities. Using the multi-influence factor (MIF) and weighted overlay techniques in Arc GIS, the eight layers were created and given set scores and weights based on their propensity to hold water. The results can be paragraphed as three zones visible on spatial exposure (a) 28% good groundwater potential zone, (b) moderate (33%), and (c) poor (39%) were evaluated. Thus, the groundwater potential model is a predominant tool for assessing sustainable groundwater resources.

Keywords Geology · Geomorphology · Groundwater potential zones · Multi-influence factor · Medchal Mandal

17.1 Introduction

One of the priceless natural resources that safeguard trade, socioeconomic growth, and human health is groundwater [4, 7, 9, 19, 56, 59]. It is a significant source of water supply for an Indian agricultural, residential, and industrial zone. Numerous hydro geomorphologic processes have influenced the occurrence, mobility, availability, storage, and transmission of groundwater in the aquifer zone [3, 34, 36]. It is necessary to field authenticate hydro-geomorphology, geomorphology, geology, and geological structures to identify potential groundwater zones [6, 12, 13, 17, 33, 35, 62].

Due to geological formation and seasonal variations, the supply of groundwater differs from one place to another [32, 45]. Identification of groundwater prospects is aided by an integrated investigation of the hydro-geomorphology, lineaments, and geomorphology of the research area utilizing geospatial tools [1, 57, 63]. Lineaments give primary baseline data that regulate the occurrence and movement of groundwater [46, 55]. Initiation of remote sensing, appropriate arrangement of high-speed computers, and GIS bid for assessment of hydro-geomorphic themes that will aid in the identification of groundwater potential zone in the Medchal area [38, 40]. Many research scholars, professors, and scientists have used spatial techniques to determine potential groundwater zones with good results [16].

The combination of remote sensing (RS) and geographic information systems (GIS) has shown to be an effective method for identifying groundwater potential zones because it requires less time and labor. Considering several important parameters, including lithology, lineament/drainage density, slope, geomorphology, rainfall, land use patterns, and soils, provides reliability and lowers the possibility of human error. Combining primary and secondary data sets is an advantage of RS and GIS.

In the past, researchers have utilized various methods, such as multi-influence factor analysis (MIF), to identify potential groundwater zones [5, 29, 54].

Sub hub of Hyderabad of Medchal Mandal population has risen from the decade, the water supply and increased demands of groundwater resources. To solve this problem, there is a need for the required method. The project aims to define possible groundwater zones utilizing spatial methodologies and statistical analysis in Medchal Mandal, Telangana state, India.

17.2 Study Area

Medchal Mandal is located northeast of the Medchal-Malkajgiri District of Telangana State, India. The Mandal is situated within of 78°22′–78°45′ East longitudes and 17°42′–17°50′ North latitudes (Fig. 17.1). There are 29 villages and 18 village panchayats in Medchal Mandal. It is 602 metres above sea level on average. There are 93,102 people living in the 196.3 km^2 area as per the 2011 Census. The amount of rain annually is 835.7 mm. Between June and September, the region receives rain from the southwest monsoon zone. It begins in the northern part of Kerala State, travels through Rayalaseema, and then ends up in the Hyderabad region. As a result of the scorching summer, the study area experienced dry climate conditions. Geomorphological features have been covered in Medchal Mandal such as the denudation hill, pediment-pediplain complex and anthropogenic bodies [2, 20, 49, 51]. The denudational origin of the Pediment-Pediplain complex has been covered in Medchal Mandal, with a minor portion covered by anthropogenic bodies. Humans sculpt and transform the landscape through the physical change of shape and behavior of surface material and subsurface substances. Contamination by harmful elements dumped at a place that simultaneously pollutes soil, water, and the air is an anthropogenic area. Girmapur, Gosaiguda, Ghanpur, Ravalkole, and Kandlakoya villages have anthropogenic bodies.

The Medchal Mandal slope can be categorized as very gently sloping, flat to nearly level, gently sloping, moderately sloping, and relatively steeply sloping [43]. Medchal Mandal contains predominantly Peninsular Gneiss Complex (PGC), granite, and Alkali feldspar granite rocky terrains and some parts covered by amphibolite, hornblende, and biotite schist rocks [27, 48]. few minerals like epidote and Pyroxene is also subsequently observed. Granite aquifers' transmissivity varies from 30 to 200 m^2/day [10]. Medchal Mandal has variable water resources but their distribution over the surface is difficult because of undulation, slope gravity, and varied rainfall.

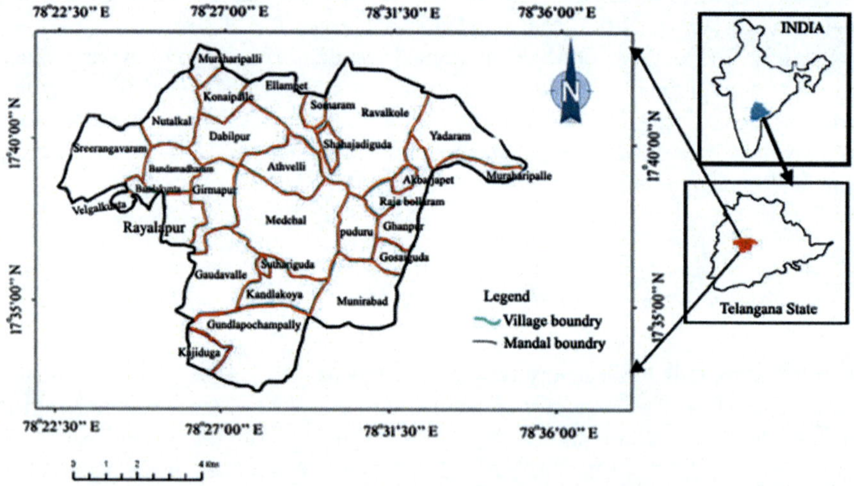

Fig. 17.1 Medchal Mandal location map, Medchal-Malkajgiri District, Telangana State

17.2.1 *Physiography and Hydrogeology*

Most of Medchal Mandal is flat and descends gradually in a northwesterly direction. The region is 420 and 640 m above mean sea level (MSL). Different lake tributaries drain rainwater from Medchal Lake and Shameerpet Lake. Undulation and lowland areas, fractures have controlled drainage, and joints in igneous rocks parallel to the sub-parallel pattern in the east, northeast, and south-eastern.

17.2.2 *Geotechnical Characteristics*

Medchal Mandal, Medchal-Malkajgiri District's terrains have been demarcated into engineering geological provinces. Granites and gneiss have low permeability, high (1000–2000 kg/cm^2) bearing capacity/ compressive strength, and 'excellent' foundation characteristics. The entire area of Medchal-Malkajgiri and Hyderabad Districts falls under seismic zone-III.

17.3 Methodology

The flow chart illustrates the process for evaluating and mapping the hydrogeomorphology in the Medchal Mandal. Survey of India toposheets (56K/6 and 56K/10 on 1:50,000 scale) and linear imaging self-scanning sensor of IRS P6 LISS III (Path 24 and Row 61). Utilised to visually comprehend many topics such as shape, size, texture, pattern, tone, and related features. Georeferencing of satellite photos was done using the WGS1984 zone 44N datum. RS data, SOI topographic maps, geological maps obtained from the GSI (Geological Society of India), and data from multiple sources were used to create several thematic maps. A base map was made using the SOI (Survey of India) toposheets. All thematic maps are created using the Medchal Mandal hydrogeomorphological map, which was created using the ARC GIS 10.2.2 program. This map is used to identify probable groundwater zones.

17.3.1 Multi-influencing Factor (MIF)

The multi-influencing factor (MIF) technique was combined with a geographic information system to map potential groundwater recharge zones. Hydrogeological components such as drainage, lineament, geology, slope, land use/land cover, rainfall, and soil influence groundwater recharge. These parameters were categorized and integrated based on weights obtained using the MIF technique.

17.3.1.1 Weightage and Assignment

Each class was weighed based on published literature. The results covered the good-to-poor ratio and were associated with potential zones.

17.3.1.2 Weighted Overlay Analysis

Analysis of weighted overlays uses GIS to overlay thematic maps of weighted data to discover the best groundwater potential zones.

17.3.1.3 Lineament Map

Lineaments are naturally occurring linear surface components immediately understood from satellite data. Lineament can be displayed as a single continuous line or a series of broken lines. The alignment of ponds and straight stream segments has revealed lines. Scale weights are assigned to each feature class and are projected in Table 17.1. These facts significantly impacted the research area's groundwater recharge. The layer's weight parameters are analyzed by GIS software, which then reveals the potential groundwater zone. The Weightage and Assignment stage connected the Liner Density stage for weightage overlay and all layers' weight conversion findings of potential groundwater zones were then displayed [14, 23]. A weight of 1 was given to the significant influencing factor, while a weight of 0 was given to the minor influencing factor (Table 17.2).

Table 17.1 Calculation of scale weights that affect groundwater recharge weights

S. no	Parameter	Feature class	Scale weight
1	Geomorphology	River, Tank	24
		Pediment	20
		Pediplain	15
		Pediment inselberg complex	10
		Denudational hill	4
2	Geology	Predominantly granite and alkali feldspar granite	20
3	Slope	Level to nearly level	10
		Very gently sloping	8
		Gently sloping	6
		Moderately sloping	4
		Moderately steeply sloping	12
4	Hydro-geomorphology	Pediplain moderately weathered	24
		Pediplain shallow weathered	18
		Pediment	15
		Pediment Inselberg complex	10
		Denudation hill	4
5	Lineament density (km^2)	<0.6	4
		0.6–0.8	8
		0.8–1.0	10
		1.0–1.2	15
		>1.2	20

Table 17.2 Shows influences, relative impacts, and weights given to each potential component

Factors	Major effect (A)	Minor effect (B)	Relative weight (A + B)	Assigned weight for each influential factor
Geomor phology	4	0	4	24
Geology	3	0.5	3	20
Slope	1	0.5	2.5	12
Hydro geomor phology	2	0	2	24
Lineaments Density	2	0.5	2.5	20
Total			Sum: 14	100

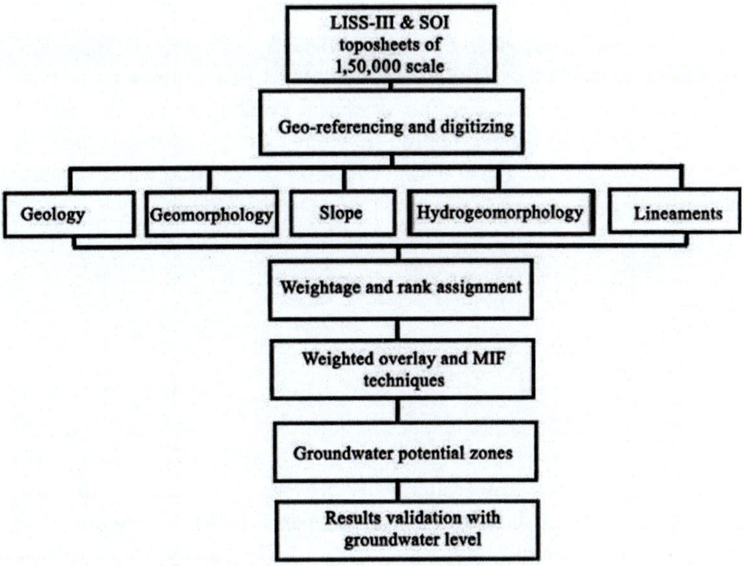

Fig. 17.2 Flowchart representing the processing method of groundwater potential map

17.3.1.4 Slope Map

The slope map was created using a digital elevation model (DEM) from the National Remote Sensing Centre's Cartosat-I DEM (NRSC). The geology map, geomorphology map, slope map, lineaments map, drainage density map, and contour map coincide with this hydro-geomorphologic map [11, 39]. The state of recharge and the types of rocks, landforms, and exploration spots are well explained by hydrogeomorphic maps and drainage networks. The final output map of the recharge potential zone was created using thematic layers with a 30 m resolution and a standard coordinate system of UTM WGS-84. The results were then verified using historical drill yield information. Figure 17.2 illustrates the study's methodology.

17.4 Results and Discussion

17.4.1 Geomorphology

The Medchal Mandal's geomorphic unit comprises the Pediment-Inselberg Complex, Pediplain, Pediment, and River Tanks. Because there are so many little inselbergs in the Pediment-Inselberg Complex, it might be challenging to tell them apart from regular pediments. The Pediment-Inselberg Complex region's low aquifer content needs better groundwater potentials. As a result, it was given little weight. Pediplain has an extensive, multi-concave, rock-cut erosion surface, in this region groundwater flow is moderate; hence we assigned 20% weightage. Pediments represent a mature stage of the erosion cycle and have good aquifer content, so we allocated 27% of weightage. Denudation hill have negligible water content so it assigned very low value of weightage [21, 50]. River and tank regions have good water resources as a weightage distribution we given 10 weightage (Fig. 17.3a).

17.4.2 Slope

Slope and undulations are essential aspects of groundwater recharge. Low land areas, lakes, and tributaries are complementary geomorphologic structures for collecting groundwater. Very gently sloping covered Mandal 42% (Table 17.3) of the area, runoff is high in this region low groundwater prospect is low. Levels to nearly level and gently sloping both are covered 20% of the part, gently sloping not suitable for underground water. Moderately sloping and moderately steeply sloping are occupied 12 and 8% of the Medchal Mandal, both land features are moderate to good for groundwater filling (Fig. 17.3b). When assessing the possibility for groundwater, the dominance of the very gently sloping slope class is unfavorable, hence they were given little weight. It was given high weight because of the good groundwater potential of gently sloping and level to the almost level slope.

17.4.3 Geology

Grey and pink granite are transitory and gradational in both lateral and vertical directions. The origin is the exact equivalent in the two rock types [8, 37, 44, 58, 61]. Medchal Mandal is covered by pink granite in a large portion and grey granite moderate in a field survey observed that weathered granite rocks present in Ghanpur, Goudavalle, and Pudoor villages (Fig. 17.3c). Therefore, appropriate weights are given to various rock units in the research region based on the availability of groundwater. Because of their resistance to weathering, faulting, and jointing in the area, predominantly granite [18] and alkali feldspar granite were given low values.

17.4.4 Lineaments

Lineament is either a single continuous line or is shown as discontinuous line segments [42]. Lineaments have been exposed in the alignment of ponds, straight stream segments, and vegetation structure [15, 28, 31]. Dykes create linear ridges on the surface. This region has good groundwater prospects [24, 41, 53]. The research area's lineament density map's thematic layers showed five distinct lineament density classes: very low (0.1–0.6 km/km^2), low (0.6–0.8 km/km^2), moderate (0.8–1.0 km/

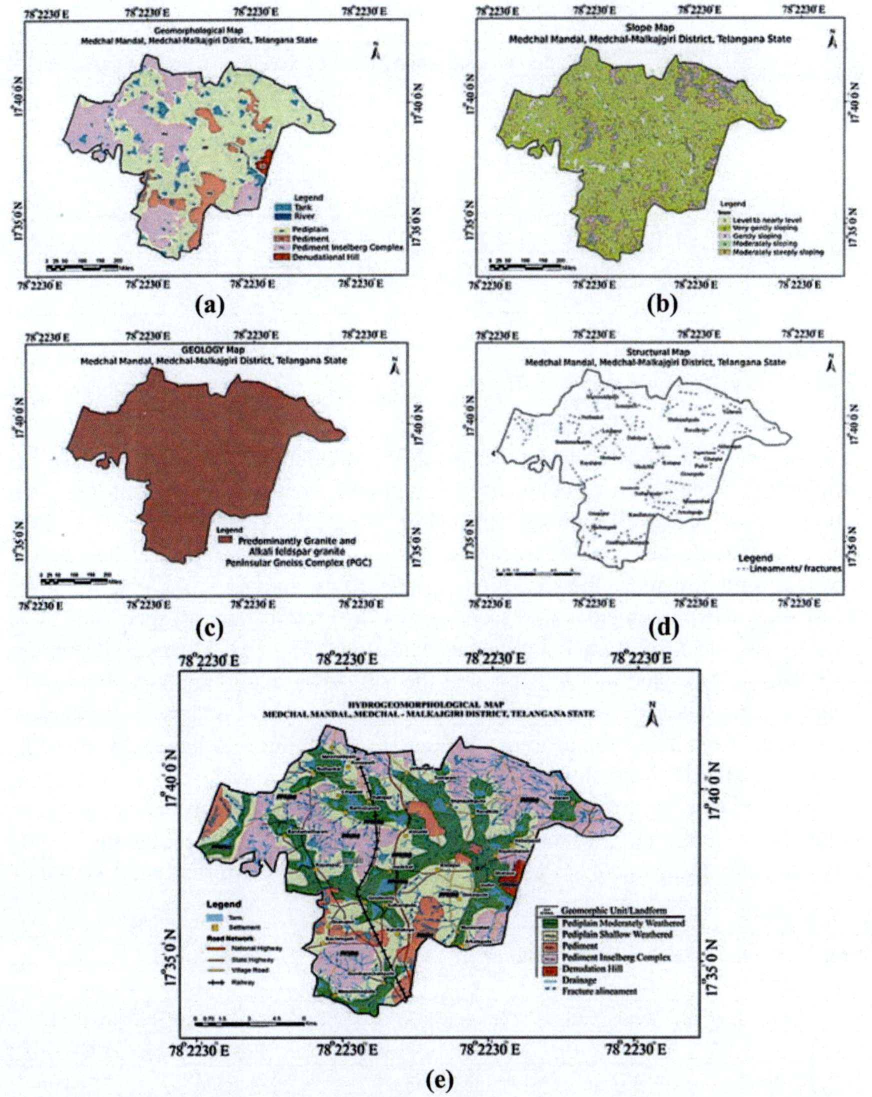

Fig. 17.3 **a** Geomorphological map, **b** slope map, **c** geological map, **d** lineament map, **e** hydrogeomorphological map with groundwater prospects of Medchal Mandal, Medchal-Malkajgiri District, Telangana State

km^2), and high (1.0–1.2 km/km^2) (Very high). Lineaments in Medchal Mandal are well exposed near the villages such as Medchal, Gaudavalli, Gosaiguda, Ghanpur, Sreerangavaram, and Nuthankal (Fig. 17.3d).

Table 17.3 Medchal Mandal slope classification as per percentage

S. no	Classification of slope	Min to max limit of the slope (%)
1	Level to nearly level	18
2	Very gently sloping	42
3	Gently sloping	20
4	Moderately sloping	12
5	Moderately steeply sloping	08

17.4.5 Hydro-Geomorphology

The hydro-geomorphology of Medchal Mandal is divided into five major categories: pediplainmoderately weathered, pediplain shallowly weathered, pediment Inselberg Complex, pediment, and denudation hills (Fig. 17.3e). Pediplain Moderately Weathered shows good water conditions, faults, lineaments, fractures, and weathered rock materials, which increase the scope for recharge of groundwater. Table 17.4 Shows the hydrogeomorphological features, corresponding groundwater prospects and their summary statistics of Medchal Mandal, Malkajgiri-Medchal District, Telangana State. This geomorphic unit is noticed at the villages around Gundlapochampalle, Kajiguda, Kondlakoyya, Suthariguda, Gosaiguda, Ghanpur, Pudoor, Rajabollaram, Akbarjapet, Rovalkol, Yadawaram, Yellampet, Somaram, Konayepalli, Atvelli, Maisereddipalle, Nuthankal, Bandamadharam, Girmapur, Goudavelly.

Pediment inselberg complex division is not possible to restore groundwater. These geomorphological units present in villages are Gosaiguda, Raja bollaram, Gundlapochampalle, Bandamadharam, Girmapur, Ravalkole, and Murharipalle. Ghanpur

Table 17.4 Hydrogeomorphology features in Medchal Mandal, Malkajgiri-Medchal District, Telangana State

S. no	Symbology	Geomorphic unit	Lithology	Groundwater prospects	Area in km^2
1	PPM	Pediplain moderately weathered	Peninsular Gneiss Complex Granite (PGC)	Good to moderate	51.8459
2	PS	Pediplain Shallow weathered	PGC	Moderate to poor	60.6610
3	PIC	Pediment inselberg complex	PGC	Poor	56.1406
4	PD	Pediment	PGC	Poor	15.5475
5	DH	Denudation hill	PGC	Negligible	1.6184

and Maisammaguda regions have denudation hills, and they have a high rate of water flow due to slope and gravitational force [22]. Continuous water flow increases weathered soil and spreads on the surface in vast regions [60].

17.5 Discussion

The Medchal Mandal has distinct hydro-geomorphological conditions, with topography and geology controlling the groundwater regime. Medchal Mandal has 124 lakes, 20 lakes are inside the outer ring road (ORR), and 104 lakes are outside the ORR [26]. Moderate to shallow weathered Pediplains allowing rainwater percolation into the ground. These hydro-geomorphological features are suitable for increased groundwater levels [47, 52]. Probably a good density in geological structures like faults, fractures, lineaments, etc., is ideal for the storage of water structures due to inherent transmissivity, agricultural lands may be avoided so that the impounding water does not inundate fertile lands, as the artificial recharge in the immediate vicinity will further aggravate the problem [25, 30].

The study areas groundwater potential map (Fig. 17.4) depicts three distinct classifications (zones) that correspond to the area's "excellent," "moderate," and "poor" groundwater potential. The lakes and drainage network are primarily included in the good groundwater potential zone, which includes areas where low-elevation areas are ideal for storing groundwater and indicate the presence of groundwater.

A good groundwater potential zone covers approximately 52 km^2, accounting for 28% of the total area. The study area's central and northern portions are in the

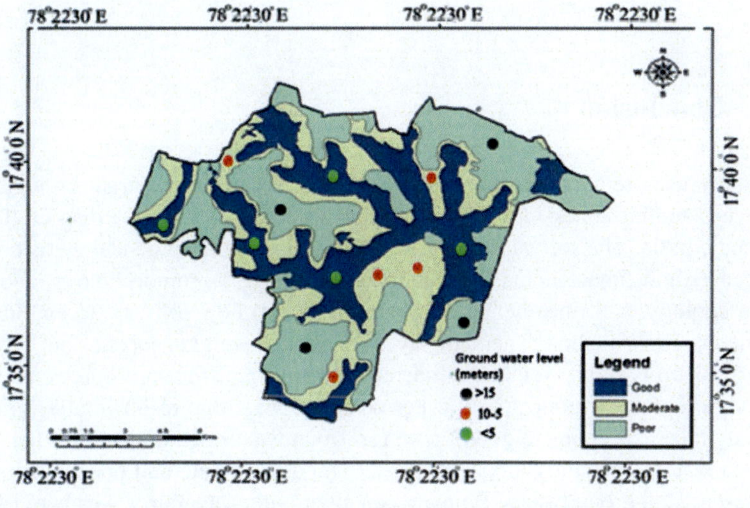

Fig. 17.4 Groundwater potential zone map of Medchal Mandal, Medchal-Malkajgiri district, Telangana

moderate groundwater potential zone; it covers 61 km^2, or about 33% of the total area. This section's hydro-geomorphic feature is deep to moderately buried under pediplain shallow weathered, implying a moderate groundwater storage capacity. However, the Pediment inselberg complex, Pediment, and Denudation Hill areas, which cover approximately 73 km^2, are located in poor groundwater potential zones. Due to the increased slope and unfavorable local geology and geomorphology, the groundwater potential in this area is lower. These prospective groundwater zones aid in the identification of artificial recharge sites in the study area.

17.6 Verification Using Data from Bore Wells

To determine the accuracy of the results of prospective groundwater zones in the Medchal Mandal, borehole data was collected from the groundwater department of Telangana state and verified by remote sensing techniques. 5.5–15 m bgl is the range of the depth of weathering. Between 1.5 and 75 m^3/day are produced via extension bores drilled to bore well depths between 20 and 45 m. The specific capacity ranges from 0.005 to 0.16 m^3/m per unit cross-section in weathered granite (mainly dry) and alluvium, while the transmissivity values in these materials range from 100 to 150 m^2/day.

Medchal Mandal's average annual borehole depth in May is 18.24 m. bgl (below ground level), seasonal November month depth is 9.12 m. bgl. The average depth of the borehole is 13.44 m. bgl. On the groundwater potential zone map, existing borehole data overlapped, revealing a positive correlation for identifying groundwater points (Fig. 17.4).

17.7 Conclusion

This study uses remote sensing and GIS techniques to demonstrate groundwater resource potential zones in Medchal Mandal, Medchal-Malkajgiri district, Telangana state, India. The software module of ARC GIS offers a practical methodology in expenses and time. Thematic maps focused on hydro-geomorphology, geology, geomorphology, and lineaments to be integrated into GIS software to employ the weighted overlay method. Thematic layers consist of weight scores as per groundwater availability, high weightage indicates good groundwater resources, and low weightage represents poor groundwater content. Thus, this result implies grading the study area as a 'poor to good' zone for groundwater occurrence. Further, with obtained results, the study area was marked good, moderate and poor location for the presence of groundwater. Being a peneplain site with fairly weathered lakes and drainage networks, the research region had 28% of the outstanding groundwater potential zone. The low land regions in the area were identified through this research, and this characteristic guarantees water availability below the groundwater table. The study also discovered that it included 61 km^2 of medium groundwater potential or 33% of the zone's total. 39% of the study region has poor groundwater, compared to 73 km^2 of the area that exhibits a poor groundwater potential zone. The groundwater potential map and the nearby drill data were thus positively correlated. The borehole yields above >75 m^3/day in areas with good and excellent groundwater potentials, whereas it yields less than 55 m^3/day in areas with moderate and poor groundwater potentials. This research opens up possibilities for additional groundwater exploration studies and could help policymakers better manage and develop groundwater resources in several promising groundwater zones in the Medchal Mandal region. The outcomes of integrated applications, techniques, and groundwater investigation are appropriate for creating and managing groundwater resources in these settlements and for sustainable management of water resources in hard-rock terrains.

References

1. Agarwal R, Garg P (2015) Remote sensing and gis based groundwater potential & recharge zones mapping using multi-criteria decision-making technique. Water Resour Manag 30(10):243–260. https://doi.org/10.1007/s11269-015-1159-8
2. Aryanto DE, Hardiman G (2018) Assessment of groundwater recharge potential zone using gis approach in purworejo regency, central java province, indonesia. E3s web conf. 31, 12002. https://doi.org/10.1051/e3sconf/20183112002
3. Balaji E, Nagaraju A, Sreedhar Y, Thejaswi A, Sharifi Z (2016) Hydrochemical characterization of groundwater in around Tirupati Area, Chittoor District, Andhra Pradesh, South India. Appl Water Sci 7(3):1203–1212. https://doi.org/10.1007/s13201-016-0448-6
4. Balaji E, Veeraswamy G, Prasad M (2018) Assessing groundwater hydrochemistry of Atmakur, SPSR Nellore District, Andhra Pradesh, India. Fresen Environ Bull 27(11):7897–7905
5. Balaji E, Veeraswamy G, Peiyue L, SiddiRaju R (2019) Deciphering groundwater potential zones using MIF technique and GIS: a study from Tirupati area, Chittoor District, Andhra Pradesh, India. HydroResearch 1:1–7. https://doi.org/10.1016/j.hydres.2019.04.001
6. El-Kadi AI, Oloufa AA, Eltahan AA, Malic HU (1994) Use of a geographic information system in site-specific groundwater modeling ground water 32:617–625. https://doi.org/10.1111/j.1745-6584.1994.tb00897.x
7. Etikala B, Adimalla N, Madhav S, Somagouni SG, Keshava Kiran Kumar PL (2021) Salinity problems in groundwater and management strategies in arid and semi-arid regions. Groundw Geochem 42–56. https://doi.org/10.1002/9781119709732.ch3
8. Elmahdy SI, Mohamed MM (2014) Groundwater potential modeling using remote sensing and GIS: a case study of the al Dhaid area, United Arab Emirates. Geocarto Int 29(4):433–450. https://doi.org/10.1080/10106049.2013.784366
9. Etikala B, Madhav S, Srinivasa Gowd S (2022) Urban water systems: an overview. Curr Direct Water Scarcity Res 6:1–19. https://doi.org/10.1016/B978-0-323-91838-1.00016-6
10. Gaur VK, Priestley KF (1997) Shear wave velocity structure beneath the Archean granites around Hyderabad, inferred from receiver function analysis. Proc Indian Acad Sci 106:1–8
11. Ghosh PK, Bandyopadhyay S, Jana NC (2016) Mapping of groundwater potential zones in hard rock terrain using geoinformatics: a case of Kumari watershed in western part of west Bengal. Model Earth Syst Environ 2(1). https://doi.org/10.1007/s40808-015-0044-z
12. Golla V, Badapalli PK, Balaji E, Sivakumar VL, Telkar SK (2021) Delineation of groundwater potential zones in semi-aridregion (Ananatapuram) using geospatial techniques. Mater Today: Proc. https://doi.org/10.1016/j.matpr.2021.02.239

13. Golla V, Etikala B, Veeranjaneyulu A, Subbarao M, Surekha A, Narasimhlu K (2018) Data sets on delineation of groundwater potential zones identified by geospatial tool in Gudur area, Nellore district, Andhra Pradesh, India. Data Brief 20:1984–1991. https://doi.org/10.1016/j.dib.2018.09.054
14. Gopinath G, Saralathan P (2004) Identification of groundwater prospective zones using irs-1d liss iii and pump test methods. J Indian Soc Remote Sens 32:329–342
15. Gumma MK, Pavelic P (2012) Mapping of groundwater potential zones across ghana using remote sensing, geographic information systems, and spatial modelling. Environ Monit Assess. https://doi.org/10.1007/s10661-012-2810-y
16. Haridass VK, Chandrasekaran VA, Kumaraswamy K, Rajendran S, Unni Krishanan K (1994) Geomorphological and lineament studies of kanjamalai using irs-1 data with special reference to groundwater potentiality. Trans Inst Indian Geogr 16(1):35–41
17. Harinarayana P, Gopalakrishna GS, Balasubramanian A (2000) Remote sensing data for groundwater development and management in keralapura watersheds of the cauvery basin, Karnataka, India. Indian Mineralog 34(2):11–17
18. Ifediegwu SI, Nnebedum DO, Nwatarali AN(2019) Identification of groundwater potential zones in the hard and soft rock terrains of Kogi State, North Central Nigeria: an integrated GIS and Remote Sensing techniques. SN Appl Sci 1:1151. https://doi.org/10.1007/s42452-019-1181-1
19. Jaiswal RKS, Mukherjee J, Krishnamurthy Saxena R (2003) Role of remote sensing and GIS techniques for generation of groundwater prospect zones towards rural development—An approach. Int J Remote Sens 24(5):993–1008. https://doi.org/10.1080/01431160210144543
20. Jothiprakash V, Marimuthu G, Muralidharan R, Senthil Kumar N (2003) Delineation of potential zones for artificial recharge using gis. J Indian Soc Remote Sens 31(1):37–47. 10.1007/bf03030750
21. Kaliraj S, Chandrasekar N, Magesh NS (2014) Identification of potential groundwater recharge zones in vaigaiupper basin, tamil nadu, using gis-based analytical hierarchical process (ahp) technique. Arab J Geosci 7:1385–1401. https://doi.org/10.1007/s12517-013-0849-x
22. Kamal R, Sahu PC (2006) Targeting groundwater in tribal-dominatedbonai area of drought-prone sundargarh district, Orissa, India-A combined geophysical and remote sensing approach. J Hum Ecol 20(2):109–115
23. Koutsoyiannis D, Montanari A, Lins HF, Cohn TA (2009) Discussion of the implications of projected climate change for freshwater resources and their management climate, hydrology and freshwater towards an interactive incorporation of hydrological experience into climate research. Hydrol Sci J des Sci Hydrol 54(2):394–405
24. Krishnamurthy JN, Venkatesa Kumar V, Jayaraman Manivel M (1996) An approach to demarcate ground water potential zones through remote sensing and a geographical information system. Int J Remote Sens 17(10):1867–1884. https://doi.org/10.1080/01431169608948744
25. Kudrna K, Šindelářová M (2006) Principles of hydrogeomorphology as a basic precondition for the solution of the territorial structure of the unitary system of agricultural, forest, and water management. J Central Euro Agricul 7(4):669–676
26. Kumar DN (2020) hydro-geochemical assessment of groundwater through statistical analysis for sustainable usage in medchal mandal, hyderabad, india. Sustain Water Resour Manag 6:119. https://doi.org/10.1007/s40899-020-00477-6
27. Lee S, Kim YS, Oh HJ (2012) Application of a weights-of-evidence method and gis to regional groundwater productivity potential mapping. J Environ Manag 96(1):91–105. https://doi.org/10.1016/j.jenvman.2011.09.016
28. Lokesh N, Gopalakrishna GS, Mahesh MJ (2007) Hydrogeomorphological studies in Kallambella watershed, Tumkur district, Karnataka state, India using remote sensing techniques and GIS. J Indian Soc Remote Sens 35(1):97–105
29. Magesh NS, Chandrasekar N, Soundranayagam JP (2012) Delineation of groundwater potential zones in Theni district, Tamil Nadu, using remote sensing, GIS and MIF techniques. Geosci Front 3(2):189–196. https://doi.org/10.1016/j.gsf.2011.10.007

30. Mohammed A, Balasubramanian A, Kondoh A, Rokhmatuloh R, Mustafa AJ (2003) Hydrogeomorphological mapping using remote sensing techniques for water resource management around palaeochannels. Geoscience and remote sensing symposium, IGARSS-APOS; 03. In: Proceedings of the 2003 IEEE international, Toulouse, France, vol 5, pp 3317–3319
31. Mondal SMD, Pandey AC, Garg RD (2008) Groundwater prospects evaluation based on hydrogeomorphological mapping using high-resolution satellite images; a case study in uttarakhand. J Indian Soc Remote Sen 36(1):69–76
32. Murthy KSR (2000) Groundwater potential in a semi-arid region of Andhra Pradesh—A geographical information system approach. Int J Remote Sens 21(9):1867–1884. https://doi.org/10.1080/014311600209788
33. Murthy KSR, Venkateswara Rao V (1999) Mapping of hydrogeomorphological features in varaha river basin using IRS data. J Indian Soc Remote Sens 27(2):155–166
34. Nagaraju A, Balaji E, Dash P (2016) Land use/land cover analysis based on various comprehensive geospatial data sets: a case study from Tirupati Area, South India. Adv Remote Sens 5(2):73–82. https://doi.org/10.4236/ars.2016.52006
35. Nagaraju A, Balaji E, Sun LH, Tejaswi A (2018) Processes controlling groundwater chemistry from Mulakalacheruvu Area, Chittoor District, Andhra Pradesh, South India: a statistical approach based on hydrochemistry. J Geol Soc India 91:425–430. https://doi.org/10.1007/s12594-018-0875-0
36. Nagaraju A, Balaji E, Tejaswi A, Sun L (2015) Quality evaluation of groundwater in Mulakalacheruvu area, Chittoor district, Andhra Pradesh, South India based on hydrogeochemistry. Fresenius Environ Bull 24(12A):4496–4503
37. Nampak J, Pradhan B, Manap MA (2014) Application of GIS based data driven evidential belief function model to predict groundwater potential zonation. J Hydrol 513:283–300. https://doi.org/10.1016/j.Jhydrol.2014.02.053
38. Naresh KD, Madhu T, Ramanaiah S (2020) Assessment of water quality in Banaganapalli, Kurnool district of Andhara Prasesh, India. Indian Ground Water 14:71–83
39. Naresh Kumar D, Madhu T (2020) Subcategories Multiple uses of land recognized by land use and land cover classification system using LISS-III and NRSC enumerate data: case study at Medchal Mandal, Hyderabad, India. Asian J Atmosp Environ 14(4):394–412. https://doi.org/10.5572/ajae.2020.14.4.394
40. Naresh Kumar D Madhu T, Rajeev Kumar G, Chaitanya Reddy E (2019) Land cover change detection of Medchal Mandal and its surroundings using SAGA software. Bull Pure Appl Sci 38(2). https://doi.org/10.5958/2320-3234.2019.00022.2
41. O'leary DW, Driedman JD, Pohn HA (1976) Lineaments, linear, lineation: some proposed new standards for old terms. Geol Soc Amer Bull 87:1462e1463
42. Obi Reddy GP, Suresh Babu R, Sambasiva Rao M (1994) Hydrogeology and hydrogeomorphological conditions of Anantapur district, using remote sensing data. Indian Geog J 69(2):128–135
43. Obi Reddy GP, Chandra K, Mouli SK, Srivastav CV, Maji AK (2000) Evaluation of groundwater potential zones using remote sensing data-a case study of Gaimukh watershed, Bhandara district Maharashtra. J Indian Soc Remote Sens 28(1):19–32
44. Pandey OP, Agrawal PK, Chetty TRK (2002) Unusual lithospheric structure beneath the Hyderabad Granite region, Eastern Dharwar Craton, South India. Phys Earth planet Int 130:59–69
45. Pankaj K, Srivastava A, Bhattacharya K (2006) Groundwater assessment through an integrated approach using remote sensing, GIS, and resistivity techniques: a case study from a hard rock terrain. Int J Remote Sens 27(20):4599–4620. https://doi.org/10.1080/01431160600554983
46. Pratap K, Ravindran KV, Prabakaran B (2000) Groundwater prospect zoning using remote sensing and geographical information system: a case study in dala-renukoot area sonbhadra district uttar pradesh. J Indian Soc Remote Sens 28(4):249–263
47. Raghu V, Mruthyunjaya Reddy K (2011) Hydrogeomorphological mapping at village level using high-resolution satellite data and impact analysis of check dams in part of Akuledu Vanka watershed, Anantapur district, Andhra Pradesh. J Indian Geophys Union 15(1):1–8

48. Ramakrishna M (1993) Tectonic evolution of the granulite terrain of South India. Mem Geol Soc India 23:35–44
49. Rani VR, Pandalai HS, Sajinkumar KS, Pradeep Kumar AP (2015) Geomorphology and its implication in urban groundwater environment: a case study from Mumbai, India. Appl Water Sci 5:137–151
50. Rao DP (2002) Remote sensing application in geomorphology. Int Soc Trop Ecol 43(1):49–59
51. Ravi Shankar MN, Mohan G (2005) A GIS based hydrogeomorphic approach for identification of site-specific artificial-recharge techniques in the Deccan Volcanic province. J Earth Syst Sci 114(5):505–514. https://doi.org/10.1007/bf02702026
52. Saraf AK, Choudhury PR, Roy B, Sarma B, Vijay S, Choudhury S (2004) GIS-based surface hydrological modeling in the identification of groundwater recharge zones. Int J Remote Sens 25(24):5759–5770
53. Selvam S, Dar FA, Magesh NS, Singaraja C, Venkatramanan S, Chung SY (2015) Application of remote sensing and GIS for delineating groundwater recharge potential zones of Kovilpatti municipality, Tamil Nadu using if technique. Earth Sci Inform. https://doi.org/10.1007/s12145-015-0242-2
54. Shaban A, Khawlie M, Abdallah C (2006) Use of remote sensing and GIS to determine recharge potential zones: the case of Occidental Lebanon. Hydrogeol J 14(4):433–443. https://doi.org/10.1007/s10040-005-0437-6
55. Shahid S, Nath SK, Roy J (2000) Groundwater potential modeling in a soft rock area using a GIS. Int J Remote Sens 21(9):1919–1924. https://doi.org/10.1080/014311600209823
56. Shankar Narayana G, Lakshmaiah N, Prakash Goud PV (1996) Hydrogeomorphological study based on remote sensing of mulug taluk, warangal district, andhra pradesh, india. Hydrol Sci J des Sci Hydrol 41(2):137
57. Shekhar S, Pandey AC (2015) Delineation of groundwater potential zone in hard rock terrain of india using remote sensing, geographical information system (GIS) and analytic hierarchy process (ahp) techniques. Geocarto Int 30(4):402–421. https://doi.org/10.1080/10106049.2014.894584
58. Sitaramayya S (1971) The pyroxene-bearing granodiorites and granites of the hyderabad area (the osmania granites). Q J Geol Min Metall Soc India 43:117–129
59. Solomon Raju P, Srinivasan A, Raghavan RV, Kousalya M (2000) Micro-tremor activity in jubilee hills area of hyderabad, andhra pradesh. J Geol Soc India 55:443–445
60. Sreedevi PD, Subrahmanya K, Ahmed S (2005) An integrated approach for delineating potential zones to explore for groundwater in the Pageru river basin, Cuddapah district, Andhra Pradesh, India. Hydrogeol J 13:534–543
61. Srinivas Rao Y, Reddy TVK, Nayudu PT (2000) Groundwater targeting in hard rock terrain using fracture pattern modeling, Niva river basin, Andhra Pradesh, India. Hydrogeol J 8:494–502
62. Venkateswara Rao B (1998) Hydromorphone-geological investigations in a typical khondalite terrain using remote sensing data. J Indian Soc Remote Sens 26(1&2):77–93
63. Venkateswaran S, Ayyandurai R (2015) Groundwater potential zoning in upper Gadilam river basin Tamil Nadu. Aquatic procedia 4:1275–1282

Chapter 18
Remote Sensing and GIS Application for Rainwater Harvesting and Groundwater Recharge to Secure Sustainable Groundwater Future of Adikavi Nannaya University, Rajamahendravaram, India

Nooka Ratnam Kinthada, Venkateswara Rao Vegala, and Murali Krishna Gurram

Abstract This study was aimed at reviewing the potential of rainwater harvesting, groundwater recharge as well as water use efficiency (WUE) within the campus of Adikavi Nannaya University (AKNU), Rajamahendravaram, India. Remote sensing data was used and geological, hydrogeological investigations coupled with rainfall analysis was carried out to evaluate the surface runoff and storm water potential. Infiltration tests were conducted to ascertain the subsurface formations which help understanding the infiltration capability of the region. A cost–benefit analysis was also done to estimate and fetch the volume of storm water, costs to be incurred and groundwater sustainability goals to be met by recommending appropriate and most efficient forms of rainwater harvesting structures in the area. The study recommends various rainwater harvesting structures such as, recharge wells, dykes, recharge trenches, recharge ponds, check dams, and recreation ponds at different locations as suitable.

Keywords Rainwater harvesting · Water conservation · Groundwater · Remote sensing · GIS

N. R. Kinthada
Department of Geosciences, Adikavi Nannaya University, Rajamahendravaram, AP 533296, India
e-mail: nooka.ratnam.kinthada@gmail.com

V. R. Vegala
Department of Geo-Engineering & RDT, Andhra University, Visakhapatnam, AP 530003, India
e-mail: vrvelagala@yahoo.com

M. K. Gurram (✉)
Department of Geo-Engineering & RDT, College of Engineering, Andhra University, Visakhapatnam, AP 530003, India
e-mail: prof.muralikrishna@andhrauniversity.edu.in

E. Balaji et al. (eds.), *Emerging Technologies for Water Supply, Conservation and Management*, Springer Water, https://doi.org/10.1007/978-3-031-35279-9_18

18.1 Introduction

As the world population continues to grow the demand for reliable water resource spirals day by day both in urban and rural areas. Various forms of socio-economical trends such as urbanization, industrialization, intense agricultural, land utilization and deforestation practices have led to overexploitation of surface as well as groundwater resource across the world. Continuous expansion of urban areas to accommodate an ever-growing population alters the natural processes like evaporation, infiltration, and runoff which ultimately affect the urban hydrological cycle [1, 2]. In India, which accounts for 15% of world population is facing a serious challenge of water shortage due to the depletion of groundwater resource. Around 55–60% of the Indian population is directly or indirectly dependent on groundwater for its livelihood [3]. The conventional water sources including stream, river, well, reservoir, etc. are inadequate to fulfill the rising water demand of populations due to lopsided rainfall. There are increasingly serious issues in different parts of the country concerning the depletion of aquifer, as groundwater tables are declining rapidly [4, 5]. This situation oblige to over exploitation of groundwater through deep tube wells, due to which the urban areas are facing the challenge of fall in groundwater table to alarming levels. Moreover, establishments like industries/institutes consume more water and at the same time due to concretization of their areas with buildings, hard surface, playgrounds and roads, affects the rainwater percolation into subsurface and thus the groundwater potential. In addition, it is expected that the severity of extreme events such as droughts and floods are likely to increase in the coming decades, therefore groundwater management need to be a major focus in a country like India [6].

Rainwater harvesting is a managed aquifer recharge technique, which aims to increase groundwater supply [7]. Rainwater harvesting is an effective option not only to recharge the groundwater aquifer but also to provide adequate storage of water for future use [8–10]. As a part of the sustainable development goals (SDGs), Government of India is more concerned about the rainwater harvesting to recharge the aquifer zones across the country. Rainwater harvesting include rooftop harvesting which is water reaping strategy through which rain water is collected from the rooftop catchments to store it in tanks or wells or aquifers. The rain water collected can be put away in sub-surface ground water repository by receiving manufactured energize strategies to meet the local needs [11].

Of late, Remote sensing and GIS technologies are effectively being used in rainwater harvesting and water conservation applications. Remote sensing (RS) coupled with geographical information systems (GIS) is now became more common for assessing the biophysical environments and identifying suitable rainwater harvesting (RWH) sites [10, 12]. Geospatial technologies are also proved to be highly effective in estimating rooftop rainwater harvesting (RRWH) quality and to simulate the implications of roofing materials and roofing surface conditions towards the urban environment. Combination of RS and GIS proves as an efficient tool in deciphering groundwater potential zones and provides accuracy by examining the influential factors such as lithology, lineament/drainage density, slope, geomorphology, rainfall, land use patterns and soils [1].

In such backdrop, this study has been taken up to see the sustainability and effectiveness of a rainwater harvesting system within the 100 acre campus area of AKNU, Rajamahendravaram, India. The study eventually recommends suitable methods for natural conservation and recharge of rainwater to improve the groundwater potential in and around the area. The final objective of the study is to recommend and develop suitable rainwater harvesting structures towards meeting the environmental sustainable goals of the university and enhancing the ground water potential.

18.1.1 Need for Rainwater Harvesting in AKNU Campus

The University in the present campus started functioning since last 2013 and a lot of development activities are at the abreast. The campus was part of agricultural land before 2013 and now being utilized for various development activities related to the university. The entire water requirements of the university including drinking, washing, toilets and greenery development is about 200m^3/day, which is met by the groundwater drawing through the five bore wells dug at various locations in the campus. Since the main aquifer zone is indicated between 180 and 200 m depth, the bore wells drilled for the purpose are ranged between 200 and 250 m depth. Existing bore well yield in the campus range between 10 and 20 m^3/h and the aquifer zone can be considered as good potential. It is noted that the groundwater usage is limited at present, and there is not much impact on yield variation during summer. However, the groundwater from the aquifer zone is also being fetched by numerous bore wells drilled in the agricultural and horticulture fields surrounding the campus. As the extent of impervious surface in the campus is expected to be increased due to the developmental activities in the form of buildings and roads, the natural infiltration of rainwater into the sub-soil will be reduced drastically and ultimately storm water increases. Though the university is drawing good amount of groundwater at present, it is likely that the requirements would increase by two to three fold in the coming years. Therefore, it is obligatory to conserve the rain water at least to the extent of utilization or the amount of percolation loss due to concretization. As per the existing subsurface geological condition, it is assessed that any water conservation measure not only improves the groundwater potential within the campus, and also in the surrounding areas.

Surface runoff/storm water generated from the roof tops of the buildings, roads and from the open areas being diverted outside through storm water drains which supposed to be good quality of water before it join with drains can be conserved and recharge aquifers artificially to improve the quantity of groundwater. As the sewerage and sullage water from various buildings is separated from the storm water drains, non-polluted storm water that joins with the drains also can be utilized for groundwater recharge with different methods of conservation.

18.2 Materials and Methods

The study was intended to harvest the rainwater in two distinctive ways importantly, surface runoff harvesting and rooftop rainwater harvesting. As required, necessary arrangements were made to collect, store the rainwater for reuse on-site, rather than allowing it to run off. The stored water will be used for various purposes, such as gardening, groundwater recharge, etc. The following sections discuss various methods of rainwater harvesting implemented as per the objectives of the study. Data representing diversified aspects of the study area such as rainfall, geology, hydrogeology, land use/land cover was collected from various primary and secondary sources. Data pertaining to surface characteristics including land use/land cover (LU/LC), terrain characteristics, soil, lithology and drainage pattern representing hydrogeological conditions have been interpreted from the high-resolution remote sensing data collected from the Google Earth of the year 2022. The data layers have been vectorized, integrated and analysed in the GIS environment using ArcGIS desktop version 10.7 software. Field checks were conducted in the campus to observe the disposition of influential parameters, conditions and also to ascertain the appropriate RWH measures to be taken. The activities include, (i) study of rainfall patterns, including daily and hourly intensity data, (ii) analyze hydrological parameters to estimate the runoff potential according to present land use, (iii) study of the surface and subsurface geological strata using secondary data to know the occurrence and distribution of ground water in various geological formations, (iv) conduct infiltration test through an experimental recharge well to understand the recharge and storage capacity of the sub-surface layers, (v) find suitable location to construct harvesting structure to conduct the infiltration test, (vi) to recommend the type and number of harvesting structures as suitable for the location, (vii) to provide designs of the suggested harvesting structures, (viii) to estimate the cost of the harvesting structures and performing cost benefit analysis, and (ix) suggest short term and long term plans for implementing rainwater harvesting scheme.

18.3 Study Area

The AKNU is located in the outskirts of Rajahmundry, East Godavari district in the state of Andhra Pradesh India. It is bounded by the geographical coordinates of N17°03′45.49″ E81°52′23.52″ in the south; N17°04′28.37″ E81°52′27.61″ in the north; N17°03′54.29″ E81°52′14.59″ in the west and N17°03′58.76″ E81°52′31.30″ in the east. The campus is spread in an area of about 100 acres consisting various buildings, open area, roads, sports complex, other ancillary units and natural drains. Figure 18.1 shows the Google earth satellite image of the location and its surroundings dated March, 2020.

The campus is abutting NH-16. A concrete road is laid from the NH at the southern end till the northern most boundary of the campus and is about 1.8 km. Surface altitude

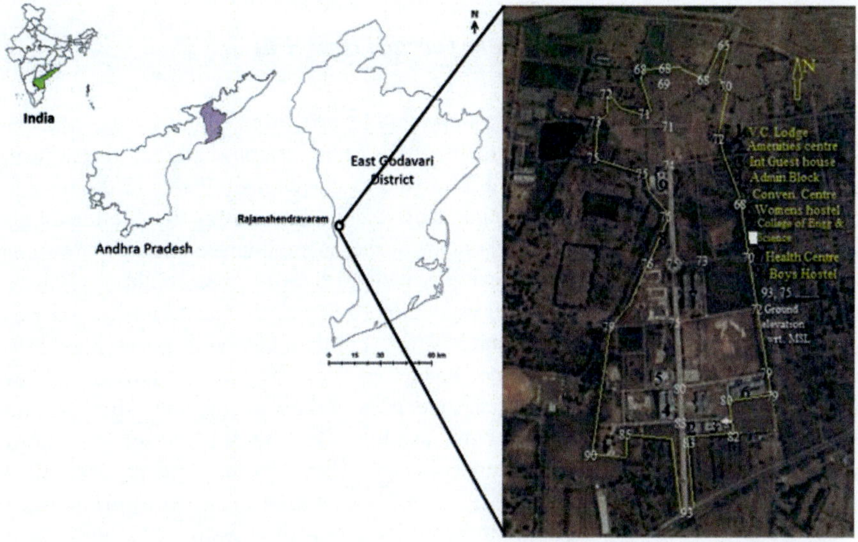

Fig. 18.1 Satellite image shows the site of investigation—ANU. Various buildings are listed on the map. Numericals in white colour area ground elevations with respect to msl

is 93 m above MSL at the southernmost point and 68 m msl on the southern point of the road. Lowest elevation is 65 m at the NE tip point near Chagalnadu canal. Most of the buildings and amenities of the campus are in the southern part of the site where the surface elevations varies between 93 and 75 m msl. In the northern part one Boy's hostel is present and mettle roads are also still to be formed. In the eastern part at the middle a big soil burrow pit of 0.41 ha. Is noticed in which storm water from the south, some part of the west and north side also getting accumulated.

18.4 Data Analysis

Spatial analysis was performed in the GIS environment by integrating various spatial information layers extracted from the satellite image which include land use/land cover, geology, hydrogeology, drainage, soil and rainfall pattern in the study area. The information layers forms a basis for identifying groundwater aquifer/prospective zones [13, 14] and for formulating a criteria for recommending groundwater recharge wells and pits in the study area.

18.4.1 Land Use/Land Cover, Terrain and Soil

LU/LC, terrain and soil are the three important factors investigated for the study as they provide key information about the pervious or impervious nature of the surface, which in a way determines the infiltration and runoff characteristics of the terrain. LU/LC map of the study area was prepared by visually interpreting the satellite image by taking into account the various elements of the image interpretation key. Various LU/LC categories deduced in the campus include built-up area (buildings, roads, pavements and concretized surface), vegetation (horticulture and scrubs), open area and waterbodies. It's noted that around 35% area comes under built-up category, 15% under vegetation, 45% remains under open area, and 5% under waterbodies. This suggests that 65% of the campus comes under impervious category and indicates good infiltration conditions. On the other hand, examination of terrain and soil conditions informs that the site of investigation is an undulating terrain with topographical variation of about 28 m within 1.8 km distance from south boundary to north boundary and slope towards north direction. Surface soil consists of gravelly soil with isolated boulders reflecting favorable infiltration conditions.

18.4.2 Geological and Hydrogeological Investigation

Study of geology and hydrogeology are essential to understand the nature of substratum and decide the type of harvesting structures as per the physical properties of the sub-surface layers in the study area. In order to ascertain the soil, rock formations and water holding potentiality, a geological as well as hydrogeological study was carried out in the study area. In addition to that, a field permeability test was conducted on the stand pipe piezometer of the boreholes, and falling head permeability tests were done on the rest.

18.4.2.1 Geology

Geological studies are based on the surface observations during the field visit and secondary data available as well from AKNU. Subsurface up to 5 m depth through shallow open wells made for the study purpose shown in Fig. 18.2a, indicates gravel and pebble bed is met within 1.5 m depth (Fig. 18.2b) and weathered rock is indicated below 3 m depth (Fig. 18.2c). At some places basalt boulders (trap rock) is noticed within 3 m to 5 m depth. Finally, an open well made up to 3 m depth lined with concrete rings and filter bed of 30 cm thick laid at the bottom (Fig. 18.2d).

Six bore wells were drilled in the campus during the last 4 years for exploring groundwater for construction and domestic utilization. Bore log data indicates that the top layer 25–30 m depth consists of soil for 3–5 m as mentioned above followed by weathered and fractured rock. Trap rock—basalt in the form of hard rock is noticed

Fig. 18.2 Subsurface lithology

Fig. 18.3 **a** Recharge Well (RW) to conserve roof water where infiltration test conducted and **b** Pipes connected to roof and recharge well

between 25/30 m and 180/200 m depth. Sandstone formation is noticed below trap rock up to 250 m which is the main aquifer zone. Geophysical Electrical Resistivity soundings conducted by Groundwater department recently also indicate the same sequential lithology.

18.4.2.2 Hydrogeology

It is the study of properties of subsurface geology and water holding and movement in it at different depths that include water table and aquifer parameters. Subsurface rock formations essentially consist of deposited soil of 3–5 m followed by weathered& fractured rock up to 30 m depth. There are no open wells in and around the site to record the shallow water table. There used to be some shallow bore wells in the surrounding agricultural lands up to 50–60 m depth. At present these bore wells are dried up and new bore are drilled to deeper depth.

Borewells yield drilled in the campus were in the range of 15 m^3 and 20 m^3/h. Deep aquifer sandstone formation beyond 180–200 m depth is found to be the only dependable groundwater source to meet the needs of the campus.

Shallow and deep aquifers are separated by about 150 thick impermeable layer of hard rock i.e., basalt. Artificial recharge of the subsurface layer at shallow depths with rainwater helps in improving the groundwater potential of shallow aquifer. There is scope for recharging deeper aquifer, if there is connectivity between them through fractures and fissures in hard rock. Shallow aquifers acts as storage system and becomes leaky aquifer. However, there is a place to store rainwater subsurface within the shallow aquifer between 5 and 30 m depth. Otherwise, the rainfall would become storm water which eventually transforms the land into concretization and gets drained out due to intense agricultural as well as land use practices manifested over the surface in the region.

18.5 Rainfall

The amount of rainfall and its intensity are the key factors in estimating the quantity of runoff and decide the number of rainwater harvesting structures that could be established to conserve rainwater. The nearest rain gauge station maintained by ITC, Rajahmundry which is about 6.0 km air distance and the monthly data of the last 16 years is presented in the following Table 18.1.

The average annual rainfall data for the last 16 years (Table 18.1) indicates a varying trend ranging between 672 mm and 1573 mm. However, the amount of rainfall is considered to be quite good. Average winter and summer rainfall together (December–February and March–May months): 153.93 mm. The average monsoon rainfall (i.e. June–November months) is 1007.08 mm.

The average annual rainfall is computed as 1173 mm out of which 86.7% occur during the monsoon season. Summer showers are very common in this area during April and May months. The rainfall other than the monsoon season is about 154mm and is about 13.3% of the annual. The annual rainfall information gives an idea about the total dependable rainfall from which the runoff quantities as well as the total quantity of water availability n the study area. However, the intensity of rainfall is the main concern to arrive at the runoff calculated from different individual units and for designing the harvesting structures instead the seasons.

18.5.1 Analysis of Rainfall Trend

Daily rainfall data analysis is required to estimate the quantity of runoff expected from each spell of rainfall. After analyzing the daily rainfall data, it has been found that the rainfall of >100 mm per/day occurred in 2 days/annum in the last 16 years and >50 mm/day occur in about 8 days in a year. The number of total rainy days in a year range between 45 and 70 days. Keeping in view of the rainfall pattern, the maximum optimum one day rainfall of 50 mm is considered. Even though the high intensity rainfall occurs in 8–10 days in a year, the amount of rainfall occurs during this period is more than 40% to the total annual rainfall. In order to catch the maximum amount of rainwater for conservation through harvesting structures,

Table 18.1 Rainfall data recorded at ITC, Rajahmundry between 1996 and 2013

Year	June	July	Aug	Sep	Oct	Nov	Dec	Jan	Feb	March	April	May	Total
1996–97	340.85	364.68	249.2	257.3	129.8	98.8	84	13	0	18	93.3	0	1573.33
1997–98	37	150.2	164.4	288.2	46.6	3.2	46	0	0	43.6	35	75.8	890
1998–99	88	195.8	318.2	324	234.7	95.9	0	0	0	0	0	68.28	1374.88
1999–00	46.9	258.8	182.6	89.3	160.72	254.75	0	0	15.2	0	15.8	49.76	1210.63
2000–01	191.88	144.7	306.15	161.02	57.25	0	0	0	0	0	0	108.8	969.8
2001–02	161.2	155.6	89.8	17.3	138.7	30.25	3.2	6	0	0	9.6	47	858.65
2002–03	89	22	267.3	92.5	90.7	0.8	0	3	3.2	84	0	1	653.5
2003–04	226	317.3	144.9	134.8	118.95	20.05	98	0	0	0	48	78	982.6
2004–05	118	243.9	270.9	147.6	92.2	0	0	0	0	0	122.2	52.2	1047
2005–06	115.08	253.24	155	457.9	475.2	7.6	0	0	0	0	46	50.4	1560.4
2006–07	136.2	103.6	312	235.6	178.4	48.8	0	0	17.6	0	63.2	12	1107.4
2007–08	230.6	171.4	286.6	257	127.2	1	0	3.2	20.8	121	29.2	12.8	1260.8
2008–09	153	186.6	272.2	85.6	46	28	0	0	0	35.2	14.4	84.4	905.4
2009–10	15.2	154.4	100.2	148	61.2	114.4	0	0	0	0	10	174.6	778
2010–11	222.2	423.2	208.6	103.4	233.6	121	99.2	0	9.6	0	25.6	18.8	1465.2
2011–12	62	103.8	338.2	42	105	0	0	0	0	13.6	0.6	6.8	672
2012–13	66.6	255.8	322	339.6	99	277.8	0	4	18.4	0	56.2	12.6	1452

50 mm rainfall intensity per day is considered to arrive at the runoff from various catchments like roof tops, roads and other concretized areas.

18.5.1.1 Optimum High Intensity Rainfall Per Hour

The daily rainfall recorded at any gauging station is 24 h duration and normally between 8 AM today and 8 AM next day. Even though the rainfall intensity is mentioned per day, actual raining hours may be very less ranging between less than an hour to 3–4 h. Storm period hourly analysis of rainfall data of East coast of India by IMD, Government of India shows the intensity varies between 40 and 60 mm/h with a return period of 20 years. But for a short return period like every year 25–30 mm/h is an idle highest intensity rainfall. As per the observed daily rainfall data the highest intensity of 30 mm/h is considered for designing the harvesting structures, so that the plan can be made to create number of harvesting structures to conserve the maximum rainfall occurred in the study area.

18.5.1.2 Optimum Low Rainfall Considered for Estimating Storm Water Potential

The concrete/asphalt surface areas require about 3–5 mm of rainfall to create surface runoff for wetting the surface and fulfilling the instantaneous evaporation. Rainfall occurred at the rate of less than 5 mm/day is not taken into account while calculating the urban runoff and also 5 mm is reduced from the rainy days that occurred more than 5 mm considered as loss due to wetting and immediate evaporation. The amount of rainfall occurred in less than 5 mm/day amounts to 5–10% of the total rainfall and the number of rainy days is 25–30% of the total rainy days.

18.6 Estimation of Surface Runoff

Surface run off of any area depends on the amount and intensity of rainfall, catchment area and its physiographic characteristics. The effective annual average rainfall that create run off from the concretized/asphalt surface in the Plant site is taken as 1056 mm after considering the above said losses (10% of average rainfall) from the total rainfall. For the purpose of harvesting studies run off from the roof tops, concretized/ road surfaces and soil surfaces in the campus are considered. The following Table 18.2 shows the estimated hourly quantity during optimum high intensity as well annual run off from a unit area of 100 M^2.

Table 18.2 Different types of land use units in the Campus area and the estimated runoff for high intensity and annual rainfall

S#	Plant unit	Rooftop area/ catchment area (M^2)	Runoff from max. hourly intensity of 30 mm rainfall (M^3)	Annual runoff from each unit 100 m^2(M^3)
1	Roof tops—concrete surface/ oped roofs with A.C. sheets	Say 100 m^2 (runoff coeff. – 0.80)	2.4	83.6
2	Asphalt surface/roads	Say 100 m^2 (runoff coeff. – 0.70)	2.1	73.15
3	Lawns/open areas around the buildings	Say 100 m^2 (runoff coeff. – 0.60)	1.8	62.70
4	Open soil covered/ agricultural land	Say 100 m^2 (runoff coeff. – 0.40)	1.2	41.80

The above table shows that the amount of runoff estimated during optimum high intensity rainfall of 30 mm/h varies between 12 and 24 L from each M^2 at the high intensity of 30 mm/h depending on the land use. Harvesting plans are to be made in various land use patterns considering the maximum hourly intensity rainfall in such a way that the entire amount of rainfall is to be conserved with suitable structures. There after the less intensity rainfall runoff automatically taken care by these structures. Total quantity of rainwater that could be collected from the above mentioned units can be arrived taking the above runoff factors and the each unit area in the plant site. However, intake capacities of subsurface formations is arrived from the infiltration test conducted through test wells-Recharge wells constructed for the purpose and the procedure is described below.

18.7 Infiltration Capability of Sub-surface Formations

Recharge structure design is mainly based on the intake capacity of the sub-surface geological formation and the intensity of rainfall. The optimum highest intensity rainfall of 30 mm/h is considered to quantify the surface run off from a given catchment area. In order to determine the infiltration rate of the sub surface geological formation, an infiltration test was conducted by making a sample recharge structures so that the optimum number of harvesting structures requirement can be planned. A recharge well is constructed near to the administrative building as shown in Fig. 18.3a, b. The method of construction and conducting infiltration test is described below.

18.7.1 *Infiltration Test (Slug Test)*

A Recharge wells is constructed with dimensions of 1.0 m diameter and 2.50 m depth and are lined with cement rings and one feet filter bed at the bottom as shown in Fig. 18.3a, b. The geological formations that come across the well section is:

0 – 1.50 m – sandy loam with gravel

1.5 – 2.5 m weathered and fractured rock and small trap rock boulders

The well is constructed in such a way that the percolation of rainwater is only through the bottom of the well.

- Cementation between each ring is essential to avoid mud flow into the well.
- 30 cm of gravel + sand filter is placed at the bottom of the well to filter the sediment come into the well along with the storm water from roof tops.
- There is no water table till the depth of excavation i.e., 2.5 m from ground level.
- Total induced water column in the well act as pressure head on the aquifer material present down below and there will be high quantity of percolation when the head is maximum and reduces with lowering of head.
- Slug test was conducted during the Ist week of July, 2020. For the purpose of the testing, the model well is filled with water to the brim level. Water levels are measured at 1 min interval up to 5 min, later 5 min interval up to 35, 15 min interval up to 140 min till the end of test.
- Infiltration rate is calculated taking into consideration of change of water level with time, diameter of the well and finally quantity of percolation of rainwater into the subsurface formations with respect to hydraulic head is arrived and is listed in the Table 18.3.

The well was filled with water through a water supply pipe line up to 200 cm height. Water level drawdown is measured at regular interval till it reached from 200 to 91 cm which took about 140 min. The amount of water percolated at head of 200 cm is about 1206 l/h (lph) and at a head of 100cm the rate of percolation is 161 lph. The above data is presented in the form of graphs (i) time in minutes Versus Hydraulic head decline in meters (Fig. 18.4a) and (ii) Hydraulic head in meters Versus infiltration rate in litres per minute (Fig. 18.4b).

The Fig. 18.4a, b representing the infiltration rate of applied water with respect to hydraulic head. As the head is more rate of percolation is more and vice versa. Permeability is 5×10^{-4} m/s at head of 200 cm and the same reduced to 1.11×10^{-4} m/s when the head reduced to 91 cm. This shows that the permeability not only varies with lithological material but also with hydraulic head.

Table 18.3 Infiltration test data of a test well at administrative block, ANU, Rajahmundry

Recharge structure-recharge well-RW-1		
Time in min	Water level-centimeters (Hydraulic head)	Infiltration rate (litres per hour)
0	200	
1	197	1206
2	194	1206
3	191	1206
4	189	804
5	187	804
10	180	562.8
15	173	562.8
20	167	482.4
25	160	562.8
30	155	402
35	150	402
50	141	241.2
65	133	214.4
80	122	294.8
95	114	214.4
110	107	187.6
125	101	160.8
140	91	160.8

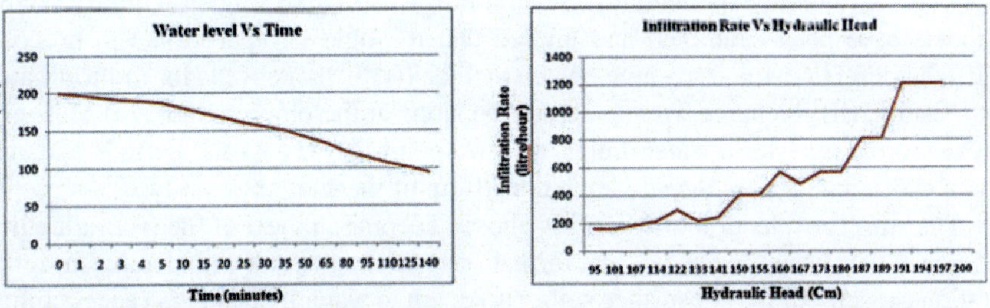

Fig. 18.4 **a** Infiltration test results-water level decline versus time. **b** Infiltration test results-hydraulic head versus infiltration rate

18.7.2 *Infiltration Test Results*

Infiltration test shows that the rate of percolation is about 1,206 lph for a given diameter of the well i.e., 1.0 m, when the hydraulic head is at 200 cm and for the same well and through the same soil, the rate of percolation is reduced to 161 lph when the head is reduced to 91 m and in terms of permeability it is 5×10^{-4} m/

s and 1.11 × 10^{-4} m/s respectively which is good infiltration rate. The experiment clearly shows that hydraulic head plays an important role in percolation of the applied water through a recharge structure for a given type of soil. In general permeability of 10^{-3}–10^{-5} m/s can be considered as good permeable zone.

It can be concluded that the infiltration through a recharge well of 1.0 m diameter and 2.5 m depth will be about 2.5 m^3/h considering the instant storage capacity of the well is about 1.5 m^3, as well the percolation capacity of the formation with respect to the water column (head) is another 1.2 m^3. At present there is no indication of shallow water table in the site of investigation.

As per the Table 18.3, total quantity of storm water expected during the optimum high intensity rainfall (30 mm/h) from the roof tops and concrete surface will be about 2.4 m^3/h/100 m^2. In order to harvest the total storm water with in the high intensity period, the number rainwater harvesting structures to be required can be arrived based on the land use, catchment area of each unit and type of harvesting structure chosen.

18.8 Proposed Rain Water Harvesting Structures

Subsurface lithological information from the borehole data and surface geological observations indicated that good permeable formation like weathered and fractured and bouldary rock layers are present up to 25/30 m below the ground level. Infiltration test conducted in the campus area indicates that the weathered rock has good permeable nature and the experiment also indicated that the increase of head increase the percolation rate exponentially. Several rain water harvesting structure (RWHS) models have been examined and mulled that it would be appropriate to propose different RWHS in the study area, which include (i) Recharge wells for rooftop water harvesting, (ii) Recharge Trench across the slope at the play ground, (iii) Masonry dykes across two storm water drains, (iv) two ponds in the east and north is part and (v) a check dam across the stream at the NE tip of the site.

The sites for selection of RWHS is chosen keeping in view of the (i) availability of space, (ii) subsurface geological formations and their extent with depth, (iii) infiltration capacity of the subsurface soils, (iv) depth to water table and (v) accessibility of roof rainwater through down take pipes. The sites suitable for these RWHS have been selected and are marked in the following satellite images.

18.8.1 Recharge Wells

As per the water infiltration capacity of the subsoil, each recharge well take care of storm water from roof top of 250 m^2 even during the high intensity rainfall of 30 mm/h. Taking it as a norm to the ANU Campus number of Recharge wells required are arrived. At present roof water from down take pipes for each building are connected to storm water drains. Now these are to be connected with the Recharge wells. Recharge wells are proposed at all the buildings in the campus. Numbers of structures required are listed in Table 18.4. Locations at each building are shown in Figs. 18.5, 18.6, 18.7, 18.8, 18.9, 18.10, 18.11, 18.12 and 18.13 in two plates below.

Figures 18.5, 18.6, 18.7 and 18.8 shows administrative building in two blocks. Roof water down take pipes are towards garden area developed in between these two. Recharge wells are located at 8 places marked as RW. Connecting 3 down take pipes to one recharge well photo is shown as inset in the satellite image. Rainwater harvesting structures—Recharge wells are marked at 4 places on the back side of the Convention centre. Presently there are no down take pipes and roof water is let out openly. Connectivity of these open let out pipes and diversion to the RW is shown in

Table 18.4 Rainwater harvesting structures—recharge wells at each building

S. no	Building	Roof area (m^2)	No. recharge wells
1	Admin building	2010	8
2	Convocation centre	2238	4+ trench
3	Science college	2892	12
4	Central library	1439	4
5	Women's hostel	2264	8
6	Boy's hostel	1062	4
7	International guest house	751	3
8	Amenities centre	480	2
9	V.C. Lodge	240	1
10	Health centre	240	1
			47

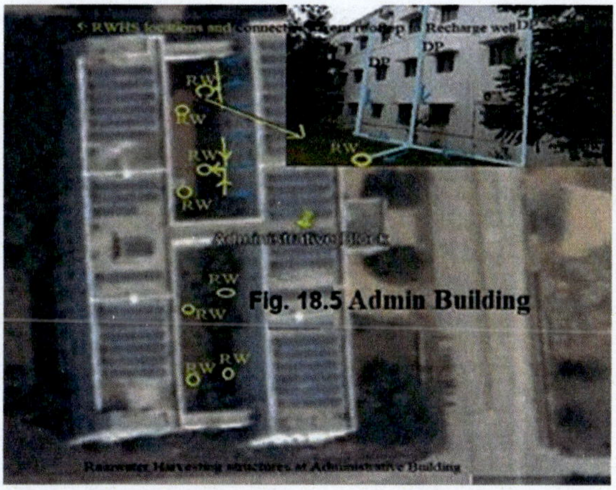

Fig. 18.5 Rooftop RWHS at admin building

inset photo in Fig. 18.6. To conserve the storm water from the open area around the centre a trench is proposed in the north-east part of the plot. A total of twelve RWs proposed in the Science & Technology building as shown in Fig. 18.7. Connecting the roof water down take pipe to RW photo is shown as inset. Four RWs are located around the boy's hostel to conserve the roof water as shown in Fig. 18.8.

Storm water from the women's hostel building rooftop drains inside the vacant place through down take pipes and let into storm water drains (Fig. 18.9). Inset photo shows the down take pipes and soil exposed area where RWs are proposed and three down take pipes are connected to each recharge well disconnecting them from storm water drains. Six numbers of RWs inside and two numbers outside the building are proposed based on the catchment area and down take pipe locations. For the storm water collection from the roof top of the Central library, four numbers of RWs are proposed-two on the north side and the other two on the south as shown in Fig. 18.10.

Fig. 18.6 Rooftop RWHS at convention center

Fig. 18.7 Rooftop RWHS at science & technology building

Fig. 18.8 Rooftop RWHS at boys hostel

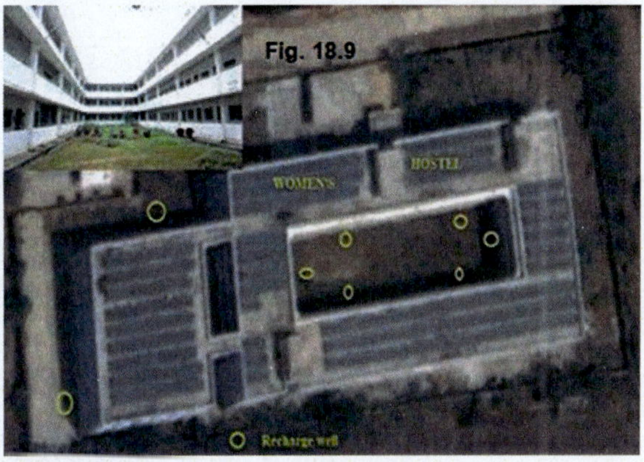

Fig. 18.9 Proposed sites for 'recharge wells (RW)' at women's hostel

Connectivity of down take pipes is shown with blue line which is the connectivity pipe direction to the RWs. Roof top area of the VC Residence is 240 m^2 for which one recharge well is sufficient. Location of RW and connectivity is shown in Fig. 18.11. Medical centre roof area is 240 m^2 for which one recharge well is sufficient and its location is shown in Fig. 18.12. International guest house and Amenities centre are located adjacent to each other. Five numbers of RWs are required for these two buildings and the RWHS locations are shown in Fig. 18.13.

18.8.2 Dykes

These are the masonry walls proposed across two storm water drains near to the north boundary as shown in Fig. 18.14. The structure consists of a masonry wall of 2–3 m width, as per the site condition, 1.5 m below the bed level of drain and 1–1.5 m above bed level and 60 cm thick wall. Small surplus weir of 50 cm × 30 cm is to be

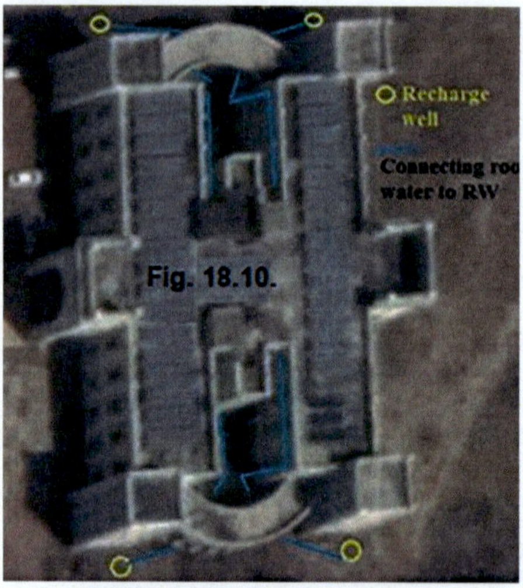

Fig. 18.10 Proposed sites for 'recharge wells (RW)' at central library

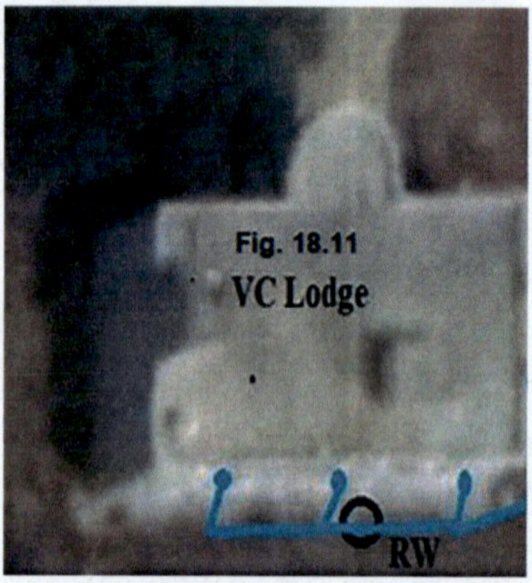

Fig. 18.11 Proposed sites for 'recharge wells (RW)' at VC lodge

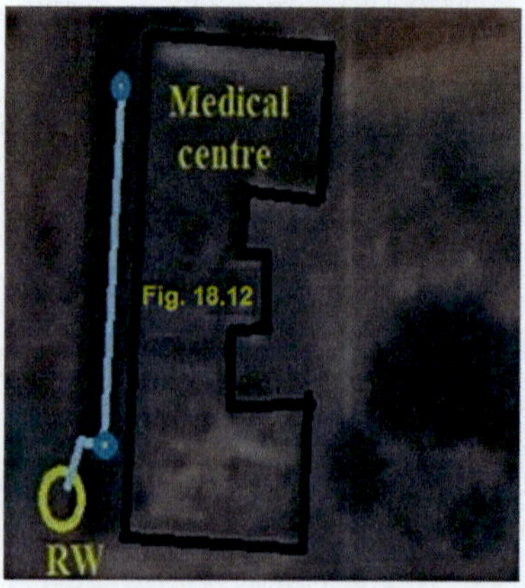

Fig. 18.12 Proposed sites for 'recharge wells (RW)' at Medical center

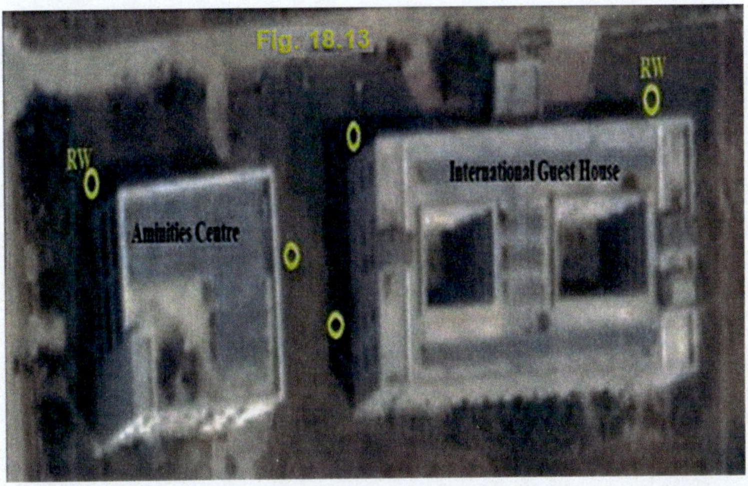

Fig. 18.13 Proposed sites for 'recharge wells (RW)' at Amenities center and International guest house

provided at the top level of the wall. Downstream side of the wall masonry apron or stone pitching is to be provided to avoid s scouring. Dykes are proposed across two drains at two places near the north-west border (Fig. 18.14). Water stored upstream of the dyke extend 80–100 m upstream as per the bed level. Thus the stored water recharges the subsurface layers through the drain bed in due course of time.

Fig. 18.14 Proposed 'dyke' structure across the local drain

Fig. 18.15 Proposed 'recharge trenches' north and western boarders of the sports complex

18.8.3 Recharge Trench

It is a trench with 2 m width and 1 m depth along the northern border of the playground which is across the ground slope. Ground slope is towards north and north-west. Trench length should be 210 m along the north border and 50 m along the west border as shown in Fig. 18.15. Trench depth may be decided based on the slope of the north border and should be filled with permeable material like—bottom 50 cm with boulders of <150 and >40 mm size, above boulders layer 30 cm thick gravel may be 20 mm chips and top 20 cm fill with coarse sand. Catchment area for this trench is 4 ha of the playground. There is scope to harvest about 3.5 ha m of rainwater and recharge groundwater.

18.8.4 Recharge Ponds

There is a soil borrow pit measuring 0.41 hec. to the north side of the Science and Engineering building. Most of the storm water from the southern part of the campus joining this pit through natural slope as shown with blue arrows in Fig. 18.16. Inset in the Fig. 18.16 shows the pit depth varies between 2 and 3 m and can hold about 1.0 ha M quantity of water. Still there is scope to borrow some soil from the pit

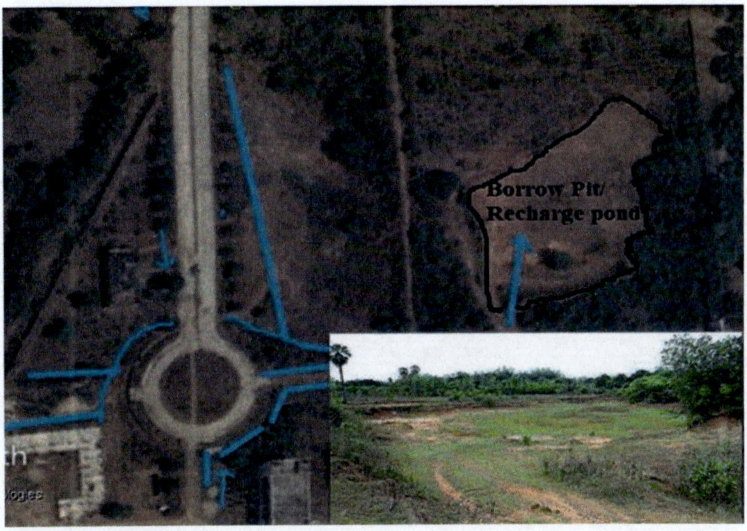

Fig. 18.16 Converted borrow pit to recharge pond

if required. Some protection measures may be taken to avoid accidents. In future it may require deepening and regular shaping and can be converted as recreation centre. Another recharge pond is suggested in the north eastern part of the campus where excess storm water from south, west and east passes through. It was told that lot of water stagnation occur at this place during rainy season. Hence, a pond P1 measuring 30 m × 20 m × 2 m depth is suggested to hold excess storm water from the upstream side as shown in Fig. 18.17. Inset photo P1 is the location of pond and the road laid at this place act as dyke and divert surface run off through a culvert further downstream (inset photo above P1) where another pond P2 is suggested by constructing a check dam at the NE tip point as shown in Fig. 18.17.

18.8.5 Check Dams

The NE point through which all the excess storm water exit consists of low lying area look like stream bed. Its width measuring 15 m at NE tip and 35 m width 100 m upstream and downstream of pond P1. Stream bed area marked as P2 is lower by 3 m in a span of 100 downstream. Check dam of 25 m length with concrete structure keying on either side of the stream banks, dam foundation 2 m below the bed level and 3 m above the bed level can hold lot of water. Stream bed is also to be deepened by 1.5–2 m depth from the bed level at places. Downstream of the check dam there should be concrete apron to avoid scouring during heavy storm water flows (Fig. 18.17).

Fig. 18.17 Proposed rainwater harvesting ponds in North–East corner (P1 and P2) of the campus

18.8.6 Recreation Pond

A recreation pond is developed in the campus about 200 m north west side of Convention centre. It is also adjacent to the west side compound wall and is made as a regular square shape measuring 80 m × 80 m is shown in Fig. 18.18. It is deepened by 1–1.5 m depth and 1 m bund over natural surface around. In total the depth of the pond varies between 2 and 2.5 m. There is small tank (tank in Fig. 18.18) outside the campus in the upstream and abutting the west side compound wall of the campus. At present excess water from the tank is allowed into the recreation pond as there is a natural slope towards the pond. However, the runoff from the upstream side in the campus includes VC lodge and other open ground flow towards the pond and the slope direction is marked on the image. Storm water is allowed through Inlet pipes into the pond is shown with circles in the inlet photo in Fig. 18.18. There is an open well in the north east corner of the pond and utilizing groundwater for construction of the buildings in the adjacent area. During rainy season the pond get filled with storm water and sometimes overflow it. But from winter onwards small quantity of water may present in small ditches here and there. Using the pond for recreation purpose there should be continuous supply of makeup water and for that purpose there should be at least one bore well. Another remedy to make the pond perennial is to reduce percolation by clay puddling by about 9″–12″ thick covering the bottom.

A Recharge trench is proposed adjacent to the approach road to pond to conserve storm water from the open area around Convention centre and large open area to the north of convention centre as shown in Fig. 18.8. Recharge trench should 2 m

Fig. 18.18 Proposed rainwater harvesting ponds near to convention centre of the campus

width and 1m depth filled with permeable material like- boulders and pebbles at the bottom, gravel as second layer above the boulders and coarse sand as top layer. Thus the total storm water from the upper catchment between VC lodge and pond area can be conserved through the above said procedure.

18.9 Cost–Benefit Analysis and Recommendations

The project cost consists of construction of rainwater harvesting structures and annual maintenance. Constructing the rainwater harvesting structures involves one time investment. However, maintenance is required once in two years to desalinize the wells and trenches.

18.9.1 Cost

- AKNU has taken up construction of Recharge wells as pilot project of eight (8 no's) at Administrative building and twelve (12 no's) at Science & Technology building with roof water connectivity.
- Each well of 1.0 m diameter, 3 m depth, lined with concrete rings. Bottom 0.3 m filled with permeable material as filter. Three roof top pipes exists to the bottom level of the building are connected to one outlet pipe and diverting roof water to one

recharge well. Cost of each well including excavation, material used for construction and finally finishing at the RW connectivity. Total cost of each Recharge well arrangement is Rs. 10,000/-

- Twenty wells have been constructed at the cost of Rs. 2,00,000/- and it is It is only one time investment.

18.9.2 Benefit

The estimated annual rainwater from the roof tops of the two buildings will be about 4,900 m^3.

$$\text{Cost of industrial waterRs. } 40/\text{- } m^3$$

$$\text{Cost of water conserved from two buildings per year} = 4,900 \times 40 = \text{Rs.}1,96,000/\text{-}$$

As per the above statistics, cost of the harvesting structure is returned within a year and more over, the precious resource of good quality of water is saved instead leaving it into surface drains. Application of Remote sensing and GIS facilitated reducing the cost drastically in terms of providing crucial information layers for the study, integration, analysis and optimally locating RWHS which is highly efficient and may not be possible to employ using any other technique available. The statistics show that the rainwater harvesting is highly cost beneficial.

18.10 Recommendations

- Rainwater harvesting structures—Recharge wells may be constructed as per the designs. Care should be taken to avoid polluted water entering the storm water drains that are connected to the RWHS.
- Rainwater harvesting project may be implemented stage wise, so that the practical difficulties aroused in the first stage can be overcome during second stage. As the pilot project is implemented and found it is effective.
- Total number of Recharge wells proposed is 47 for roof water conservation from all buildings. Recharge wells construction may be completed this year at all the buildings.
- Establish two observation bore wells (i) at the circle area near science building, and (ii) at the proposed pond P1 in the north east corner to collect base line data like water levels in the bore holes and quality of groundwater before completing all the RWHS, so that the effect of the rainwater harvesting can be assessed in due course of time.

18.11 Conclusions

The overall quality of rainwater was quite satisfactory and implies that the system could be sustained during critical periods as well as normal periods. Additionally, the system is cost effective as large amounts of money can be saved per year. Energy conservation and related reduced emissions are crucial parts of this system. The small and medium residential and commercial construction can adopt this system as sustainable option of providing water. It is almost the only way to upgrade one's household water supply without waiting for the development of community system. The system could become a good alternative source of water supply to cope up with the ever-increasing demand and should be accepted. The developed system has the potential to counter the ever increasing water crisis and can be widely adopted and implemented at community levels. The analysis results shows that the present RWH system is having the storage 53,96,816 L/year and construction cost of Rs. 5 lakh respectively and is reasonably well in comparison with conventional water sources. The developed system, when implemented by considering all the technical aspects, adequately meets the water requirements of rural and urban areas. The study has successfully demonstrated the effective application of Remote sensing and GIS especially in identifying RWH and groundwater recharge locations based on land use/land cover, slope, soil, runoff, and drainage, along with built-up characteristics in the study area.

References

1. Balaji E, Veeraswamy G, Peiyue L, SiddiRaju R (2019) Deciphering groundwater potential zones using MIF technique and GIS: a study from Tirupati area, Chittoor District, Andhra Pradesh, India. HydroResearch 1:1–7. https://doi.org/10.1016/j.hydres.2019.04.001
2. Balaji E, Sughosh M, Somagouni SG (2022) Urban water systems: an overview. Curr Direct Water Scarcity Res 6:1–19. https://doi.org/10.1016/B978-0-323-91838-1.00016-6
3. Villholth KG (2006) Groundwater assessment and management: implications and opportunities of globalization. Hydrogeol J 14(3):330–339
4. Rodell M, Velicogna I, Famiglietti JS (2009) Satellite-based estimates of groundwater depletion in India. Nature 460:999–1002
5. Shah T et al (2007) Groundwater: a global assessment of scale and significance. In: Molden D (ed) Water for food, water for life: a comprehensive assessment of water management in agriculture. Colombo: IWMI, London, Earthscan, pp 395–423
6. Gupta SK, Deshpande RD (2004) Water for India in 2050: first-order assessmentof available options. Curr Sci 86(9):1216–1224
7. Glendenninga CJ, Vervoortb RW (2010) Hydrological impacts of rainwater harvesting (RWH) in a case study catchment: the Arvari River, Rajasthan, India. Part 1: Field-scale impacts. Agricul Water Manag 98(2):331–342
8. Hajani E, Rahman A (2014) Reliability and cost analysis of a rainwater harvesting system in Peri-Urban regions of greater Sydney, Australia. Water 6:945–960. https://doi.org/10.3390/w6040945. ISSN:2073-4441
9. Hammerstrom J, Younos T (2014) Single-family home rainwater-harvesting system demonstration project for storm water-runoff control and utility-water saving. The Cabell Brand Center Special Report No. 100-2014

10. Mishra SS (2014) Rainfall analysis and design of water harvesting structure in water scarce Himalayan Hilly Regions. Int J Civ Struct Eng 5(1):29–41
11. Sabale PD, Yadav SJ, Bangale C, Kharat A, Patil C, Waghule S (2018) Review on design of rooftop rainwater harvesting in Nimgaon Village-A case study of Junnar Tahsil. Int J Eng Res Technol (IJERT) 7(04):494–499
12. Adham A, Riksen M, Ouessar M, Ritsema CJ (2016) A methodology to assess and evaluate rainwater harvesting techniques in (semi-) arid regions. Water 8(5):198
13. Golla V, Badapalli PK, Etikal B, Sivakumar VL, Telkar SK (2021) Delineation of groundwater potential zones in semi-aridregion (Ananatapuram) using geospatial techniques. Mater Today: Proc. https://doi.org/10.1016/j.matpr.2021.02.2393
14. Golla V, Etikala B, Veeranjaneyulu A, Subbarao M, Surekha A, Narasimhlu K (2018) Data sets on delineation of groundwater potential zones identified by geospatial tool in Gudur area, Nellore district, Andhra Pradesh, India. Data Brief 20:1984–1991. https://doi.org/10.1016/j.dib.2018.09.054